HANDBOOK OF CERAMIC GRINDING AND POLISHING

HANDBOOK OF CERAMIC GRINDING AND POLISHING

Edited by

Ioan D. Marinescu
The University of Toledo
Toledo, Ohio

Hans K. Tonshoff
University of Hannover
Hannover, Germany

Ichiro Inasaki
Keio University
Yokohama, Japan

NOYES PUBLICATIONS
Park Ridge, New Jersey, U.S.A.

WILLIAM ANDREW PUBLISHING, LLC
Norwich, New York, U.S.A.

Library of Congress Catalog Card Number: 98-17520
ISBN: 0-8155-1424-7
Printed in the United States

Published in the United States of America by
Noyes Publications/William Andrew Publishing, LLC
13 Eaton Avenue, Norwich, New York 13815

10 9 8 7 6 5 4 3 2 1

Library of Congress Cataloging-in-Publication Data

Handbook of ceramic grinding and polishing : properties, processes, technology, tools and typology / edited by Ioan D. Marinescu, Hans Kurt Tonshoff, and Ichiro Inasaki.
p. cm.
Includes bibliographical references and index.
ISBN 0-8155-1424-7
1. Grinding and polishing. 2. Ceramic materials.
I. Marinescu, Ioan D. II. Tonshoff, H. K. (Hans Kurt)
III. Inasaki, Ichiro.
TJ1280.C36 1998
666--dc21 98-17520
CIP

MATERIALS SCIENCE AND PROCESS TECHNOLOGY SERIES

Series Editors

Rointan F. Bunshah, University of California, Los Angeles
Gary E. McGuire, Microelectronics Center of North Carolina
Stephen M. Rossnagel, IBM Thomas J. Watson Research Center

Electronic Materials and Process Technology

CHARACTERIZATION OF SEMICONDUCTOR MATERIALS, Volume 1: edited by Gary E. McGuire

CHEMICAL VAPOR DEPOSITION FOR MICROELECTRONICS: by Arthur Sherman

CHEMICAL VAPOR DEPOSITION OF TUNGSTEN AND TUNGSTEN SILICIDES: by John E. J. Schmitz

CHEMISTRY OF SUPERCONDUCTOR MATERIALS: edited by Terrell A. Vanderah

CONTACTS TO SEMICONDUCTORS: edited by Leonard J. Brillson

DIAMOND CHEMICAL VAPOR DEPOSITION: by Huimin Liu and David S. Dandy

DIAMOND FILMS AND COATINGS: edited by Robert F. Davis

DIFFUSION PHENOMENA IN THIN FILMS AND MICROELECTRONIC MATERIALS: edited by Devendra Gupta and Paul S. Ho

ELECTROCHEMISTRY OF SEMICONDUCTORS AND ELECTRONICS: edited by John McHardy and Frank Ludwig

ELECTRODEPOSITION: by Jack W. Dini

HANDBOOK OF CARBON, GRAPHITE, DIAMONDS AND FULLERENES: by Hugh O. Pierson

HANDBOOK OF CHEMICAL VAPOR DEPOSITION, Second Edition: by Hugh O. Pierson

HANDBOOK OF COMPOUND SEMICONDUCTORS: edited by Paul H. Holloway and Gary E. McGuire

HANDBOOK OF CONTAMINATION CONTROL IN MICROELECTRONICS: edited by Donald L. Tolliver

HANDBOOK OF DEPOSITION TECHNOLOGIES FOR FILMS AND COATINGS, Second Edition: edited by Rointan F. Bunshah

HANDBOOK OF ION BEAM PROCESSING TECHNOLOGY: edited by Jerome J. Cuomo, Stephen M. Rossnagel, and Harold R. Kaufman

HANDBOOK OF MAGNETO-OPTICAL DATA RECORDING: edited by Terry McDaniel and Randall H. Victora

HANDBOOK OF MULTILEVEL METALLIZATION FOR INTEGRATED CIRCUITS: edited by Syd R. Wilson, Clarence J. Tracy, and John L. Freeman, Jr.

HANDBOOK OF PLASMA PROCESSING TECHNOLOGY: edited by Stephen M. Rossnagel, Jerome J. Cuomo, and William D. Westwood

HANDBOOK OF POLYMER COATINGS FOR ELECTRONICS, 2nd Edition: by James Licari and Laura A. Hughes

HANDBOOK OF REFRACTORY CARBIDES AND NITRIDES: by Hugh O. Pierson

HANDBOOK OF SEMICONDUCTOR SILICON TECHNOLOGY: edited by William C. O'Mara, Robert B. Herring, and Lee P. Hunt

HANDBOOK OF SEMICONDUCTOR WAFER CLEANING TECHNOLOGY: edited by Werner Kern

HANDBOOK OF SPUTTER DEPOSITION TECHNOLOGY: by Kiyotaka Wasa and Shigeru Hayakawa

HANDBOOK OF THIN FILM DEPOSITION PROCESSES AND TECHNIQUES: edited by Klaus K. Schuegraf

HANDBOOK OF VACUUM ARC SCIENCE AND TECHNOLOGY: edited by Raymond L. Boxman, Philip J. Martin, and David M. Sanders

HANDBOOK OF VLSI MICROLITHOGRAPHY: edited by William B. Glendinning and John N. Helbert

HIGH DENSITY PLASMA SOURCES: edited by Oleg A. Popov

HYBRID MICROCIRCUIT TECHNOLOGY HANDBOOK, Second Edition: by James J. Licari and Leonard R. Enlow

IONIZED-CLUSTER BEAM DEPOSITION AND EPITAXY: by Toshinori Takagi

MOLECULAR BEAM EPITAXY: edited by Robin F. C. Farrow

SEMICONDUCTOR MATERIALS AND PROCESS TECHNOLOGY HANDBOOK: edited by Gary E. McGuire

ULTRA-FINE PARTICLES: edited by Chikara Hayashi, R. Ueda and A. Tasaki

Ceramic and Other Materials—Processing and Technology

ADVANCED CERAMIC PROCESSING AND TECHNOLOGY, Volume 1: edited by Jon G. P. Binner

CEMENTED TUNGSTEN CARBIDES: by Gopal S. Upadhyaya

CERAMIC CUTTING TOOLS: edited by E. Dow Whitney

CERAMIC FILMS AND COATINGS: edited by John B. Wachtman and Richard A. Haber

CORROSION OF GLASS, CERAMICS AND CERAMIC SUPERCONDUCTORS: edited by David E. Clark and Bruce K. Zoitos

FIBER REINFORCED CERAMIC COMPOSITES: edited by K. S. Mazdiyasni

FRICTION AND WEAR TRANSITIONS OF MATERIALS: by Peter J. Blau

HANDBOOK OF CERAMIC GRINDING AND POLISHING: edited by Ioan D. Marinescu, Hans K. Tonshoff, and Ichiro Inasaki

HANDBOOK OF INDUSTRIAL REFRACTORIES TECHNOLOGY: by Stephen C. Carniglia and Gordon L. Barna

SHOCK WAVES FOR INDUSTRIAL APPLICATIONS: edited by Lawrence E. Murr

SOL-GEL TECHNOLOGY FOR THIN FILMS, FIBERS, PREFORMS, ELECTRONICS AND SPECIALTY SHAPES: edited by Lisa C. Klein

SOL-GEL SILICA: by Larry L. Hench

SPECIAL MELTING AND PROCESSING TECHNOLOGIES: edited by G. K. Bhat

SUPERCRITICAL FLUID CLEANING: edited by John McHardy and Samuel P. Sawan

Preface

Ceramics have been identified as one of the primary fields where processing improvements and more advanced products can be anticipated. Such products will have an increased "technological knowledge content" and will have to be manufactured using processing technology that is more advanced and better controlled. Advancements in ceramic processing and manufacturing technology will be necessary for the commercialization of new processing technology; these innovations may lead to eliminating expensive steps, improving productivity, and increasing product reliability.

Most of the industrialized countries of the world have invested heavily in the manufacturing (processing) of new ceramic materials; this has led to the production of lower priced ceramics with better properties. This successful development is useful, but it is not good enough for the expected boom in ceramic materials in industry. The main problem concerning ceramic use in industry is that machining of ceramics is still very expensive. This prohibits the replacement of metal parts with ceramic parts in the following industries: automotive, aerospace, and nearly all other industries that use machined parts.

Silicon nitride, silicon carbide, and zirconia are a few among many advanced structural ceramics with high strength at elevated temperatures, resistance to chemical degradation, wear resistance, and low density. Improvements in processing advanced materials will benefit the following: electronics, telecommunications, optical systems, sensors, catalysts, bone replacements, heat exchanges, and heat engines. The main applications for these advanced materials are listed in Table 1.[1]

According to the Office of Technology Assessment (1988), the market for structural ceramics will double every 5 to 7 years and increase to five billion dollars by the year 2000. However, if this estimate is to be met then the following challenges must be addressed:

- The technical feasibility of replacing metal engine parts with ceramics must be demonstrated;
- Ceramic material and component reliability must be improved so that more accurate design predictions can be made;
- Improved nondestructive evaluation must be developed; and
- Fabrication cost must be reduced.[2]

In some of these areas, progress has been realized. Large national programs have been developed in the United States, Germany, and Japan to research the methodology of replacement of metallic engine parts. There are several benefits to replacing metal parts with ceramic parts: ceramic parts are lighter, ceramics can withstand higher temperatures, and the wear rate of ceramics is much lower than in the case of steel and other metals. The metal parts which have been replaced so far by ceramic parts are valves, injection pins, bearings, and seals.

For all these parts, nondestructive control is a challenge. Ceramics are brittle materials; therefore, surface and subsurface damages are induced during machining operations. To detect these damages new techniques must be developed; these techniques are usually expensive from the point of view of equipment utilization and time consumption. New approaches have been developed including new sensors and acoustic emission techniques.

The greatest challenge of increasing the use of ceramic parts is reducing the cost of machining operations. Ceramic parts cannot be obtained at a very near final shape because the sintering process is at high temperatures, which makes it difficult to control the variation of the parts' dimensions with the temperature. As a result, mechanical operation (machining) is necessary to obtain the final shape and dimensional accuracy. Ceramics are hard, brittle, and have high abrasivity (some of them are used as abrasives); therefore, the only efficient tool to machine these ceramics is made from diamond. Diamond tools are expensive. It is necessary to work with very low depth of cut be-

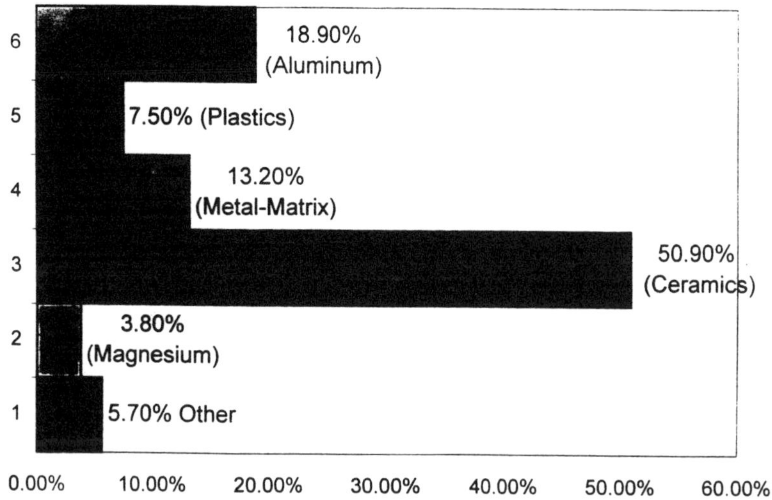

Figure 1: Will Materials Improve the Efficiency of Engines by the Year 2000? (Source: Ward Auto World)

cause of the damages introduced by any mechanical operation. This induces very low productivity; increasing productivity is a big challenge. One of the most important applications of ceramics is in engine components. Ceramics allow the production of lighter components, greater fuel efficiency, better wear capability, and better engine miles in light trucks and cars.

Figure 1 predicts which material will improve the efficiency of engines by the year 2000. The predicted trend of higher use of ceramics will bring about greater focus on CAN and diamond wheels for optimizing the processing of highly hardened ceramic materials.[3]

The main objective of this book is to present abrasive machining of ceramic materials in order to obtain a better understanding of ceramic materials, machining processes, and their applications. This book will not discuss the manufacturing of ceramics, as many books have been written in this field. It is important to have an idea about the general processing of ceramic materials -- from starting materials to final product -- this is presented in Figure 2.

Chapter 1 presents properties of ceramics with emphasis on wear mechanism, selection criteria, microstructure reinforcement of ceramics, and

Table 1: Current and Future Market Advanced Ceramics [Wortendke1989]

* Aerospace -advanced electronics -bearings -combustors -fuel cells -fuel systems and valves -high temperature auxiliary power units, low weight components for rotary equipment such as starters -seals -structures -thermal protection systems -turbine engine components * Automotive -advanced reciprocating engines -advanced rotary regenerators -catalytic converters -drivetrain components -electronic substrates -fixed boundary recuperators, -iuel injector components -low heat rejection diesels -turbines -turbocharger rotors -valves and valve seats -waterpump seals * Bioceramics -artificial teeth bones and joints -heart valves	*Chemical Process Industry -catalysts and igniters -mechanical seals -nozzles -radiant tubes and burners -recuperators -reformers -refractories -valve components *Defense -armor -bearings -engine combustor sections -gun barrel liners -improved armor -optics -rocket nozzles -SDI (optical/heat transfer properties) -submarine shaft seals -tank power trains *Electric Power Generation -bearings -ceramic gas turbines -cogeneration -filters (gas clean-up) -fuel cells (solid oxide) -high temperature -components -reactor pump seals * Electronics -advanced multilayer integrated package -electro-optic packaging	-multilayer capacitors -optical wave guides -pressure and gas sensors -semiconductor packages -substrates -superconductors *Environmental -advanced components and systems for environmentally harsh processes -filters and scrubbers -incinerator liners and afterburners -radiant burners and boilers -wastewater treatment *Metals Processing -boats -burners -crucibles and ladies -cutting tools and dies -insulation -molten metal filters -seeded gel abrasives for metal/ceramic finishing *Oil Industry -bearings -blast sleeves -flow control valves -pumps -refinery heaters

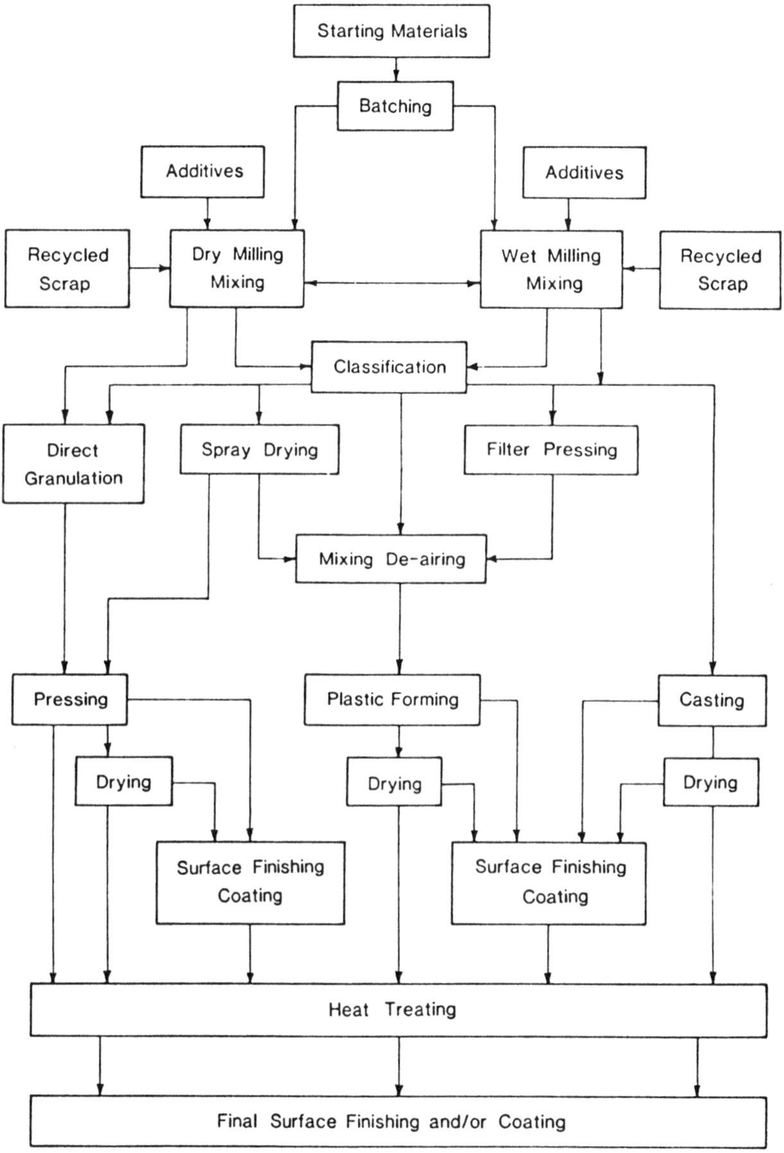

Figure 2: General Processing Flow Diagram Illustrating Different Possible Processing Paths from Starting Materials to Final Product.[4]

fundamental properties.

In order to understand the machining process of brittle materials, it is necessary to have some knowledge of the mechanics of these materials. Chapter 2 introduces the basics of deformation and fracture mechanics of ceramics, crack initiation and propagation, and indentation processes.

Chapter 3 gives a general view of abrasive processes: typology, and tribology of abrasive processes, single and multi point scratch test, and the general model of abrasive processes. Very importantly, at the end of Chapter 3, surface topography and surface integrity of ceramic materials are presented.

The most familiar process for the machining of ceramics is grinding. Chapter 4 presents grinding beginning with fundamentals of grinding, grinding tools, and conditioning of the grinding tools. Both the wear mechanism of grinding wheels and cooling lubrication are fundamental in establishing efficiency and productivity of grinding of ceramics. In today's world, environmental aspects are very important, and in Chapter 4 some of these aspects of grinding of ceramics are presented. The properties of ground ceramic surfaces and the main types of grinding machines are also presented.

In Chapter 5, two abrasive processes: honing and superfinishing of ceramic materials are discussed. For honing and superfinishing, typology of process, tools, machines, and technologies are presented, and this methodology will be used for all the operations.

The final chapter, Chapter 6, is dedicated to loose abrasive processes: lapping and polishing. A new type of lapping, ultrasonic lapping is also presented.

After reading this book, it will be much easier to understand the advantages of ceramic materials, and the reality that abrasive machining is not an easy operation. Ceramics are part of a category of materials named "difficult to machine materials."

This book is a synthesis of the knowledge in this field from the United States, Germany, Japan, and the United Kingdom, countries with an important contribution to the field of machining of ceramic materials.

Toledo, Ohio
March, 1999

Ioan D. Marinescu

REFERENCES

1. David W. Richerson, *Modern Ceramic Engineering*, Marcel Dekker, New York,1992
2. *NIST Special Publication 834*, Ceramic Machining: Assessment of Current Practice and Research Needs in the United States, NIST, 1992
3. Kevin V. G. Bevan, *Grinding Partnership 2000*, Finer Points, Vol. 9 No. 1, 1997
4. James S. Reed, *Principles of Ceramic Processing,* John Wiley & Sons, Inc., NewYork, 1995

Contributors

Toshiro Doi
Saitama University
Urawa
Saitama, Japan

Ichiro Inasaki
Keio University
Yokohama, Japan

Toshio Kasai
Saitama University
Urawa
Saitama, Japan

Fritz Klocke
Institut fur Gesteinshuttenkunde der RWTH Aachen
Aachen, Germany

Yin Ling
Tinjin University
Tinjin, China

Ioan D. Marinescu
The University of Toledo
Toledo, Ohio

Hans R. Meyer
Ernst Winter & Sohn
Norderstedt, Germany

Mariana Pruteanu
The University of Toledo
Toledo, Ohio

Brian Rowe
Liverpool John Moores University
Liverpool, England

Junji Shibata
Shibaura Institute of Technology
Omiya City
Saitama, Japan

Gunther Spur
Technical University of Berlin
Berlin, Germany

Rainer Telle
Institut fur Gesteinshuttenkunde der RWTH Aachen
Aachen, Germany

Hans K. Tonshoff
Institute for Production Engineering & Machine Tools
University of Hannover
Hannover, Germany

Hans G. Wobker
Institute for Production Engineering & Machine Tools
University of Hannover
Hannover, Germany

NOTICE

Contents

1

Properties of Ceramics

Rainer Telle

1.0 INTRODUCTION

Although ceramic materials for technical application have been known for more than two hundred years, especially-designed synthetic ceramics, unlike traditional materials in composition, microstructure, and properties, have been developed since approximately 1970. Whereas silicate ceramics and refractory materials are basically derived from natural minerals and manufactured by comparatively simple processing steps, this new class of materials, the "advanced", "high-tech," or in Japanese terms "fine" ceramics require an entirely different fabrication route starting from chemically well-defined, fine, highly-purified, and artificial raw materials. These materials have been created for distinct applications in which other conventional materials like metals or polymers have failed. Due to the large variety of chemical, electrical, biological, and mechanical properties that ceramics presently exhibit, there is almost no social and industrial application without ceramics (Table 1.1). In the electronic and manufacturing industries, as well as in technologies that require materials sustaining extremely high temperatures and corrosive environments, high-tech ceramics play the role of key materials; novel technologies, processes, and machines are finally made possible only by means of especially tailored ceramics.

Table 1.1: Classes of Ceramics and Fields of Application

Materials Group	Property	Application
	Compressive Strength	Bricks
Traditional	Density + Strength	Ceramic Hollow Ware
Ceramics	Density + Wear Resistance	Structural Clay Products
	Heat and Corrosion Resistance	Refractories
Structural	Hardness	Grinding Grits and Disks
Ceramics	Strength + Toughness	Engineering Ceramics
	Biocompatibility, Bioactivity	Bioceramics
	Nuclear Properties	Nuclear Ceramics
	Corrosion Resistance Catalytic Properties	Chemoceramics
Functional Ceramics	Electric Resistivity Dielectric Properties	Electroceramics
	Magnetic Susceptibility	Magnetoceramics
	Diaphaneity, Anisotropic Optical Properties	Optoceramics

Surprisingly, this development was initiated by metal scientists or - more precisely - by powder metallurgists rather than by traditional ceramists. The reason for this is that the manufacturing route used for the production of metallic parts by powder molding and compaction followed by subsequent consolidation by a heat treatment, i.e. sintering, was investigated fundamentally since the turn of the century for steel, refractory metals, and since 1920, for hard metals which could not be casted or molded otherwise. With regard to natural multicomponent raw materials and comparatively simple chemical systems, the basic understanding of these originally "ceramic" processing procedures was much easier than in the case of traditional ceramics. Thus, the break through in the science of sintering was achieved in 1970 to 1980 yielding knowledge on the reproducible production of high-performance powder and metallurgically-prepared parts. Being easily transferred to ceramics of "simple" composition, the foundation for the development of tailored microstructures with as-desired properties was created. The simultaneous development of high-toughness zirconia and highly wear-resistant silicon nitride ceramics indicated a promising way to overcome the most important disadvantage of traditional ceramics: their brittleness. The capability of the entire control of residual porosity together with the so-called transformation toughening by zirconia as well as the science of phase relationships in multicomponent systems that yielded the opportunity to synthesize silicon nitride -based high-temperature materials initiated a world wide boom in ceramic research and development. Figure 1.1 shows one of the many predictions for future markets and turn-over opportunities related to the various branches of application. To further the collaboration between industry and research institutes, large investments in ceramic development and research programs by industrial countries have been implemented. As a consquence of these efforts, a novel understand of matter was achieved in the field of fracture mechanics yielding insights in toughening phenomena and reinorcing strategies for static and dynamic load. Models for the prediction of the long-term behavior of complex parts have been derived, and the term "fatigue" was described in respect to brittle fracture originating from microstructural defects which have been quantified by means of statistics. High-resolution electron transmission microscopy gave information about the internal structure of grain boundaries and thus enhanced the development of creep resistant high-temperature silicon nitride based monoliths. Micro- and nanoscaled molding techniques brought about new possibilities to manufacture electrically and electronically active ceramics: ubiquitous components of modern electronic devices. Additionally, the invention of the ceramic high-temperature superconductors contributed to the tremendous increase in materials research.

Not in all cases, however, have ceramics been able to meet the sometimes extraordinarily high demands of the applying industry. The progress in understanding the particular influence of the manufacturing procedures to the microstructure and mechanical properties was slower than expected. The market did not develop as projected due to the lack of reliability of the ceramic

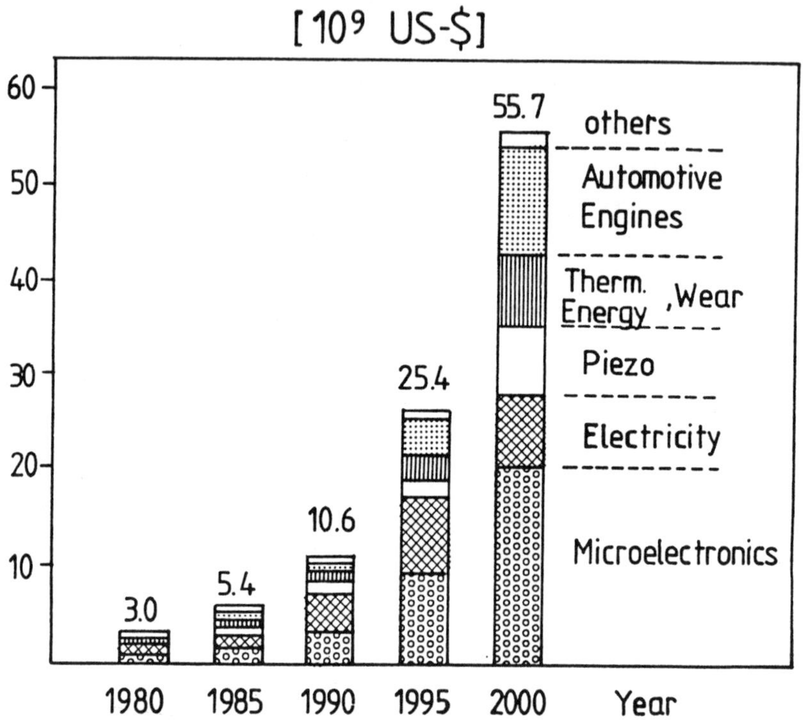

Figure 1.1: Market Forecast for High-performance Ceramics. (Courtesy, Hoechst 1988)

parts and due to problems in its acceptance by construction engineers. Furthermore, the request for high quality products led to high-cost raw materials and products which had some time to compete with metals or even with polymers. Thus, some strategic investments by big companies came too early and turned out risky, especially in Europe, but the competition with Japan and the United States, as the two most important providers of advanced ceramics, was severe. Imports from Japan where part development and production was strongly supported and funded by the government, were sometimes preferred to imports from the European providers.

Today (1994-96), the worldwide economic problems govern the entire market. The exponential increase in market demand for high-tech ceramics is stopped, even in Japan (Figure 1.2). New machining techniques for shaping sintered parts to final dimensions, however, have significantly lowered the costs of structural parts. On the other hand, more accurate analyses of the mechanical properties being really requested for ceramics in automotive engines show a clearly lower level of performance being necessary than aimed at before, hence the dramatically decreasing costs in raw materials, processing, and final machining. Together with new fields of application (e.g. tools for semiconductor fabrication, Figure 1.2) these facts bring about new prospects for high-tech ceramics in the near future, because they are still what they have been designed for: key materials of a modern technology.

One field of application that developed with an increasing intensity, as was predicted, is related to the excellent wear behavior of ceramics: the application as cutting tools and grinding grits. In the last decades, ceramic grinding and cutting tools initiated a strong impact to the manufacturing technology of metals. New turning and milling machines were developed; these required high hardness and toughness materials that were capable to work at very high feed rates, speed, and therefore at high temperatures yielding smooth surfaces free of damage. Numerically controlled manufacturing techniques, the strong increase in process reliability, and quality reproducibility were made possible by especially developed alumina and silicon nitride ceramics. The most important step towards high performance ceramics was the basic understanding of fracture initiating mechanisms and strategies to minimize the material-inherent brittleness.

Functional ceramics in the sense of components of electronic or electric devices such as capacitors, piezo ceramics, chip carriers, insulating housings, spark plugs, etc., are prepared by thin film techniques or extrusion processes, sometimes followed by glazing, yielding suitable surface roughness and sufficient accuracy in final dimensions. Grinding and polishing operations are usually not requested as an additional finishing step. Therefore, this class of ceramics will not be treated further in the following paragraphs. Structural ceramics, however, which have to sustain external loads and to fit into a mechanically active construction consisting of a large variety of different materials, e.g. an engine, must strictly meet the desired final dimensions and surface qualities to guarantee the requested properties in serv-

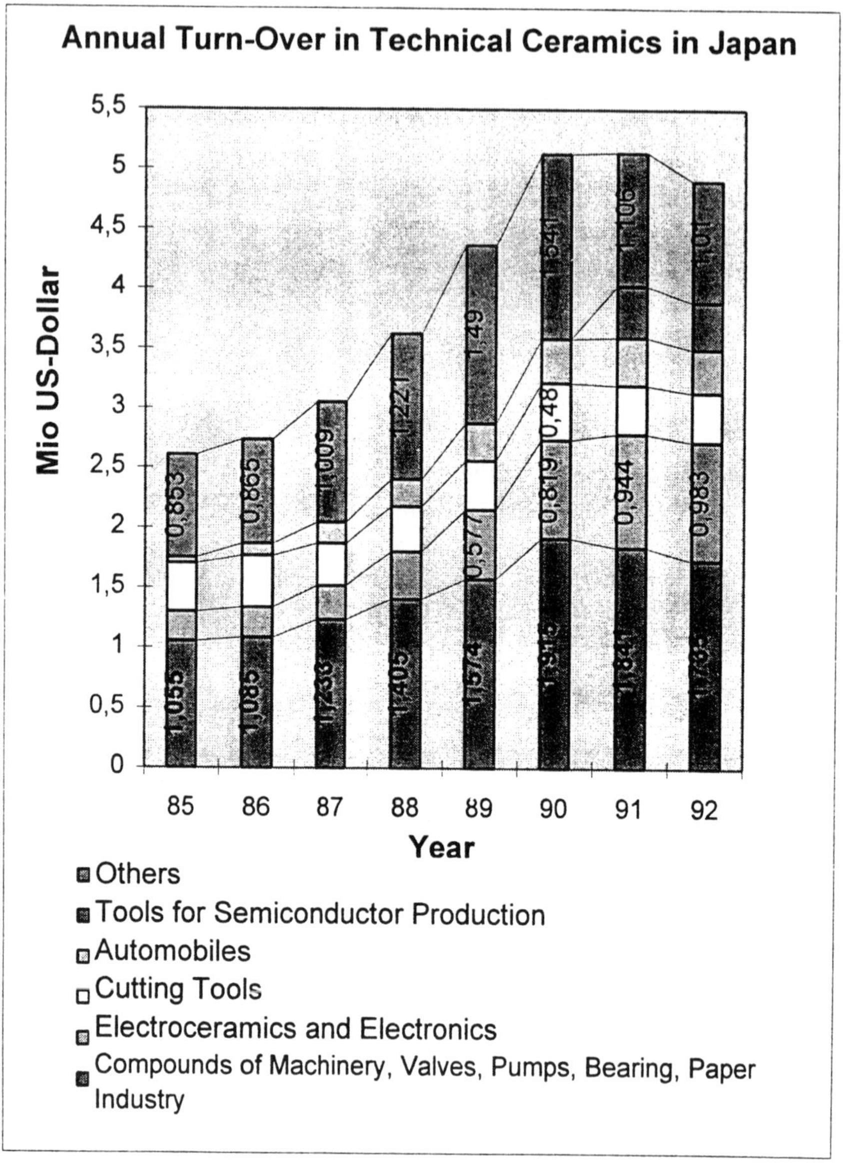

Figure 1.2: Fine Ceramic Market Development in Japan as it is. (Courtesy, Hitachi Metals Inc., 1996)

ice with sufficient reliability and life time.

Since hardness, stiffness (Young's modulus), toughness, and strength are the most important mechanical properties of structural ceramics determining the wear resistance, the goal of this article is to introduce one to the fundamentals of material-inherent properties as well as of wear mechanisms and reinforcing strategies which have been applied to technical ceramics. This is of a particular importance because grinding and polishing (*i.e.*, mechanical material removal during shaping) of ceramics which have been especially optimized to resist material removal (*i.e.*, wear in service) is accordingly difficult. These conflicting properties, ease in machining and simultaneous resistance in service, are surprisingly not yet regarded by the material developers nor by the manufacturing engineers.

Additionally, a basic understanding will be developed to enable the reader to choose suitable material combinations for appropriate applications and to understand the difficulties in manufacturing but also the risks and origins of failure during service live. Besides parts of structural ceramics, grinding grits or small cutting tools suffer basically from the same problems and can therefore be strengthened by the same methods. Another goal of this article is, however, to show the chances and the limits of a future materials development.

2.0 WEAR MECHANISMS OF CERAMICS MATERIALS

Because of their partially covalent and partially ionic chemical bonding, ceramics are extremely hard and corrosion resistant and therefore excellent wear resistant materials at both room temperature and high temperatures. One important limiting factor is, however, their inherent brittleness. High stiffness, high hardness, and consequently the brittleness, are based upon the little deformability of the crystal lattice in contrast to metals and polymers. At low temperatures, strain energy in the vicinity of a crack tip cannot be released by dislocation movement or creep. In comparison to metals, the activation energy for the movement of dislocations is so high that the ultimate fracture strength is by far exceeded. As the crystal structures of ceramic possesses lower symmetries compared to metals, even an increase of temperature closest to a melting point does not result in the activation of more than two or three dislocation slip systems. Therefore, the plastic deformability remains poor which means that the brittleness and also the high hardness persists to high temperatures. Talking in terms of stress-strain relationships, the linear elastic range of the stress-strain curve is terminated by immediate catastrophic fracture releasing the entire stored elastic strain energy (Figure 1.3). This is in particular the case if the stored elastic strain energy exceeds the work of fracture required for the formation of a new crack surface or if at a tip of a preexisting crack or microstructure inhomogeneity tensile stresses are accumulated in the order of the theoretical strength of the material.

One measure of work of fracture is related to the critical stress intensity factor K_{Ic} also denoted to as fracture toughness. The critical stress intensity factor describes a particular stress intensity at a tip of a crack which is required to make a crack propagate. The ultimate fractures strength σ_c is thus a very important mechanical parameter which describes the critical tensile or bending stress which is required to initiate the crack. For brittle materials like glasses and most of the non-reinforced ceramics, fracture strength and fracture toughness are linked by the so-called Griffith-equation:

$$\sigma_c = Y \frac{K_{IC}}{\sqrt{\pi a}} \tag{1.1}$$

This equation **Y** means a geometry factor which describes the shape and the position of a microstructural inhomogeneity, e.g., a crack or a pore, and ***a*** the maximum elongation of this particular inhomogeneity, e.g., the crack length or the pore diameter. As it will be shown later, this fracture mechanical equation does not only correlate the basic mechanical parameters but also shows the direction of a further improvement of properties and thus inherently contains the basic understanding of fracture statistics. Due to the fact that geometrical parameters are involved in the description of fracture initiation, the materials strength cannot be described as a single constant being valid for a certain product but is, instead, a function of the probability of the spatial and size distribution of supercritical microstructural defects.

Since the Young's modulus is given by the stress-strain relationship, another mechanical property is still missing which is very important for

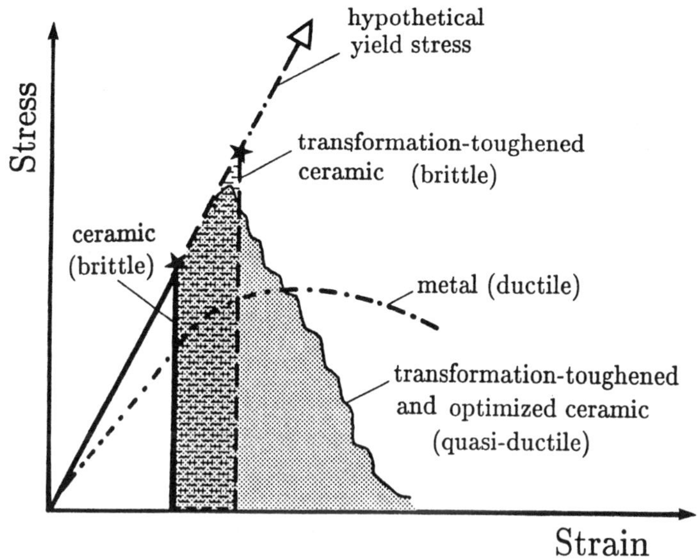

Figure 1.3: Stress-strain Curve for Ceramics and Metals

cutting tools and grinding materials: the hardness. The hardness is defined as the resistance of a material against the penetration by a testing device. From the viewpoint of physics, the hardness is related to the lattice properties of crystals and can be therefore derived solely from interatomic forces. In practice however, the hardness is a combined property which involves microstructural characteristics such as porosity, grain size, grain boundaries, dislocation movement, cleavage fracture, and other geometry- and temperature dependent bulk properties.

In the case of cutting tools and grinding grits, these mechanical properties cannot be discussed at room temperature alone. Due to the very small surface area being in contact with the material machined, very high temperatures may develop at the interface between the work material and the cutting material. Accordingly, the temperature dependence of strength, fracture toughness, hardness, and Young's modulus have to be discussed as well as other thermo-physical properties such as thermal expansion and thermal shock behavior. Additionally, at the contact between ceramics and metals, chemical reaction may be initiated under the high contact pressure and the high temperatures.

Although many theories in fracture mechanics have been developed to describe the service behavior of brittle materials, the prediction of the wear properties from the static mechanical properties is not easy since the interaction between wear couples is manifold. Usually, tool and work material is not simply in contact with each other but a third medium such as cooling agents, lubricants, abrasive additives, chips of the work piece, hard material, and certain atmospheres may form an environment which contributes strongly to the particular wear mechanisms. Taking this third medium into account, one can distinguished four basic wearing effects:

- surface fatigue,
- abrasion,
- adhesion, and
- tribochemical reactions.

Figure 1.4 summarizes schematically the basic interactions, mechanisms, and effects that can be observed in wear couples. Material removal by formation of adhesive bridges between tool and work material, crack formation by delamination, and opening of grain boundaries are visualized for the case of sliding wear in Figure 1.5. From both figures, it becomes evident that chemical interactions contribute to the wear behavior in addition to the mechanical interaction. In the following paragraphs, the particular wear mechanisms are described in detail.

Abrasion

The term "abrasion" comprises all groove-forming mechanisms on the surface of a material by micro chipping and micro ploughing. This mechanism is a consequence of a high ratio of the hardness of the tool material and the work material. An estimation of this hardness ratio must, however, consider the dramatic decrease with temperature in ceramic materials while metals may reveal an increasing hardness by work hardening effects. Additionally, the dynamic hardness of a metal may be considered higher than the hardness measured by indentation techniques due to incorporated carbide particles. Generally, the ratio of tool material hardness to work material hardness should not be less than 1.5 to 1.7.

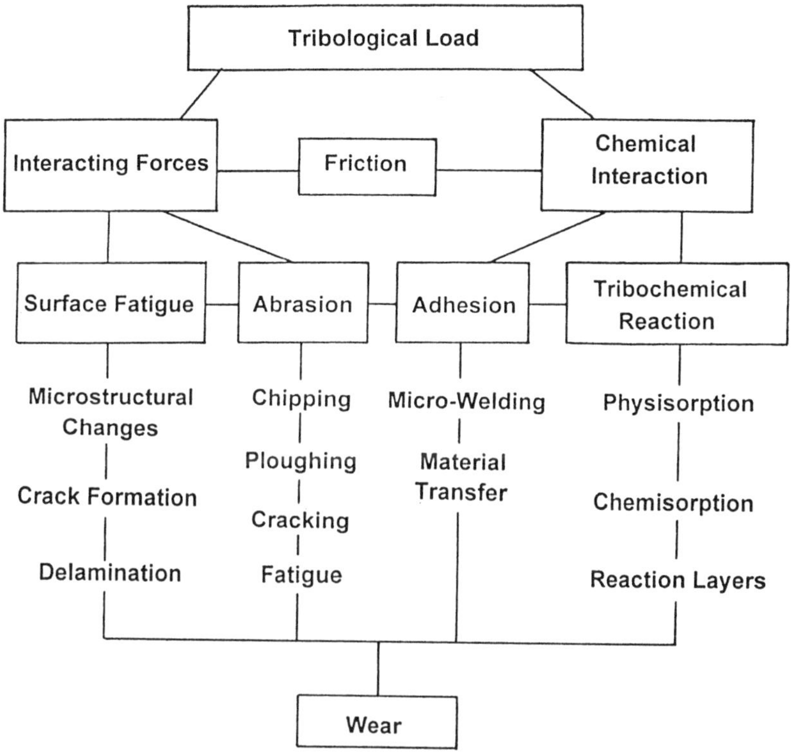

Figure 1.4: Principal Mechanisms and Effects of Wear

Although grooving is an evidence for plastic deformation, the pull-out of chips and particles from the microstructure of both cutting tool and work material must also be considered. Accordingly, the local fracture toughness must be taken into account. Model wear tests performed on a large variety of material couplings indicate that a correlation of the wear amount, or wear rate, to both hardness and fracture toughness is generally possible. Several empirical formulae have been developed from pin-on-disc tests relating the wear resistance to fracture toughness times hardness of several exponents. Table 1.2 shows some more important empirical formulae that have been proved to fit well with experimental results. The Evans-Wilshaw-Equation

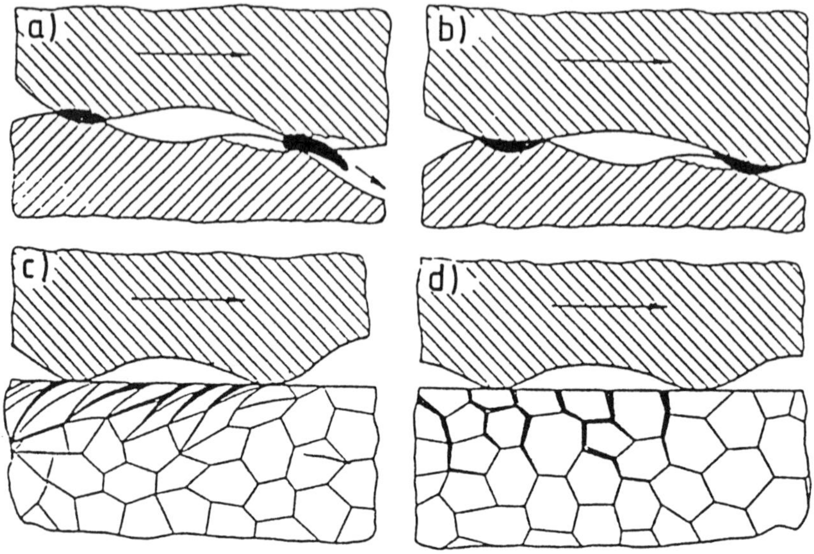

Figure 1.5: Surface Effects of Sliding Wear (after Zum Gahr)
a), b): Material; removal by adhesive bridges and their chip-off
c): Crack formation by delamination (grain boundary sliding and cleaving)
d): Crack formation by grain boundary opening

is accepted most for ceramic-ceramic pairs. It is evident from this expression that the high hardness must always be combined with a high fracture toughness to yield suitable wear properties. Surprisingly, the infl-uence of fracture toughness is more important than the hardness as can be concluded from the particular exponents.

Surface Fatigue

The term "surface fatigue" covers the combination of wear mechanisms, operating within a surface layer of several micrometers in thickness, that are caused by tangential shear stresses at the material surface as well as by iterative impacts. The surface fatigue is characterized by crack formation

Table 1.2: Empirical Relations Between Wear Resistance Factor *R* and Mechanical Properties

$R = W^{-1} \sim K_{IC}^{2}\ H^{3/2}$	Hombogen
$R = W^{-1} \sim K_{IC}^{3/4}\ H^{1/2}$	Evans & Wilshaw
$R = W^{-1} \sim K_{IC}^{2}\ H^{-1/2}$	Zum Gahr
$R = W^{-1} \sim K^{4/3}\ H^{-1/9}$	Ruff & Wiederhorn

R = inverse volume loss *W*

along the grain boundaries or cleavage planes starting at the surface and progressing continuously to greater depth by subcritical crack growth. This wear mechanism is especially detrimental since the ultimate depth of the cracks cannot be estimated by looking at the surface of the material. Upon service, however, they can grow slowly to more than 100 μm extension becoming the rupture-initiating failure of the part by reaching the critical length as given by the Griffith Equation. Tool failure by surface fatigue is a characteristic for cycling compressive and tensile loading as observed by, e.g., intermittent cutting operations or by reverse sliding of seals. A similar effect may cause the pull-out of ceramic grinding grains if the particle interface to the binder is slowly and steadily subjected to cycling loads and debonds.

Subcritical crack growth by repeated impact may be supported by iterative thermal shock. In case of the grinding operation, for instance, the temperature during the milliseconds of cutting action may give rise to a strong temperature increase at both the cutting tip and the work material surface area in contact. Local stresses may develop due to the accordingly introduced thermal gradients, due to an isotropy effects, or due to differences in thermal expansion of the various compounds. Since crack growth is the basic mechanism of this wear effect, a high fracture toughness, a high thermal conductivity, and the low thermal expansion coefficient of the ceramic material is requested.

Adhesion

Adhesion comprises the chemical interaction between the wear materials. Depending on the affinity between cutting tool and work material, a local joining or even welding of both materials may occur. The binding forces may become so high that chips may be pulled out or chipped off from the work material, e.g., the metal debris of the work material may adhere at the ceramic cutting tool. This effect is also known as material transfer and is responsible for the fact that the cutting tool is not in contact any more with the work material. Figure 1.6 shows several models to explain the effect of adhesion. Besides clamping as a mechanical effect, diffusion of atoms and ions, electron transfer, or dielectric polarization effects are considered to be responsible for the development of chemical bonding.

Tribo-chemical Reactions

Tribo-chemical reactions between wear couples, e.g., tool material and work material may occur if both materials are not in a thermodynamic equilibrium, especially at higher temperatures. In the contact area, a new reaction product is formed which is usually removed together with the chip or adheres at the cutting tool material. These reactions may also be caused by environmental materials like lubricants or atmospheric gases. The chemical wear becomes visible by very smooth and lustering surfaces or by deformation of built-up cutting edges. To avoid tribo-chemical reactions, appropriate tool material selections may be recommended as well as lower cutting powers to avoid the generating of high temperatures.

Combined Wear Mechanisms

Of course, the above-mentioned mechanisms do not occur separately but in combination with each other where they are not acting additively but multiplicatively. Environmental material like lubricants, gases, or tribo-chem-

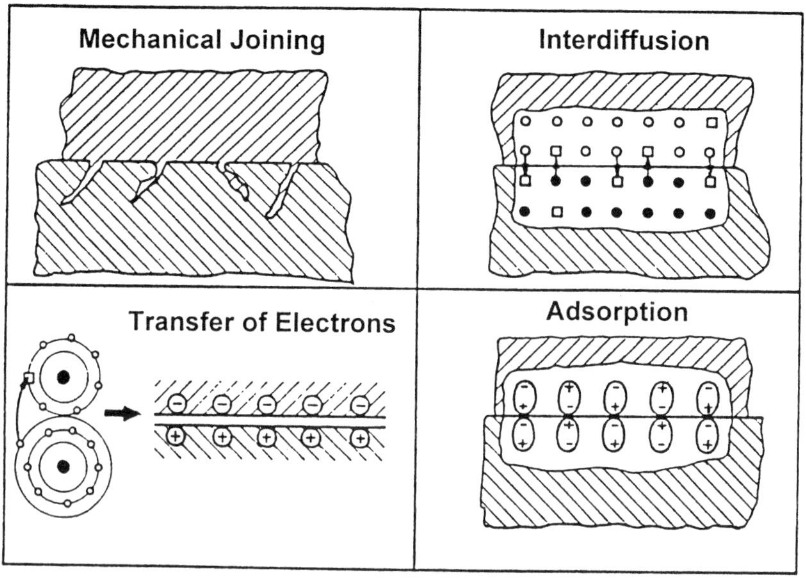

Figure 1.6: Possible Reasons for Adhesion (After Zum Gahr)

ical reaction products may infiltrate surface cracks opened by dynamic fatigue, possibly initiating new stress corrosion mechanisms and therefore enhancing the subcritical crack growth. Similarly, abrasion may be drastically accelerated if the surface of the material is partially dissolved by chemical attack or if the grain boundaries are weakened. Furthermore, surface fatigue may contribute to enhanced abrasion by weakening the grain boundary strength by a cycling load that facilitates the pull-out of single particles.

The combination of adhesion and tribo-chemical reaction causes even more severe wear problems. Figure 1.7 shows an example where both materials have adhesive contact at the apices of the surface roughness while including reactive environmental material in the adjacent concave surface areas. Chemical reactions may now result in the formation of a passivation layer on both surfaces preventing a further chemical attack. Together with the material removal by adhesive contact, however, this passivation layer may be destroyed whenever it is newly formed. Consequently, the concave structures are filled with debris acting as very small abrasive particles enlarging the concave structures by interactive microgrinding effects. This synergetic wear

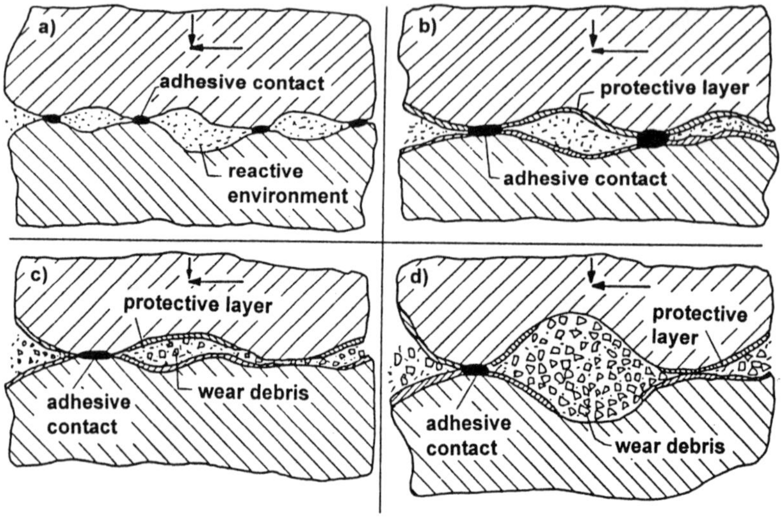

Figure 1.7: Synergetic Effects of Combined Adhesion and Tribo-chemical Reaction (after Zum Gahr)

a) adhesion
b) formation of passivation layer
c) formation of debris by adhesive pull-out
d) removal of passivation layer by wear debris

mechanisms result in a very fast pull-out of the protruding hard material grains.

3.0 FUNDAMENTAL PROPERTIES AND SELECTION CRITERIA

For ruling out wear resistant materials for special applications, the specific mechanical properties such as hardness, toughness, strength, thermal conductivity, oxidation resistance, and chemical inertness against the work material must be considered as functions of temperature in service. For this, hardness, thermal conductivity, oxidation resistance, and chemical inertness are considered intrinsic properties that can be assigned to a particular chemical compound; they follow the known rules of mixtures if another compound is added to form a composite material. Fracture toughness, fracture strength, and consequently also thermal shock resistance are basically influenced by the microstructure and can therefore be modified by certain optimization techniques. This chapter is devoted to the intrinsic properties whereas the improvement strategies will be addressed in the chapter: Reinforcing mechanisms.

Hardness

It has been shown already that, besides fracture toughness, hardness is the property determining the resistance against abrasive wear. Figure 1.8 shows the temperature dependence of hardness for some important ceramic materials in relation to diamond and cubic boron nitride (CBN). Due to its perfect covalent bonding, diamond is the hardest natural and synthetic material known. Theoretically, other compounds have been predicted by calculation of interatomic forces having a hardness superior to that of diamond. Compounds like C_3N_4 are, however, not stable under technically available pressures and are therefore only hypothetical candidates for hard materials but nevertheless investigated as coatings on silicon nitride substrates. Although diamond is a high-cost product, cutting tools made of polycrystalline diamond or grinding grits consisting of diamond particles are widely used for grinding, milling, and machining treatments of ceramics as well as metals. Because of its metastability under normal pressure, diamond has the disadvantage of transforming to the stable graphite phase at temperatures above 500-600 °C. Upon transformation from the cubic to the hexagonal modification with a weakly bonded layered structure, diamond undergoes a lattice softening which causes a dramatic decrease of hardness. A similar behavior is observed of cubic boron nitride which is also a high-pressure compound with the same structure like diamond. It also turns to hexagonal boron nitride (hBN) graphite structure and shows, therefore, the same decrease in hardness but at much lower temperatures. The material ranking at the third order is boron carbide, B_4C, which does not undergo phase transformation. It is followed by silicon carbide, SiC, silicon nitride, Si_3N_4, and finally by a series of transition metal borides and carbides which have,

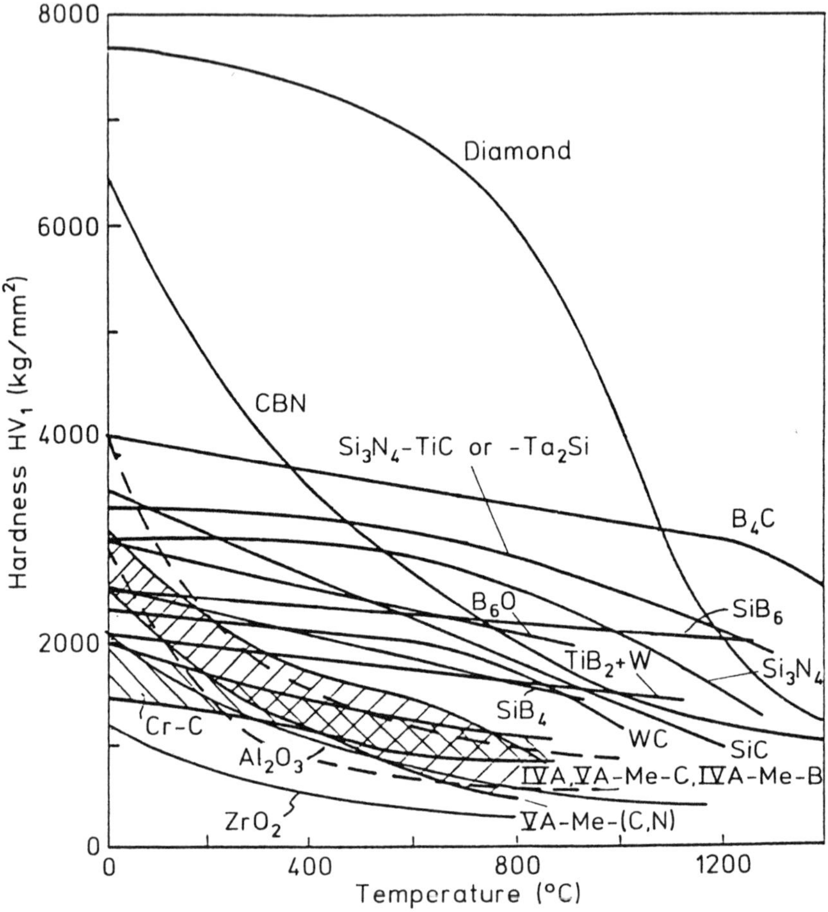

Figure 1.8: Temperature Dependence of Hardness of Ceramics

however, only 20 to 25 % of the hardness of diamond and 50% of the hardness of boron carbide. The first oxide ceramic of interest is boron suboxide (B_6O) which is technically unimportant up to now followed by alumina (Al_2O_3), and spinels ($MgAl_2O_4$) ranging at 2000 kg/mm^2 and less (Figure 1.8). In comparison to these materials, zirconia (ZrO_2) is rather soft with the hardness of 800 to 1100 kg/mm^2 at room temperature and the strength further decreases upon heating. Zirconia, however, is a very important compound in oxide ceramic composites being responsible for a strong increase in fracture toughness as will be shown later. Another grinding and polishing material, silica (SiO_2) starts with a hardness on the order of 600 kg/mm^2 but shows a transient sudden increase in hardness at 573 °C to 1500 kg/mm^2 due to the reversible transformation to a high temperature structure.

Fracture Toughness

The inherent fracture toughness of single phases in terms of therefore isolated particles depends strongly on the crystal lattice and the interatomic forces determining the bonding of the particular cleavage planes. Particular lattice planes being densely occupied by an electrostatically equivalent amount of anions and cations (typical for rock salt) or weakly bonded planes separating low-energy substructural units with internally saturated bonds (typical for clays and micas) may be preferential paths for cleavage. Alumina, for instance, shows a preferential cleavage along the rhombohedral planes, whereas the other planes exhibit an irregular conchoidal rupture surface like glass. This can be observed sometimes upon grinding of alumina ceramics if single particles are partially pulled out. Boron carbide, silicon carbide, zirconia and silicon nitride particles usually show a conchoidal fracture surface where, e.g., high-temperature superconductors exhibit a pronounced cleavage along the basal plane. In grinding tools, single diamond grains fail sometimes by perfect cleavage along the pyramidal (111) plane whereas CBN fractures preferentially along the rhombohedral (101) plane. In the case of polycrystalline ceramics, however, the fracture toughness is strongly affected by the micro-structure, i.e., by grain size, grain shape, intergranular phases, and residual stresses influencing the crack propagation. Additionally, the measures of fracture toughness are extremely dependent on the testing procedure.

In Table 1.3, some data of the critical stress intensity factor $_{K.C.}$ are listed for single phase materials. Extraordinarily high values are reported for CBN, diamond, tungsten carbide, and titanium carbide. In case of the superhard materials, this data have been calculated or measured by indirect methods since appropriate test samples of diamond or cubic boron nitride are not available. In comparison to hard metals (Co-bonded WC), the fracture toughness of the pure hard materials such as SiC, Si_3N_4, and Al_2O_3 are rather poor. As mentioned before, an increase of fracture toughness can be obtained by a tailored microstructural design in multicomponent ceramics where a doubling of the values is not unusual. This is, for example, evident in case of alumina ceramics which are reinforced with zirconia or in the case of SiC-TiB_2 composites (Table 1.3).

Fracture Strength

The temperature dependence of the fracture strength is documented for some hard materials in Figure 1.9. The values reported here are only valid for some particular microstructures. In general, room-temperature strength and high-temperature strength depend very strongly on the size distribution of microstructural inhomogeneities and intergranual phases. In the case of silicon nitride ceramics, for instance, materials have been developed with a room-temperature strength exceeding 1000 MPa and maintaining this value up to 800 to 1000°C followed by a drop to 600-800 MPa at higher temperatures. Other silicon nitride materials may start with the lesser strength of 800 MPa but may maintain this value up to more than 1200°C. This depends not only on grain size and related flaws but also on the volume, fraction and glass transformation temperature of the intergranual phases. As shown in Figure 1.9, the strength of ceramics remains usually constant up to temperatures of 600 to 800°C followed by a more-or-less decrease. It should be mentioned that in case of nonoxide ceramics, the values reported in Figures 1.8 and 1.9 are only valid for non-oxidizing atmospheres.

Certainly, oxidation of the ceramics along the grain boundaries is followed by a strong stress corrosion-induced crack propagation during service and therefore by a significant decrease in residual strength at temperatures above 600°C. Besides silicon nitride and sialons, silicon carbide be-

Table 1.3: Mechanical Properties of Ceramics

Materials	Fracture Toughness K_{Ic} [Mpam$^{1/2}$]	Bending Strength σ_b [Mpa]	Hardness HV1 [Gpa]	Young's modulus E [Gpa]
cubic Boron nitride CBN	15 - 17 4.5 - 6	700 - 1000	70 - 90	700 - 720
Diamond	9 - 10	1000 - 2000	80 - 100	910 - 1050
Boron carbide B_4C	2.5 - 3.6	250 - 450 (600)	25 - 45	420 - 450
B_4C + SiC	2.5 - 4.5	300 - 600 (1200)	25 - 35	420 - 450
B_4C+ TiB_2	3.5 - 5.5	450 - 800 (1100)	25 - 35	430 - 450
Silicon carbide SiC				
hot-pressed	2.5 - 4.0	300 - 450 (800)	18 - 25	410 - 450
SiC + TiC	5.5 - 6.5	450 - 650 (800)	18 - 20	
SiC + TiB_2	5.5 - 8.0	450 - 800 (900)	18 - 25	420 - 480
Tungsten carbide WC (without binder)	5.5 - 7.5	400-500	20 - 28	700 - 720
Titanium carbide TiC (without binder)	3.5 - 4.5	240 - 280 (400)	25 - 30	450 - 470
Titanium diboride TiB_2	4.5 - 6.0	450-600	22 - 30	550 - 570
Silicon nitride				
sintered: SSN	5 - 5.5	400 - 650 (1100)	15 - 18	150 - 320
hot-pressed: HPSN	5 - 8	700 - 1000 (1300)	15 - 20	250 - 340
gas-pressure sintered: GPSN	6 - 9	600 - 900 (1200)	15 - 20	280 - 340
reaction-bonded: RBSN	2 - 4	250 - 300	10 - 12	80 - 220
SiAlON				
$Si_3N_4 + Al_2O_3 + Y_2O_3$	3 - 8.5	600 - 900	12 - 16	200 - 320
Si_3N_4 + MgO	3 - 6	400 - 600	12 - 16	200 - 300
Alumina Al_2O_3	2.5 - 4.5	250 - 400 (600)	18 - 20	300 - 330
ZTA: $Al_2O_3 + ZrO_2$	4.0 - 9 (15)	400 - 800 (1200)	15 - 18	200 - 250
Zirconia ZrO_2				
cub. ZrO_2	2.8 - 3	240 - 260	10 - 15	100 - 200
MgPSZ	6 - 9	500 - 700	9 - 11	200 - 210
Y - TZP	4.5 - 9.0 (20)	400 - 1200 (2000)	8 - 12	200 - 210

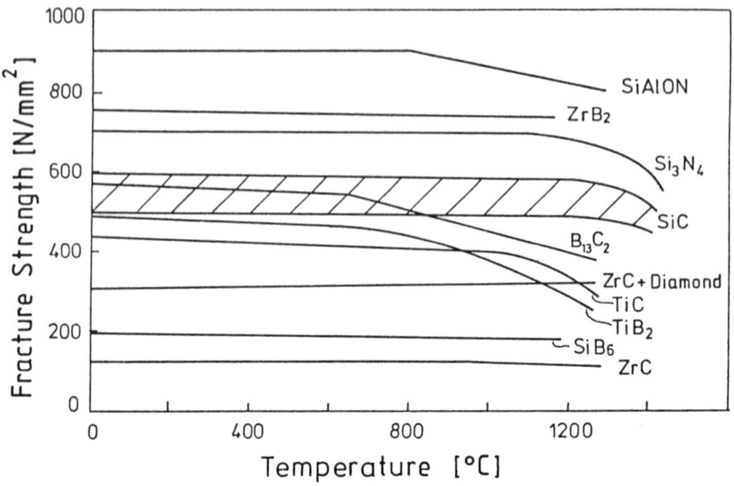

Figure 1.9: Temperature Dependence of Fracture Strength

haves best since it exhibits the best oxidation resistance maintaining its strength to temperatures exceeding 1000 °C due to the lack of glassy intergranual layers and by building up a passivation layer of silica which may even close surface cracks. The disadvantage of the glassy phase-containing materials is furthermore the plastic deformation by creep which is a significant risk of failure in silicon nitride-based tool materials.

Thermal Conductivity

The thermal conductivity plays a significant role especially in case of the selection of cutting tools, wear parts being in sliding contact, or structural parts being subjected to thermal cycling. In contrast to metals with an excellent temperature conduction, the contact temperature of, e.g., ceramic cutting tools may increase to more than 1200 °C at the cutting edge and may therefore create stresses combined with a risk of thermally-induced fracture.

In case of diamonds as grinding grits, the diamond acts as an thermal sink due to its extraordinarily high thermal conductivity.

In the case of ceramic tools, intermittent cutting, repeated impact, and grooving action of single particles results in the most critical thermal loading followed by thermal shock and thermal fatigue failure. In case of ceramic work materials, storing of heat in small surface volumes may create a local temperature increase close to or even above the melting point; this, in cooperation with the multiaxial stresses makes the observed plastic deformation understandable. Therefore, the thermal conductivity is a very important factor for the applicability of a wear resistant. From Figure 1.10, it is obvious that diamond is the material with the highest thermal conductivity due to its perfect covalent bonding. It is followed by high purity aluminum nitride (AlN) which is developed for electronic substrates and heat sinks but is unsuitable as a hard material. The next in ranking is high-purity silicon carbide followed by transition metal carbides and borides which possess transport properties like metals. As can be seen from this diagram, thermal conductivity is not a pure function of the crystal structure of the particular compounds but is also influenced by impurities and, in case of polycrystalline materials, by grain boundaries acting as barriers for phonon transport. In the same way, pores may scatter phonons and therefore cause a strong decrease in thermal conductivity. Compared to the metallically and covalently bonded borides, carbides, and nitrides, the thermal conductivity of the oxides is little. Zirconia acts even as a more-or-less perfect insulator which limits its application as wear-resistant material, although its fracture toughness is excellent.

Oxidation Resistance

For the selection of ceramics in an oxidizing high-temperature environment and, as hard materials, for working or even simply contacting metals, the oxidation resistance is an important criterion. Unfortunately, the data reported in the literature about the oxidation velocity, in particular about the rate constant, are very unreliable and show a large scattering range. This fact can be explained by either the variations in microstructures or by the particular additives and dopants which may change the chemical behavior of the

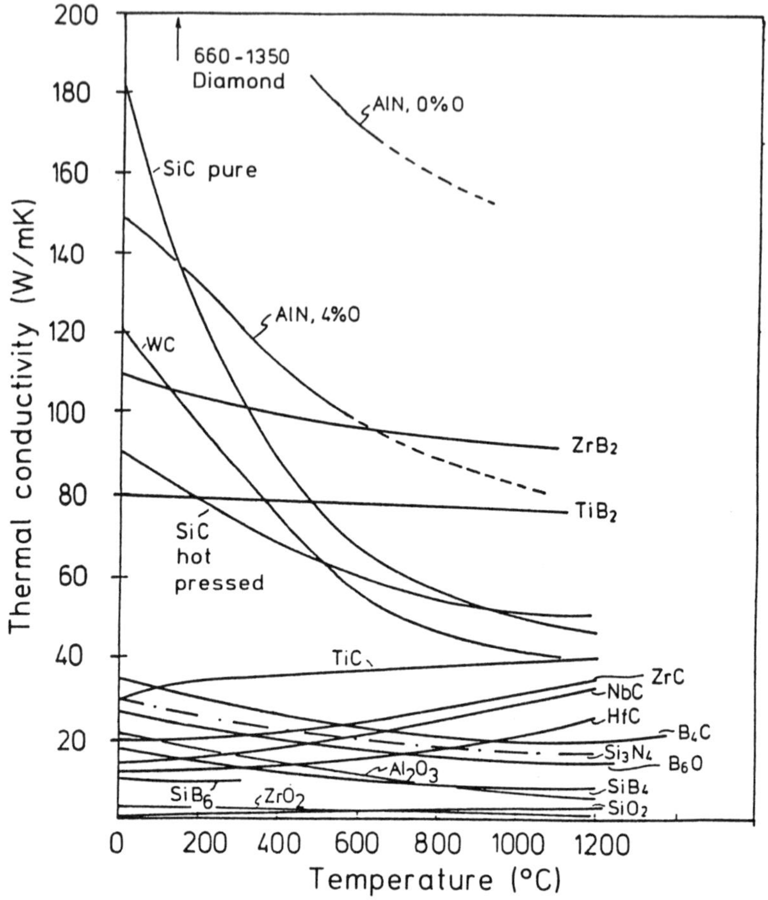

Figure 1.10: Thermal Conductivity of Ceramics Compounds. Note the Little Conductivity of Oxides Compared to Borides, Carbides, and Nitrides.

ceramics significantly. Furthermore, the characterization treatments are usually different and yield values that are not comparable. Accordingly, Figure 1.11 represents only tendencies of the oxidation resistance of various ceramics. The most insensitive materials are silicon compounds such as silicon borides, molybdenum disilicide ($MoSi_2$), silicon carbide, and silicon nitride. These materials form a relatively dense silica layer on the surface exposed to air preventing oxygen diffusion to the bulk material beneath. The stability of this oxidation layer being proved efficient in laboratory experiments is, however, in question if it is exposed to aggressive atmospheres containing alkaline volatiles, exposed to alkaline solutions, or removed by abrasive or impact wear. In this case, the material is consumed by continuous oxidation and removal of the newly formed oxidation layer. Another example may show that the weight gain as a measure for oxidation resistance must be evaluated with great care. Boron carbide, (not presented in Figure 1.11) does not exhibit a significant weight gain or weight loss when exposed to air. In reality, the weight gain by the formation of boron oxide layers and the weight loss due to the evaporation of boron oxide balance each other causing the recorded weight to remain almost constant up to 1000°C until all the boron carbide is transformed to boron oxide. Transition metal borides and carbides undergo severe oxidation during the very first time of exposure followed by more or less parabolic time dependence.

Chemical Reactivity

In order to minimize chemical wear at higher temperatures, reactions between materials in contact with each other and work material must be generally excluded or at least run very slowly. This means that the materials are in a thermodynamic equilibrium throughout the expected temperature range or both materials possess a very low enthalpy of formation which is a measure of the stability of compounds. In Figure 1.12, the enthalpy of formation of some more important structural materials are shown. Accordingly, zirconia and alumina are the most stable compounds followed by titanium nitride, titanium carbide, silicon nitride when compared with hard metals: WC and Co which, as an element, has a zero enthalpy of formation by definition. According to this "figure of merit", the hard metals are less suited for a series

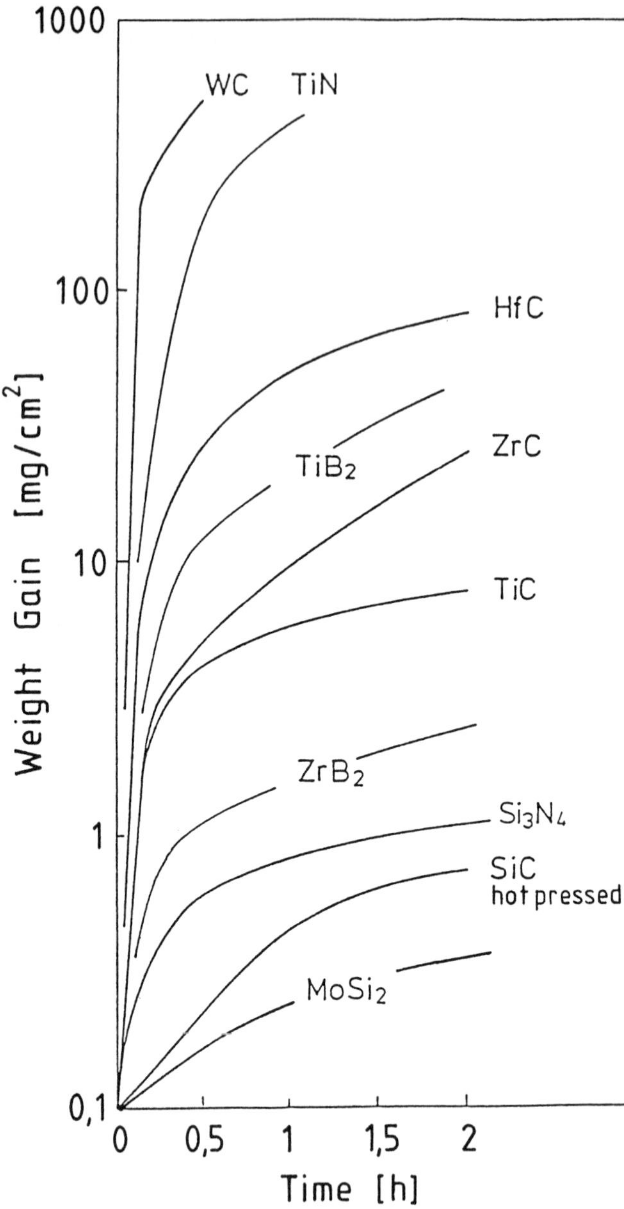

Figure 1.11: Time Dependent Oxidation Behavior of Non-oxides Exposed at 1000°C to Air

of applications due to their high chemical wear. If the enthalpy of formation of the contact materials are known, chemical reactions can be predicted by thermodynamic calculations. Many phase diagrams have already been assessed by experimental investigations together with computer-aided thermodynamic modeling. As an example, in Figure 1.13 the ternary system B-C-Fe is presented. The dashed line shows the considered contact between a boron carbide-based cutting tool during machining steel or a comparable wear-resistant boron carbide coating a steel. This line crosses more then six multicomponent equilibria involving iron borides and iron carbides. This means that boron carbide is not stable together with iron even below 1000°C but decomposes to form FeB+C, FeB+Fe_3C+C, or other borides being rich in iron. From the same phase diagram, it can be predicted that diamond grinding grits are partially dissolved in the iron work material forming cementite if the concentration of carbon gets high enough. This is a well-known problem from drilling steel-reinforced concrete with diamond tools.

Alumina ceramics, silicon nitride, titanium nitride, and titanium carbide react minimally with steel whereas zirconia may undergo strong chemical wear during sliding contact in spite of its high chemical stability. This is because zirconia is an excellent oxygen ion conductor and suffers therefore from a rapid oxygen transfer to the steel, i.e., zirconia is reduced and iron is oxidized. This reaction may continue to a certain critical oxygen deficiency above which zirconia degrades and forms metallic zirconium. An additional risk for cutting steel with zirconia is the adhesive interaction with manganese sulfide which may result in a build-up of the cutting edges. In contact with metals under nitrogen atmosphere, zirconia may accommodate N instead of O in the crystal lattice and therefore change its transformation behavior.

Thermal Shock Resistance

The resistance against thermal shock describes the tolerance of a material against temperature-induced stresses in the microstructure. Both the isotropy of the thermal expansion of non-cubic crystalline phases as well as the variation of thermal expansion between different phases may give rise to a stress accumulation if the material is loaded with temperature gradients. The critical parameter for thermal shock is usually given as temperature diff-

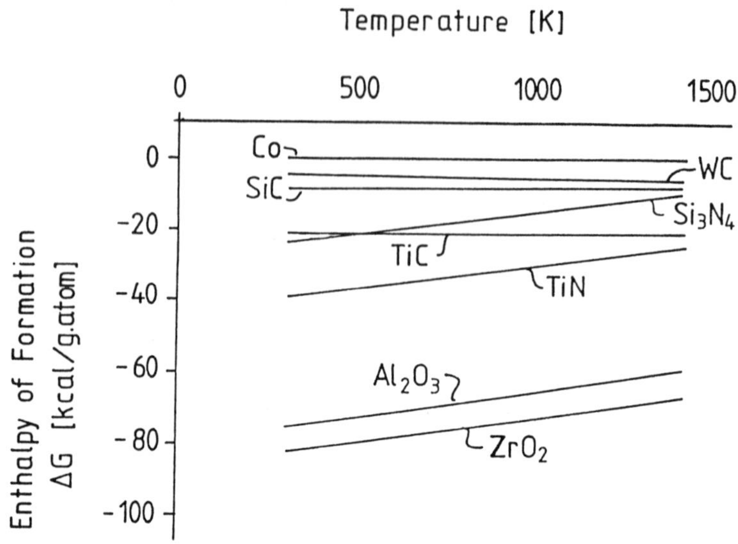

Figure 1.12: Enthalpy of Formation of Various Hard Materials

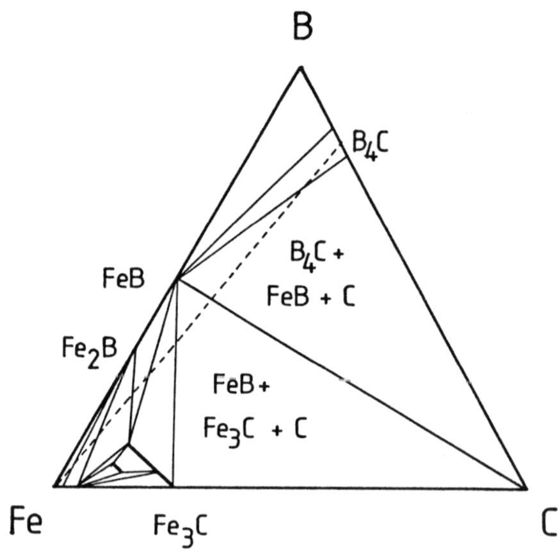

Figure 1.13: Isothermal Section of the B-C-Fe System; the dashed line represents the contact between Boron Carbide and Steel

erence ΔT_c by which a material can be quenched without significant mechanical damage, e.g., decrease of fracture strength σ_c. For some materials, so-called R-factors are available which give figures of merits for ranking the particular materials for certain applications. The R-factors involve elastic properties like Young's modulus (E) Poisson ratio (ν), the thermal expansion coefficient (α), the thermal conductivity (λ), and the fracture toughness (K_{Ic}). The R-factors can be calculated according to the following equations:

$$R_1 = \frac{\sigma_b(1-\nu)}{\alpha E} \tag{1.2}$$

$$R_2 = \frac{\sigma_b(1-\nu)\lambda}{\alpha E} \tag{1.3}$$

$$R_3 = \frac{K_{IC}\lambda\Delta T}{\alpha E} \tag{1.4}$$

Since most of the parameters are microstructure-dependent, they are more or less the critical values for ranking. The critical temperature difference ΔT_c is measured by quenching experiments using bending bars and subsequent testing of the fracture strength. In this case, additional information is given by the residual fracture strength σ_r even if ΔT_c is exceeded. The resistance against thermal shock is higher the larger the R-factors have been calculated or the higher the temperature difference is tolerated without decrease of strength. Table 1.4 shows calculated R-values for a series of ceramics and hard materials; this yields evidence that diamond is again the best material if the thermal conductivity is considered. As described before, fracture strength and toughness of diamond or related superhard materials are not adequately known, so these figures of merit give only a tendency. Metallically-behaving, hard materials like transition metal borides and carbides have a comparatively poor figure of merit due to their high thermal expansion and in spite of their high thermal conductivity.

Table 1.4: Thermal Properties of Ceramics

	Thermal Expansion Coefficient	Thermal Conductivity at RT	Thermal Shock Parameters		
Material	$\alpha_{RT-1000}$ [10^{-6} K^{-1}]	λ [Wm^{-1} K^{-1}]	R_1 [K]	R_2 [Wm^{-1}]	$R_3 / \Delta T$ [$Wm^{-1/2}$]
Diamond	0,8 - 1,5	1000 - 2000	10^4	10^6	10^4
Boron Carbide B_4C	4,2 - 4,5	40 - 45	180	7600	53
Silicon Carbide SiC	4,5 - 4,8	60 - 150	195	19500	211
Tungsten Carbide WC	5,2 - 6,0	100 - 120	41	4500	181
Titanium Carbide TiC	7,4 - 7,7	33 - 35	58	2030	39
Titanium diboride TiB_2	6,5 - 8,5[1)]	65 - 120	95	9500	129
Silicon Nitride Si_3N_4	3,2 - 3,5	32 - 35	500	17000	195
Alumina Al_2O_3	7,2 - 8,6	27	129	3500	43
Zirconia ZrO_2	8 - 10	1,7 - 2,2	220	400	7

Wear Resistance

If the relation between wear resistance and the factor $K_{Ic}^{3/4}$ x $H^{1/2}$ (compare Table 1.2) is considered, an adequate measure for ruling out wear resistant materials (and accordingly difficult to machine) and a ranking order of hard materials can be established led by diamond and cubic boron nitride (Table 1.5). The higher capability of the superhard materials is, of course, due to the extraordinary hardness and additionally to the assigned but not accurately proved high fracture resistance. They are followed by boron carbide and tungsten carbide, the first because of its high hardness, the latter because of its comparatively high toughness. The advantage of composite materials is based upon the improved fracture toughness whereas, on the other hand, the hardness can be significantly lowered by adding a softer second phase. This is, for instance, the case for Al_2O_3-ZrO_2 composites.

Consequently, these materials must be improved according to their final application as structural wear resistant material (15-18% ZrO_2) or cutting material (which means higher alumina content). Again, it must be noted that both fracture toughness and hardness depend strongly on the microstructure of the component and the manufacturing parameters. Furthermore, Table 1.5 does not imply any temperature-dependent variations of the mechanical parameters given. In most cases, information about a high-temperature fracture toughness is not available anyway. Like the other ranking methods, this is also a very crude way to rule out materials.

Table 1.5: Abrasive Wear Resistance of Ceramics

Material	Wear Factor $K_k^{3/4} \cdot H^{1/2}$
CBN	59
Diamond	52
B_4C	15 - 28
WC	24
TiC	15
SiAlON	14
SiC	14
Si_3N_4	12
Si_3N_4 + TiC	21
Al_2O_3 + TiC	17
SiC + HfO_2	15
Al_2O_3 + ZrO_2	15
Al_2O_3	9

Comparison Between Ceramics and Other Hard Materials

A comparison between the properties of hard materials as given in the preceeding tables and figures may result in the conclusion that, excluding diamond and cubic boron nitride for certain cutting and grinding operations, alumina, zirconia, silicon nitride, and silicon carbide are basically the most suitable wear resistant materials. These materials can be readily improved by suitable combinations of materials such as alumina reinforced with titanium carbide or zirconia particles due to their high hardness combined with a high fracture toughness. Figure 1.14 shows a comparison between ceramics and metallic materials concerning hardness and fracture strength. It is evident that new developments in the field of hard metals, e.g., iron-bonded TiB_2, binderless tungsten carbide, and new coatings have filled the gap between super hard materials and tool steels with regard to toughness and hardness.

4.0 MICROSTRUCTURAL REINFORCEMENT OF CERAMICS

Fundamentals

Low-Temperature Behavior

In the class of structural ceramics, oxides, carbides, nitride, and borides are utilized. They consist of very hard crystallites that are usually held together by an amorphous grain-boundary phase. Below the softening temperasure of the glassy phase, these materials behave in a linear-elastic manner; above, softening of the glass phase, inelastic deformation and grain boundary sliding may occur. Below the softening temperature, mechanically stresssed ceramics may fail by the following mechanisms:

(a) rapid low-strain fracture,
(b) subcritical crack propagation including stress corrosion, and
(c) mechanical fatigue.

All of these failure modes are triggered by process-induced microstructural flaws or defects that result from shaping or load application. Because of the linear-elastic behavior, the rapid low-strain fracture of ceramics can be modeled with methods of linear-elastic fracture mechanics. From the Griffith

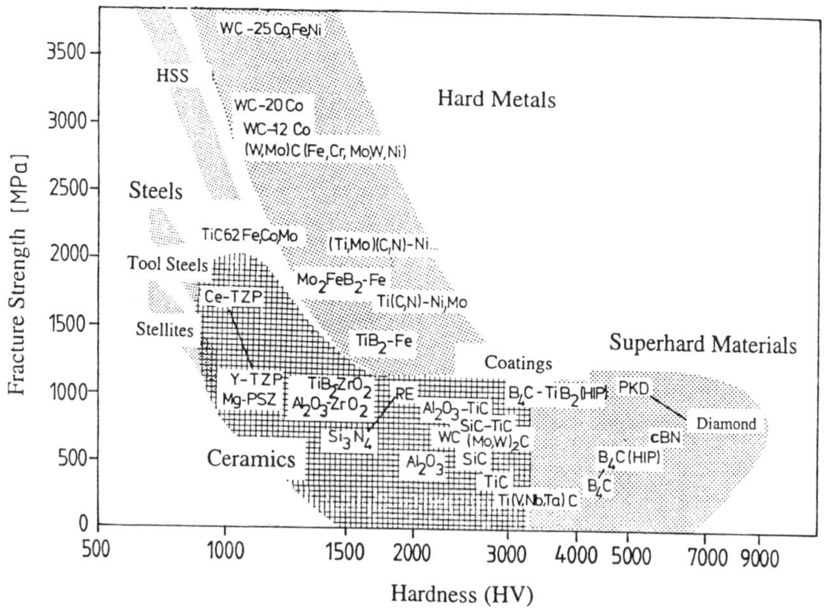

Figure 1.14: Comparison Between Metals, Hard Metals, Ceramics and Superhard Materials in Respect to Hardness and Strength

Equation (1.1), it follows that the strength must increase either when the fracture toughness is increased or when the size of the fracture-initiating flaw is reduced. From these two possibilities, the basic trends toward developing improved structural ceramics is pursued. The enhancement of fracture toughness is often referred to as "microstructural strengthening," and the reduction of flaw size is referred to as "defect-controlling manufacturing methods."

In metals and alloys, strength and toughness tend to move in opposite directions, i.e., increasing the toughness decreases the strength. Of course, it is possible to overcome this trend in metals by hardening the grain boundaries or by increasing the purity. On the other hand, in ceramics an increase in toughness is translated directly into an improvement in strength [Equation (1.1)]. The toughness-enhancing measures have the additional advantage that in the previously elastic ceramic, minor inelastic processes (quasi-ductility) are possible. Differing from the case of metals, the source of this deformation is not dislocation motion but rather phase transformation (transformation-induced toughening) or the formation of microcracks (micro-crack-induced toughening).

The second approach to improving the strength of ceramics is to reduce the flaw size. This is accomplished chiefly by the consistent improvement of all steps in ceramic processing. It is difficult to eliminate or to reduce existing defects in size during subsequent processing steps (e.g., sintering). As a rule, such defects tend to increase in size. Controlled-defect processing requires fine powders with a favorable particle size and shape distribution, a uniform spatial distribution of additives within the powder, optimization of the compaction process, and the possibility of eliminating, by destructive or nondestructive methods, bodies that contain large defects.

Another interesting consequence of Equation (1.1) is that the strength of ceramics can only be described by statistical methods and, furthermore, is dependent on specimen size. For the most part, the largest defect in a given specimen is also the one that triggers failure, yet in two specimen of identical appearance, the size of the largest defect is never the same. Therefore, it follows from Equation (1.1) that the strength of the two specimen cannot be exactly the same, so that statements about the strength of ceramics can only be related by a distribution function. An example of such a function is the Weibull distribution, as shown in Figure 1.15 and Equation (1.2). Here the probability, P, of failure is plotted against the applied stress, σ. At low loads, the probability of failure is small; at high loads, it approaches unity. The data points show that the Weibull distribution is a good approximation.

The origin of the previously-mentioned size dependence of the fracture strength is similarly based. Because of the higher specimen volume, the probability of finding a large fracture-triggering flaw is higher in a large specimen than in a small one. Hence, the failure probability of a large specimen is higher for a given stress. For an expected failure probability, the permissible stress is lower. These facts are expressed in the Weibull distribution in the form:

$$P = 1 - \exp\left[-\frac{V}{V_o}\left(\frac{\sigma}{\sigma_o}\right)\right] \qquad \textbf{(1.5)}$$

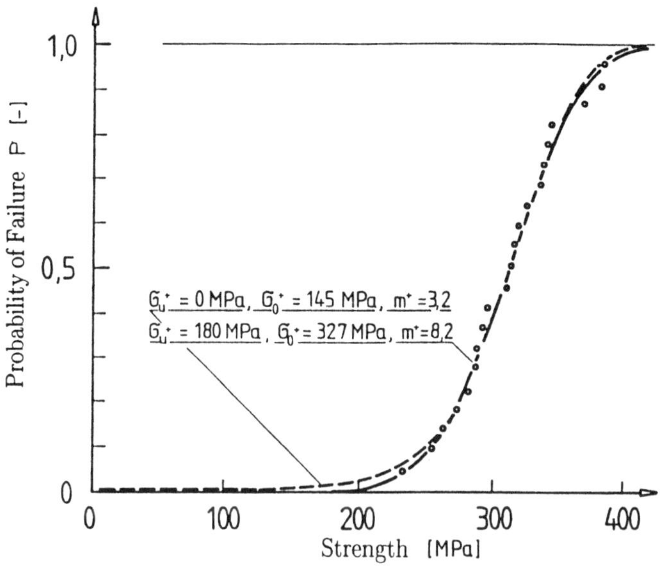

Figure 1.15: Probability of Failure of a Ceramic. Points: Measurement, Dashed Line Two-parameter Weibull-function, Continuous Line: Three-parameter Weibull Function(after Danzer, 1988)

The failure probability P is highly sensitive to the Weibull exponent *m*, especially in the technologically interesting range of low stresses. It has been shown that the Weibull exponent is a descriptor of the size distribution of flaws and, as such, can be determined ceramographically. When in a specimen of volume V_o the stress reaches the characteristic value σ_o, the probability of failure becomes l/e. The characteristic volume V_o is usually set equal to 1 mm^3.

The Weibull distribution is frequently used to predict the failure probability in mass-produced components. The risks inherent in this procedure can be assessed with the following example.[1]

Consider a part for which a failure rate of 10-5, i.e., one part in 100,000, is acceptable on economic grounds. To keep costs down, the Weibull distribution is held to a limited sample of usually less than 100 parts. With 100 samples the Weibull distribution is defined in the range P = 0.01 (1%) to 0.99 (99%) so that the experimental distribution has to be extrapolated over three orders of magnitude to obtain valid information on the stress to which the part can be subjected. While the Weibull distribution of Equation 1.5 is a possible function to describe the strength of brittle materials, it is by no means the only one. This can be seen in Figure 1.16 where the curves of Figure 1.15 are re-plotted on a semi-logarithmic scale.[1] It is obvious that the three-parameter Weibull distribution, although it fits the experimental data too, deviates considerably in the extrapolated range. Therefore, extrapolation makes sense only when additional information concerning size and distribution of defects is available. Such information can be extracted from process data, nondestructive and destructive testing (e.g., ultrasonic, x-ray, and/or proof tests), and from ceramographic characterization methods. The importance of proof tests is visualized in Figure 1.17 showing the applied stress distribution loading a ceramic part during service and the strength distribution of a manufacturing batch of this particular part. Ideally, both Gaussian distributions should be clearly separated from each in order to prevent the weakest ceramic part from rupture due to the highest possible load. This is accomplished either by shifting the medium fracture strength to values as high as possible, still running the risk of providing parts with a certain probability of an extremely low strength, or by cutting off the low-strength part from the distribution by subjecting each part to a certain test load above the maximum applied stress and providing only those parts which have survived.

The low-temperature failure modes of subcritical crack propagation and fatigue are also triggered by the aforementioned defects, so that the familiar statistical problems are encountered. Obviously, this mechanism is affected by the volume fraction, distribution, and composition of the grain boundary phase as described before.

High-Temperature Behavior

At temperatures above the softening temperature of the grain boundary phase, permanent inelastic deformation (creep) occurs by grain boundany sliding or creep pore formation and rupture. Therefore, the strength of the material is time-dependent and because of the temperature dependence of grain boundary sliding, also temperature-dependent. These basic mechanisms

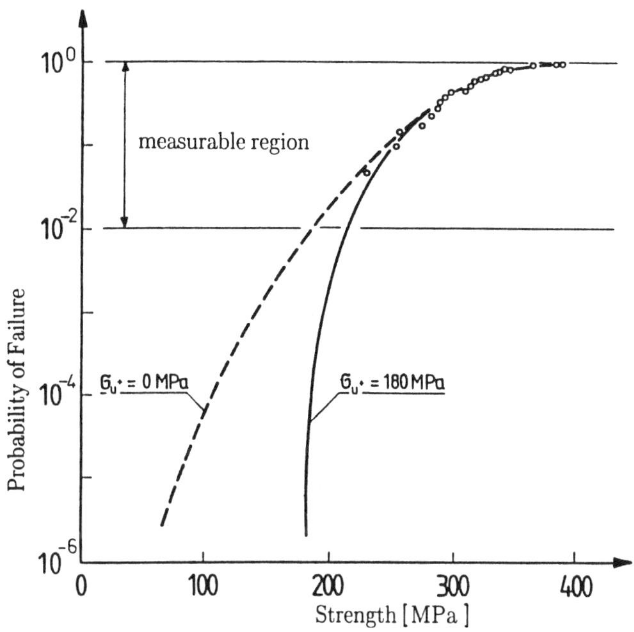

Figure 1.16: Extrapolation of the Weibull Distributions of Figure 1.15 for Small Failure Probabilities(after Danzer, 1988)

of failure are schematically summarized in Figure 1.18. Additional causes of failure at high temperatures are oxidation and corrosion as already mentioned. The failure mechanisms discussed in the previous section are still operative at high temperatures, but they are of less significance. The high-temperature properties of ceramics can be expected to improve with all measures that inhibit grain boundary sliding. Primarily, this requires the utilization of materials that contain little or no glassy phase in the grain boundaries. By a judicious choice of composition and with the help of a suitable annealing procedure, the glass can be crystallized, resulting in a significant reduction of grain boundary sliding. This is the usual procedure for the high-temperature strengthening of silicon nitride ceramics which need a liquid phase to aid sintering. If the densification is terminated, the especially doped glassy phases are crystallized by an annealing treatment. A further possibility is the creation of roughened grain boundaries that interlock as grain boundary slid-

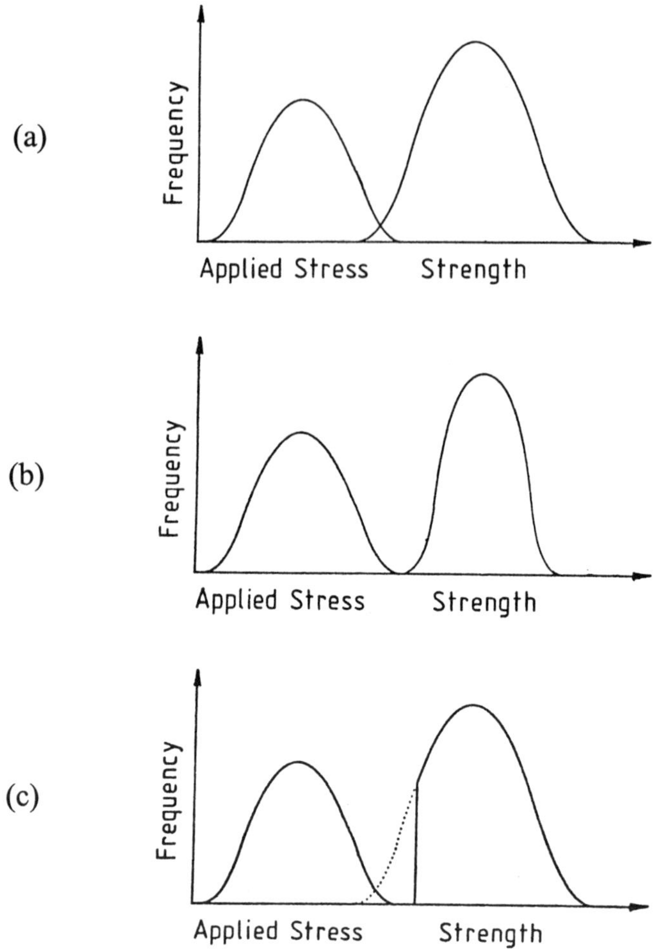

Figure 1.17: Applied Stress Profile in Service and Strength Distribution of a Manufacturing Series of Ceramic Parts

A) Risk of Failure Due to Distribution Overlap

B) Curve Separation by Improved Manufacturing (Higher Reliability)

C) Curve Overlap Truncated by Proof Testing

ing progresses or the *in-situ* precipitation of particles from supersaturated solid solutions that grow across grain boundaries.

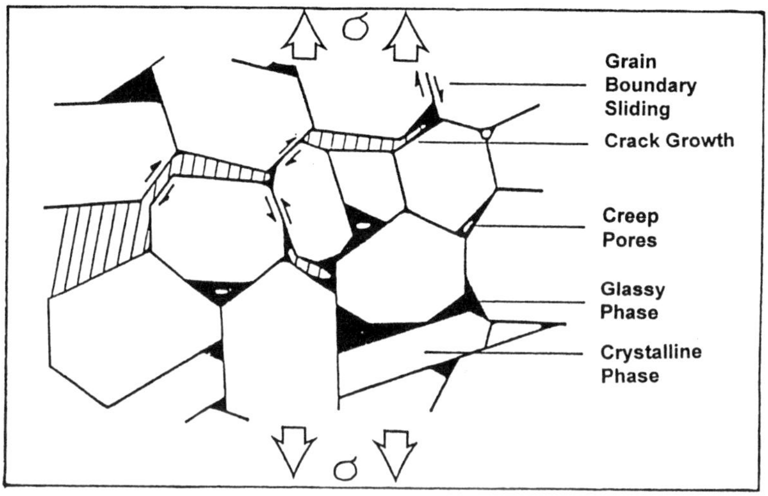

Figure 1.18: Ht-deformation of Ceramics Containing Glassy Phases

Strengthening Mechanisms and Microstructure

The fracture strength of ceramics can be improved only by mechanisms that influence crack propagation. Whereas in metals the high toughness is obtained by plastic deformation within a zone ahead of the crack tip (K_{Ic} = 20-200 MPam$^{1/2}$), the tensile stress in an unreinforced ceramic is entirely concentrated in a small volume ahead of the crack tip. One goal of reinforcement is to deflect the crack from its straight linear path, with several resulting positive effects. On the one hand, the fracture surface is increased; on the other, the crack is reoriented in the field of the applied stress. Further energy-consuming processes occur, such as crack flank attrition and crack bridging. All of these mechanisms reduce the stress intensity at the crack tip, and they are referred to as crack shielding. Another method of increasing the toughness is transformation toughening in which quasi-plastic deformation is induced in the process zone surrounding the crack tip by a martensitic transformation in an internal phase. This introduces compressive stresses and microcracks that significantly reduce the stress at the crack tip. With the latter method, it has been possible to obtain the highest toughness increments achievable to date (up to 20 Mpam$^{1/2}$). The various strengthening mechanisms are explained in greater detail in the following sections.

Dispersion Strengthening

In dispersion strengthening, the toughening effect of dispersed particles is exploited.[2] For the most part, it operates by crack deflection. A special case is whisker strengthening in which several mechanisms operate at the same time, as will be shown. The dispersed particles introduce elastic inhomogeneities into the matrix material to influence the propagation of nearby cracks. Dispersion strengthening requires that the mechanical properties of the second phase differ from those of the matrix. Thus, the elastic modulus of the particles should exceed that of the matrix, the thermal expansion should differ from that of the matrix, the strength of the particle should exceed that of the matrix, and the interfacial strength should be smaller than that of the matrix.

After sintering, the material contains misfit stresses caused by the difference in thermal contraction between the matrix and the dispersed particles during cooling. Figure 1.19 shows schematically the case of thermal expansion coefficients α of the particles that are smaller than, equal to, or greater than that of the matrix. For idealized particle morphologies and $\alpha_{particle} < \alpha_{matrix}$, one obtains radial compressive stresses and tangential stresses that are tensile. A growing crack tends to run normal to the tensile stress because it opens the crack. In a sense, the crack is attracted by the particle and spends a short time normal to the particle/matrix interface. If the particle cannot be split by the crack, the energy must be increased to make it run around the particle. In the case $\alpha_{particle} = \alpha_{matrix}$ there are ideally no stresses at the particle/matrix interface. However, the interface acts as a crack deflector because of the difference in elastic moduli between particle and matrix. When $E_{matrix} < E_{particles}$, the elastic strain is greater in the matrix than in the particle ($E_{matrix} > E_{particle}$) so that part of the crack-forming tensile stress is transmitted to the particle. Consequently, tensile stresses are concentrated at the particle/matrix interface and can lead to crack deflection.

The third case ($\alpha_{particle} > \alpha_{matrix}$) leads to radial tensile stresses and tan-

$$\sigma_{m_r} = \frac{(\alpha_p - \alpha_m)\,\Delta T}{[(1+\nu_m)/2E_m] + [(1-2\nu_p)/E_p]}$$

Selsing (1961)

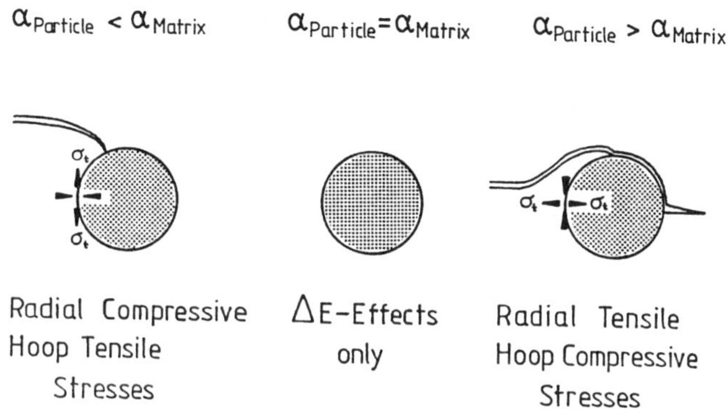

Figure 1.19: Influence of the Thermal Misfit Strain on Crack Propagation: Crack Arrest and Crack Deflection at Dispersed Particles.

gential compressive stresses during cooling. The tensile stresses are perpendicular to the particle/matrix interface and introduce a crack that moves along the interface. The energy consuming processes consist of (a) the increased crack surface in the interface and (b) the increased energy required to exit the interface and re-enter the matrix (Figure 1.19). Occasionally, this effect can be observed during grinding and polishing when in multiphase ceramics the grains of one particular phase are preferentially pulled out. Figure 1.20 shows the resulting crack path for a SiC/TiC and SiC/TiB_2 composite, respectively. If the thermal misfit stresses get close to the critical rupture stress for the grain boundaries, they may give rise to strain-induced cracking if additional loads are applied by the grinding and polishing treatment. Consequently, particular grains may be liberated by circumferential cracks and can be pulled out.

Ceramics in which the crack deflection is frequently 90° to the direction of the main crack show the additional strengthening effect of crack flank friction. When opposite sides of the fracture surface are locked together, shear forces are generated which may create additional crack systems. When

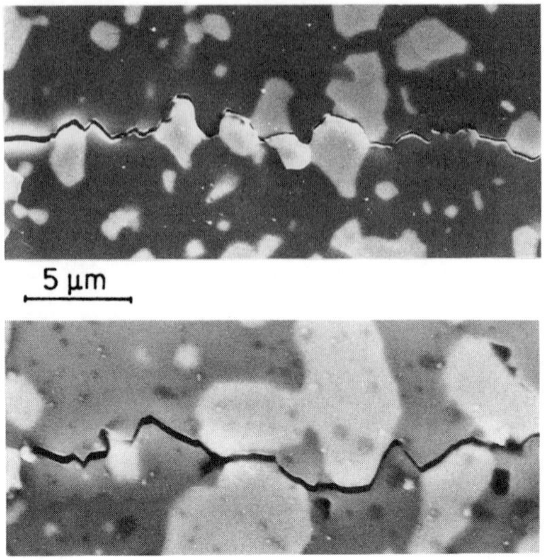

Figure 1.20: Non-planar Crack Propagation in Sic/ti (Top) and Tic/tib$_2$ Compositics

individual particles are neither bypassed nor cleaved, crack bridging occurs (Figure 1.21). Crack bridging by a highly viscous glass phase is frequently found in ceramics with sinter additives at high temperatures.

For optimal strengthening by crack deflection, geometrical factors are essential in addition to the properties of matrix and particles. Since the misfit stresses at the particle/matrix interface decrease in the matrix with R^{-3}, the volume fraction and homogeneity of the dispersion are important. To prevent catastrophic crack propagation, a minimum distance between the particles must be maintained.

Theorctical analyses of fracture in dispersion-strengthened ceramics have shown that the amount of crack deflection and effective strengthening also depends on the shape of the included particles.[3] It became evident that spherical particles already cause a definite improvement of the mechanical properties, but platelets are significantly better; the greatest strengthening effects are achieved with fibers. A precondition for maximum exploitation of the predicted behavior is that the particle orientation be random for random loading. For axial loads, they are oriented one- or two-dimensionally perpend-

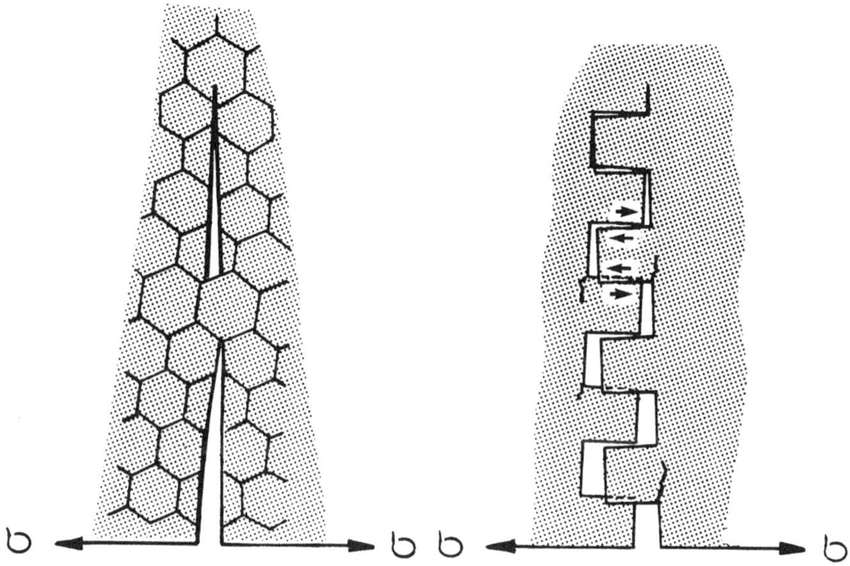

(a) (b)

Figure 1.21: Crack Flank Effects:
A) Crack Wake Friction and Interlocking Crack Surfaces
B) Crack Bridging

icular to the applied force (Figure 1.22). Since conventional processing usually leads to deformation textures, hot pressing by axial compression generally causes the powder particles to orient perpendicular to the hot-pressing direction; both random and preferred orientation are not always easily attained.

The desired ideal shape of a one-dimensionally elongated particle is found in whiskers or short fibers. By adding high-strength, single-crystal whiskers or polycrystalline fibers, improvement of high-temperature strength, fracture toughness, thermal shock resistance, and reliability of brittle ceramics

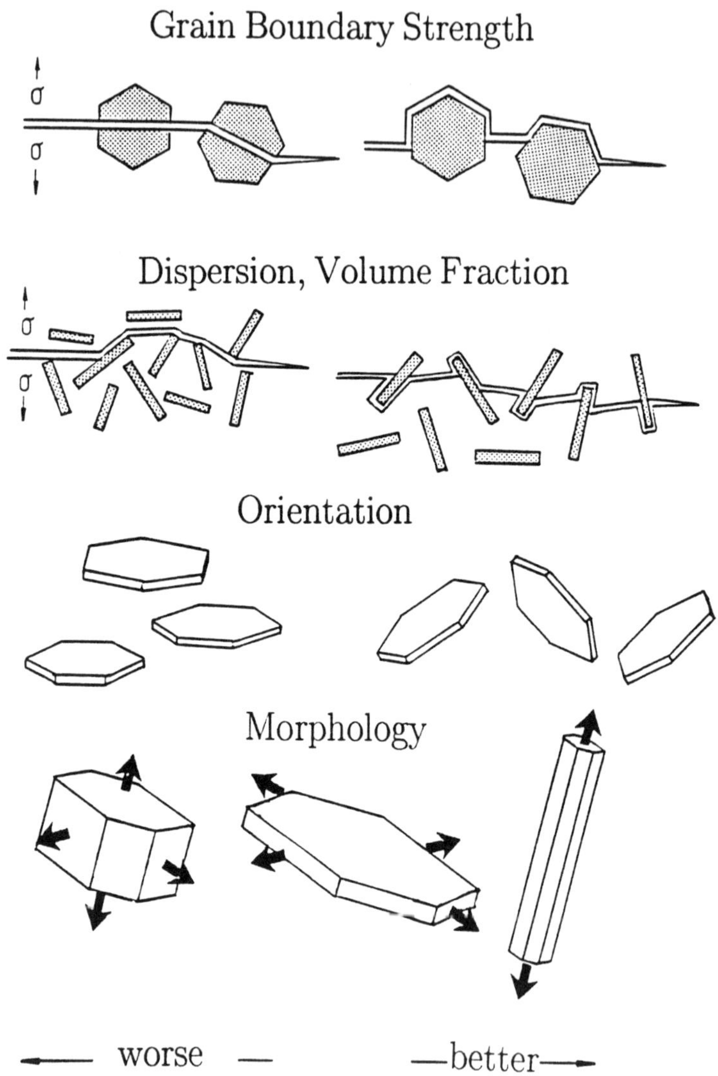

Figure 1.22: Preconditions for an Effective Microstructural Toughening of Composites.

is achieved. In addition to the aforementioned strengthening mechanisms of crack deflection, proper selection of fiber, whisker, and matrix materials can further assist with crack deflection, load transfer, pull-out, and matrix pre-stressing.

Load transfer from ceramic matrix to whisker presupposes a higher elastic modulus in the whisker: a condition that is achievable even in single-phase materials because of the elastic isotropy of the single-crystalline whisker. Because of the elastic stiffness of the whisker, a portion of the applied stresses is transferred to the fiber (Figure 1.23). This releases the stresses in the matrix.

When fibers are pulled out of the matrix during crack growth, an additional significant mechanism of high-energy consumption is involved. During fracture, the fiber tips may remain clamped in the opposing crack flanks and are first strained elastically. Ideally, the whisker does not fracture but is pulled out of the matrix. Depending upon the properties of the interface, this occurs with increasing or decreasing friction. This type of crack bridging by fibers and/or the pull-out accompanied by friction deflects a large fraction of the stresses in the crack flank region, thus relieving the stresses at the crack tip. Although the pull-out is the most effective of all strengthening mechanisms, it is rarely observed in real fracture surfaces so that its actual contribution to the strengthening of ceramics is small.

The principle of matrix pre-stressing is based on differences in the thermal expansion coefficients of fiber and matrix ($\alpha_{whisker} > \alpha_{matrix}$) and is similar to crack deflection. Because of the greater contraction of the whiskers during cooling, the whisker/matrix interface shows radial tensile stresses and tangential compressive stresses in the matrix itself. The effect of matrix pre-stressing is comparable to the behavior of pre-stressed concrete where the brittle mixture of cement and gravel is kept under compression by pre-stressed reinforcing bars.

There are numerous examples of dispersion-strengthened ceramics. This includes such material combinations as Al_2O_3-TiC, Al_2O_3-spinel, ZrO_2-Al_2O_3, SiC-TiC, and B_4C-TiB_2 (Table 1.3). A significant jump in mechanical properties has been attained by the inclusion of SiC-whiskers (Figure 1.24).[4] In matrix ceramics (e.g., Si_3N_4, ZrO_2) with an inherently high toughness, this can be improved further at the expense of overall strength. Between 1985 and 1992, silicon carbide-whisker reinforced oxides have been successfully developed for cutting tools and structural applications. However, the activities in

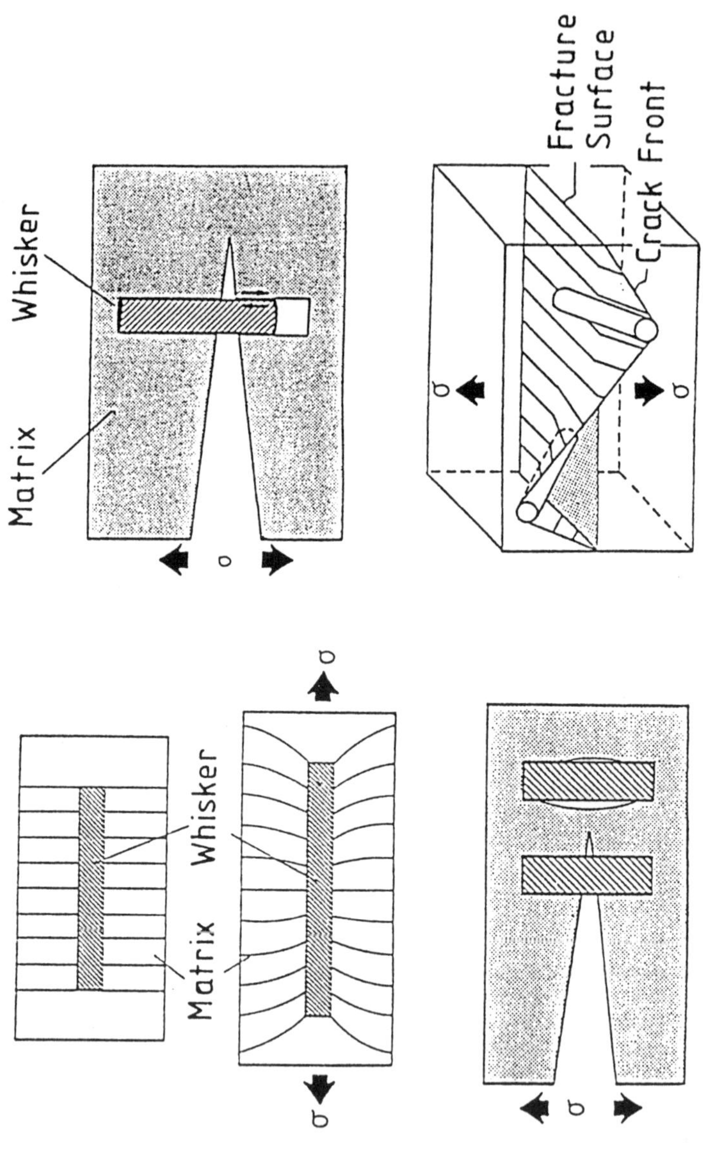

Figure 1.23: Mechanisms of Whisker-reinforcement (Courtesy, Hoffman, Karlsruhe).

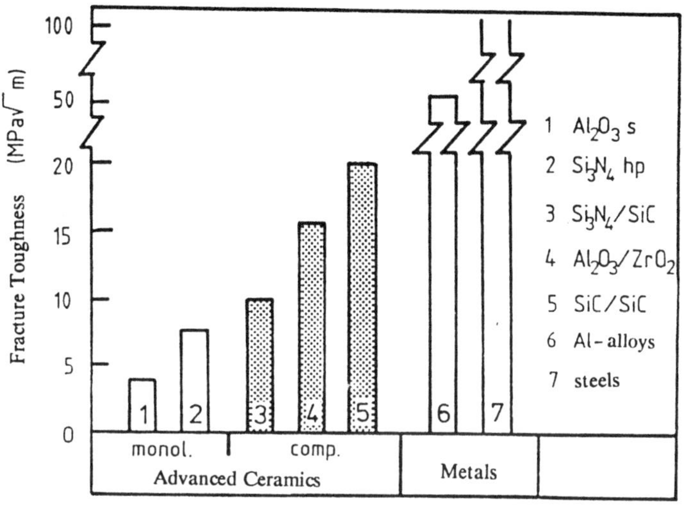

Figure 1.24: Toughness Increase by Whisker Reinforcement Compared to Non-toughened Monoliths (1 and 2) and Transformation Toughened Alumina (4) (Courtesy, Greil, Erlangen)

fiber-reinforced materials decreased generally when it became evident that ceramic fibers could be carcinogenic. In Europe, production of whisker-reinforced materials stopped but some Swedish companies continued until the beginning of the '90s to provide this high-performance materials to the world market. Besides the potential carcinogenic effects of the whiskers, the risk of introducing large defects into the ceramics with supercritical length or whisker agglomerates could cause a tremendous decrease in strength; this retarded the enthusiasm.

Microcrack Reinforcement

A further possibility of reducing the stress concentration in the crack tip is found in the deliberate introduction of a large number of small cracks. This so-called microcrack reinforcement operates by a judicious choice of phase combinations, morphology, and dispersed particles so that the particles partially separate from the matrix when subjected to an additional external load. At the matrix/particle interface, cracks on the order of tens to hundreds of nanometers are formed. Such microcracks are found in the so-called process zone along the crack flanks some distance ahead of the crack tip because the tensile stresses emanating from the crack tip extend into the undisturbed microstructure.[5]

When the load is further increased, the favorably oriented microcracks, of which there are hundreds, grow by a fraction of their original length and thereby reduce the peak concentrations. The stress concentration at the tip of a macrocrack essentially spreads to a multiplicity of small cracks. The material behaves in an elastoplastic manner because the microcrack formation is partially reversible upon load removal.

This reinforcement strategy is not without risk. Unless the misfit stresses are optimized, the presence of excessively high local stresses can lead to spontaneous formation of microcracks to the point of coalescence so that the mechanical properties of the material are severely affected. A macrocrack is enhanced by having to provide only the additional energy needed for microcrack coalescence before catastrophic failure occurs. In Figure 1.25,[5]

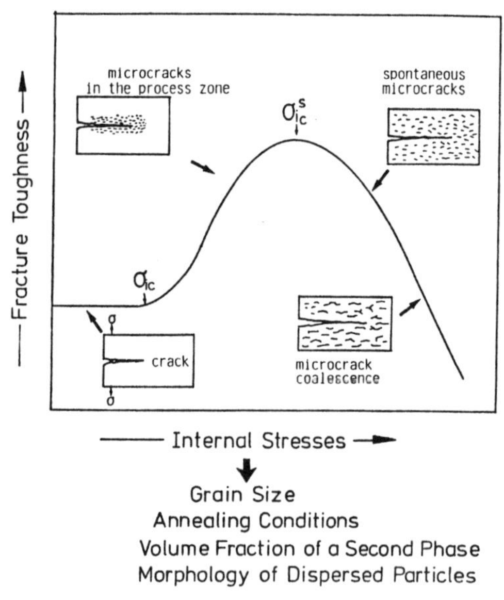

Figure 1.25: Microcrack Formation and Influence on Fracture Toughness

the dependence of fracture toughness on internal stresses is shown schematically. Below a critical stress (σ_{ic}), no microcracks are formed even under an applied load so that no interaction takes place between macrocrack and microcracks and no increase in toughness is observed; however, above σ_{ic}, the formation of the process zone around the main crack starts. By superposition of the internal and applied stresses, the formation of stress-induced microcracks away from the crack tip becomes possible. The shielding of the principal crack from the external load leads to a significant increase in toughness. The maximum toughness is reached at σ_c when the stresses are high enough to induce spontaneous microcrack formation. Further increases of toughness in this region are conceivable when existing microcracks split or branch out. Then the principal crack repeatedly changes its course and may show multiple branching. Because of the deflection of the stress field and the increase in fracture energy associated with the increased fracture surface, a high-energy consumption is still present. At higher internal stresses, the microcracks begin to interact and may accelerate the propagation of a macrocrack under an applied load. Therefore, it is the goal of materials development to find an optimum increase in toughness in the upper third of the ascending branch of the curve in Figure 1.25 by exploiting the microstructure and selecting phases with suitable thermal expansion coefficients.

Reinforcement by microcracks has proven itself especially in materials sensitive to thermal shock. In materials that are subjected to reversible mechanical stresses, it may, however, lead to subcritical crack growth, of which little is known, or to mechanical fatigue. Figure 1.26 presents a TEM-micrograph of a ZrB_2/ZrO_2 composite where radial microcracks have been originated from the transformation of tetragonal zirconia to the monoclinic modification.

Transformation Toughening

Transformation toughening is the deliberate exploitation of a phase transformation of dispersed particles in a ceramic matrix.[6][7] It is a prerequisite that the low temperature phase of the material to be transformed has a lower density, i.e., is a larger volume than the high-temperature modification. A condition for the matrix is that the stresses accompanying the trans-

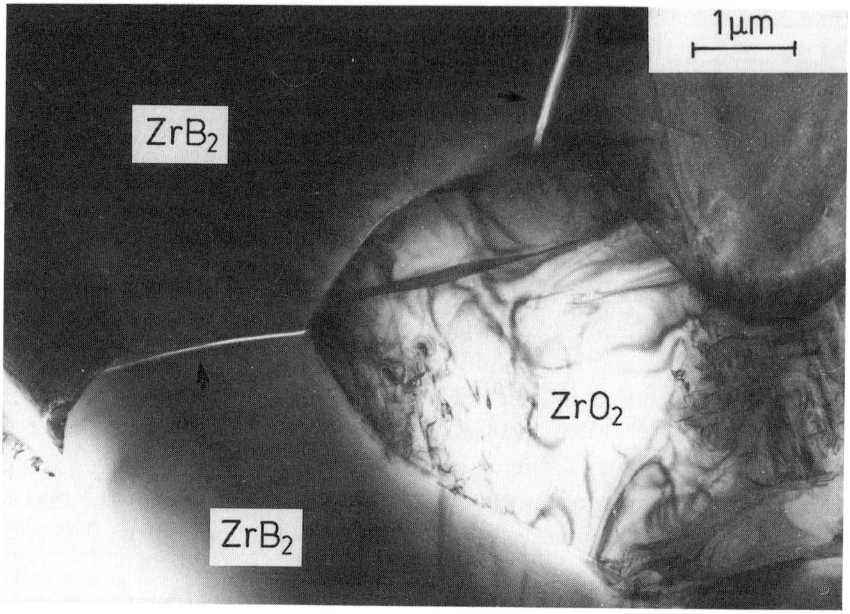

Figure 1.26: Radial Microcracks Origination from T→m transformed Zirconia; Tem-micrograph

formation be relaxed very slowly or not at all. Among the many materials that exhibit such a transformation, ZrO_2 has proven to be especially suitable.

Zirconia crystallizes in three modifications. Below the melting point near 2,700° C, the cubic (c) phase with the calcium fluoride structure exists that transforms to a tetragonally (t) distorted CaF_2 structure at 2,370°C. Below 1,170°C, a further distortion, accompanied by a volume expansion of

3% - 5%, leads to a monoclinic (m) modification. This transformation, tetragonal to monoclinic, which is mainly utilized, is martensitic, i.e., diffusionless and reversible. During the modification change, the tetragonal structure is sheared by 9° while the a and b axes are stretched. This phase transformation is sudden, but the tetragonal high-temperature modification is capable of being metastably undercooled. Therefore, t-ZrO_2 can be dispersed in a ceramic matrix such as Al_2O_3 so that the work of volume expansion has to be performed against the surrounding matrix. The activation energy of the transformation is obtained from the undercooling, and the nucleation energy increases with decreasing size of the ZrO_2 particles. The activation energy can also be gained from the stored elastic energy that results from the misfit between particle and matrix or from the strains ahead of a macrocrack.

Optimal utilization of the martensitic transformation of ZrO_2 requires that it take place in a controlled fashion. This is accomplished by control of the grain size distribution and by selection of a matrix material with a lower coefficient of thermal expansion. This generates radial tensile stresses at the particle/matrix interface that adjust the activation energy of the transformation precisely; the tetragonal modification is retained at room temperature but can be transformed under the influence of an applied stress. Furthermore, ZrO_2 can be stabilized chemically by adding MgO, CaO, or Y_2O_3. Since magnesium, calcium, and yttrium have an ionic radius near that of zirconium, they are introduced into the ZrO_2 lattice by ionic diffusion and impede the t $\rightarrow$ m transformation by creating oxygen vacancies. Such stabilizers are required especially in pure ZrO_2 single phase ceramics or in composites with matrices having a significantly lower coefficient of thermal expansion than zirconia.

Three mechanisms are responsible for the absorption of fracture energy in transformation strengthening (Figure 1.27):[6]

(a) stress-induced transformation during loading,
(b) microcrack formation, and
(c) crack deflection.

The stress-induced transformation is essentially based on a crack shielding effect resulting from the volume and shape changes of the ZrO_2. The fracture energy is consumed by the work of transformation, and tensile stresses in the region of the crack tip are removed by the compressive stresses accompanying the volume increase of the ZrO_2; in other words, the crack is partially closed again. Such process zones in which transformed monoclinic ZrO_2 is found

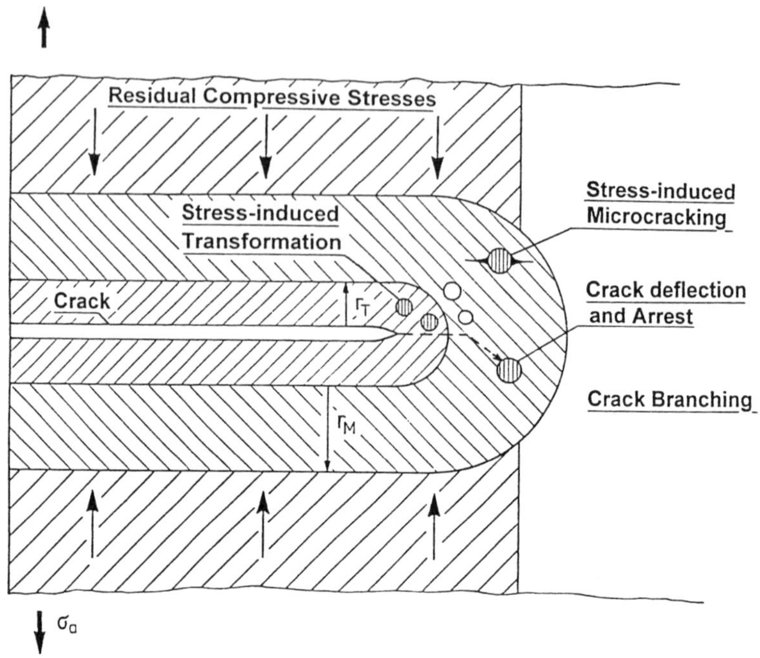

Figure 1.27: Schematic Representation of the Reinforcing Mechanisms in Zirconia Ceramics (Courtesy, Claussen, Hamburg)

can be observed by electron microscopy in the crack flanks or in the region of the crack tip.

An additional result of the stress-induced transformation is the production of permanent surface compressive stresses when the transformation is deliberately introduced in the surface region of a material by grinding or sand blasting. The resulting compressive stresses prevent crack penetration in the surface region which leads to increased strength and thermal shock resistance.

If the ZrO_2 is not stabilized in the tetragonal modification during fabrication, (spontaneously transformed) or if the strain-induced transformation already occurred, then the formation of radial microcracks is conceivable around such transformed particles as a result of the volume expansion (Figure 1.27). The energy of an externally-induced macrocrack is absorbed by the same mechanism as described in microcrack reinforcement:

the extension of the strain-induced microcracks and the associated crack branching. A reduction in fracture energy can also be attained by forcing a transcrystalline crack in the matrix to move around the ZrO_2 particle (intercrystalline fracture). For example, a tetragonal ZrO_2 crystallite, because of its thermal mismatch against Al_2O_3, generates radial tensile stresses; a monoclinic ZrO_2 generates tangential tensile stresses so that a crack bypasses a tetragonal particle and cuts through a monoclinic one in a transcrystalline way. The aforementioned strengthening mechanisms can be simultaneously active. The most significant is undoubtedly the stress-induced transformation.

Three fundamentally different types of transformation-strengthened ceramics can be accordingly deduced:

1. Partially stabilized zirconia (PSZ) in which tetragonal ZrO_2 particles are coherent with cubic ZrO_2. It is obtained by chemically doping the material with MgO forming so-called MgO-PSZ.
2. Tetragonal zirconia polycrystals (TZP) consisting of an extremely fine-grained (<1 μm) tetragonal matrix. This is chemically stabilized with Y_2O_3. The materials exhibit transverse bending strengths in excess of 2,000 MPa that have not been attained by any other ceramic. This is also the group of materials that has achieved the highest fracture toughness ($_{K.C.}$ up to 20 MPa $\sqrt{m}$) if stabilized with ceria.
3. Ceramics composites in which tetragonal or monoclinic ZrO_2 particles are dispersed: ZrO_2-strengthened Al_2O_3 (ZTA), mullite (ZTM), spinel (ZTS), cordierite, titanium diboride, and others.

Because of these properties and also because of the analogy to the crack tip plasticity observed in metals, such materials are often referred to as "ductile ceramics"; the mechanisms involved are in no way comparable to plastic deformation in metals. Examples of the improvement of mechanical properties by ZrO_2 addition are shown in Figure 1.28 and listed in Table 1.6.

High-Temperature Strengthening

In sialons, silicon nitride, and silicon carbide ceramics, high-temperature properties are impaired by the presence of intergranular phases. Above the softening temperature of the grain boundary phase, permanent in-

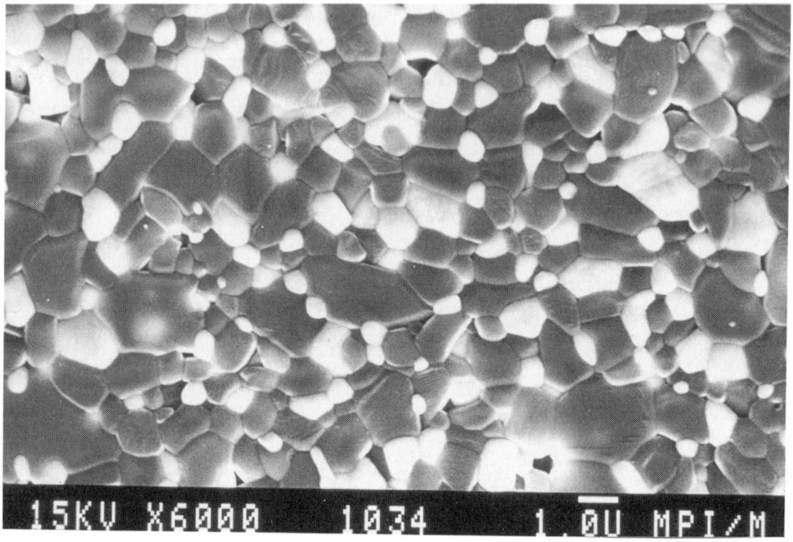

Figure 1.28: Typical Microstructure of a Zirconia-toughened Alumina Composite (Courtesy, Claussen and Schubert)

elastic deformation may occur by grain boundary sliding glass phases. As a result, the strength becomes dependent on temperature and load time and additional failure modes of oxidation and stress corrosion are encountered. A strategy to improve the creep resistance, therefore, is to avoid such glassy phases; on the other hand, they are needed in the sinter process as a liquid that transports matter quickly and supports densification. One feasible method of high-temperature strengthening, in addition to reducing the added glass orig-

Table 1.6: Properties of Transformation-toughened Ceramics

Matrix	Vol.-% $Zr0_2$	Tetragonal Fraction, Vol.-%	K_{IC} Mpa m	$\frac{K_{IC} - K_0}{K_0}$	σ_F MPa	$\frac{\sigma_F - \sigma_0}{\sigma_0}$, %	Remarks
$A1_20_3$	15	20	9,6	+ 85	500	-10	hot-pressed. (HP)
$A1_20_3$	16	100	15,0	+270	1200	+300	HP Sol-Gel-Powder
$A1_20_3$	7,5	95	6,8	+ 55	600	+ 30	sintered, Sol-Gel-P .
Spinel	17,5	50	4,6	+140	500	+150	HP
Mullite	23	30	4,8	+ 60	400	+ 35	Reaction sintered
Mullite	23	80					Reaction sintered
ZnO	36	0	3,3	+ 95	n.d.	n.d.	HP
Si_3N_4	25	30 cub./70 mon.	8,5	+ 50	950	+ 45	HP, SiAlON 10
Si_3N_4	10	30 cub/60 mon.	7,2	+ 30	700	+ 15	HP, SAl0N 1
SiC	5,5 EM2	0	6,7	+ 25	n.d.	n.d.	HP (+ Mullite)
Mg-HZ	60 % cub.	40	9,0	+260	600	+140	3,4 wt.-% MgO
Ca-PSZ	50 % cub.	50	6,6	+165	650	+165	3,3 wt.-% CaO
Y-PSZ	20 % cub.	80	6,4	+155	650	+165	2 wt-% Y_20_3

n.d. = not determined

inators and oxygen impurity to an acceptable minimum, is to crystallize such residual molten phases by devitrification subsequent to the densification process. [8]

To crystallize the glass phase, one must know its composition so that further additives can be included to change its composition to the stoichiometry of a crystalline phase. The phase to be crystallized should satisfy the following conditions:

- High symmetry,
- No anisotropic grain growth,
- High nucleation rate,
- Low growth rate,
- No segregation of impurities ahead of the crystallization front, and
- Specific volume similar to that of the glassy phase.

The devitrification process follows the ceramic densification and consists of a complicated anneal in the temperature range of maximum nucleation and minimum growth.

Devitrification of the glass phase leads to a stiffening of the structure at high temperatures. Viscous flow and/or grain boundary sliding can thereby be reduced to a tolerable level. In Figure 1.29, the high-temperature bend strength is shown for a silicon nitride ceramic with and without a crystallized glass phase. It is obvious that the drop in strength above 1000°C is arrested. Similar sintering procedures with glassy phases and subsequent annealing have been recently applied to silicon carbide (SiC) where also Al_2O_3 and Y_2O_3 are used to initiate liquid phase sintering followed by yttria-alumina-garnet formation.

Another possibility to prevent high-temperature deformation is to create an interlocking microstructure by the *in-situ* growth of elongated particles. Figure 1.30 presents a plasma etched microstructure of a silicon nitride ceramic in which the nuclei and the growing steps of silicon nitride needles in the liquid phase can be recognized.

Fracture-Producing Defects

To improve the strength of brittle ceramics, the Griffith Equation (1.1) suggests a scheme in addition to the deliberate introduction of multi-

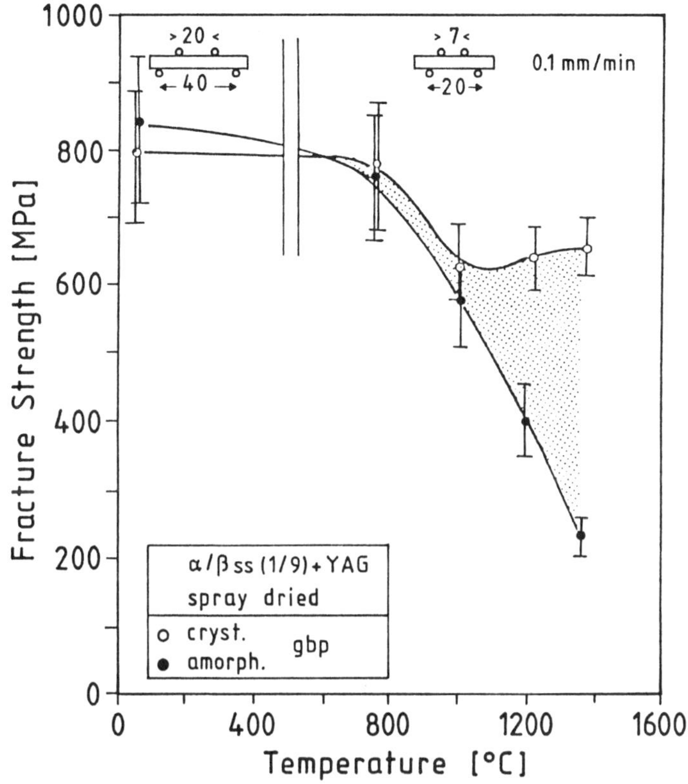

Figure 1.29: High Temperature Bending Strength of Silicon Nitride Ceramics with Amorphous and Crystalline Grain Boundary Phases (Courtesy, Greil, Erlangen)

phase materials and optimum microstructures.[9][10] As mentioned before, the critical defect size enters the equation that indicates the maximum allowable dimensions of a microstructural defect, such as pores, inclusions, agglomerates, or giant grains, before it triggers fracture under an applied load. The goal of the increase in toughness is to increase the tolerance of the material towards such structural defects. Ideally, the ceramic should be tough enough so that defects with dimensions on the order of 1 mm can readily be ignored. Figure 1.31 shows a diagram representing the Griffith Equation by combining

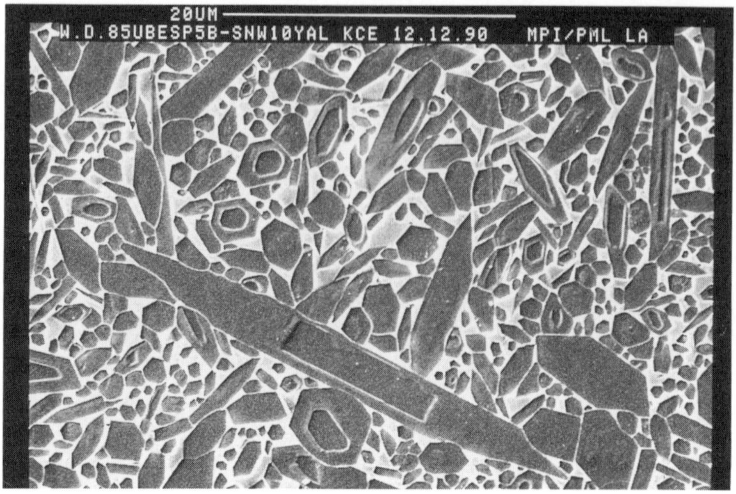

Figure 1.30: Strengthening Against Creep by the Formation of *in-situ* Grown Elongated Silicon Nitride Particles, Plasma-etched Section (Courtesy, Hoffmann, Karlsruhe)

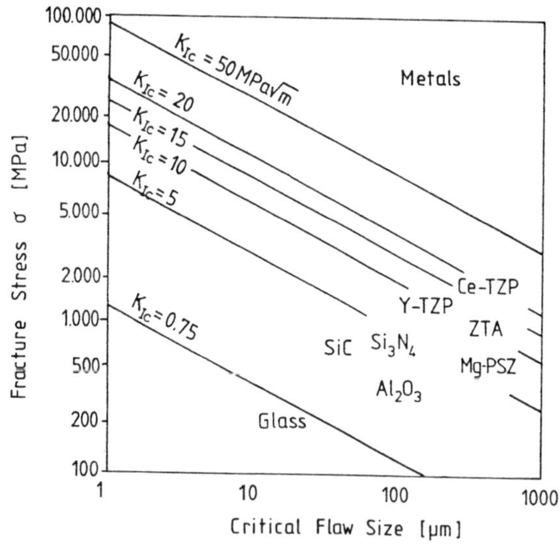

Figure 1.31: Principal Dependence of Failure Stress from Critical Flaw Size for Different Fracture Toughness Values

the ultimate rupture stress due to a critical flaw size with the value of fracture toughness of that material. In other words, the area beneath each of the "iso-toughness" lines represents the stress level tolerated by a test piece with microstructural inhomogeneities being smaller than the critical flaw size.

However, not only are the absolute dimensions of a microstructural defect important but also the defect concentration in a ceramic component. The distribution function of defect size and frequency determine the fracture statistics, i.e., the probability that a work piece will fail under a given load. In effect, one measures the bend strength or tensile strength of specimen and attempts to extrapolate from the comparatively small test volume to larger components or greater numbers of pieces.

Conditions leading to the coarsest microstructural defects are created as early as the powder production and preparation stage. Mechanical and chemical impurities caused by careless work, concentration gradients caused by the inhomogeneous distribution of sintering additives, or local density variations based on agglomerates are the consequences of improper powder preparation methods and cannot be eliminated by subsequent sintering or hot pressing. On the contrary, such inhomogeneities are coarsened during heat treatment because of the differing shrinkage rates during sintering: regions of high density in the powder or high concentration in the additive shrink faster and pull away from the less densely packed particles of the compact. This results in large pores or pore channels that are not readily removed even by hot isostatic pressing. Inclusions of organic matter, such as fibers, flakes of skin, or dust burn-off during sintering, leave a void that is also difficult to densify. Figure 1.32 (a-d) shows examples of typical fracture-causing defects.

Examples of other process-related defects that originate in a later stage of powder preparation are pressing defects: variations in density caused by an uneven pressure distribution in the die, or machining defects: grinding marks or surface rupture due to fatigue. In the sintering process, the condition of the grain boundaries and triple points as well as the grain size are effective. Figure 1.33 shows qualitatively the frequency distribution of such defects as a function of their size. It is obvious that three defect populations exist. The most unproblematic defects are the micro-structure-related ones, such as grain boundaries, triple points, and crystallites that represent the largest fraction of defects. With respect to size, they usually remain sub-critical, and the only exceptions are giant grains that may grow to several 100 μm in size. The second population, with a maximum around 10-20μm, are agglomerates,

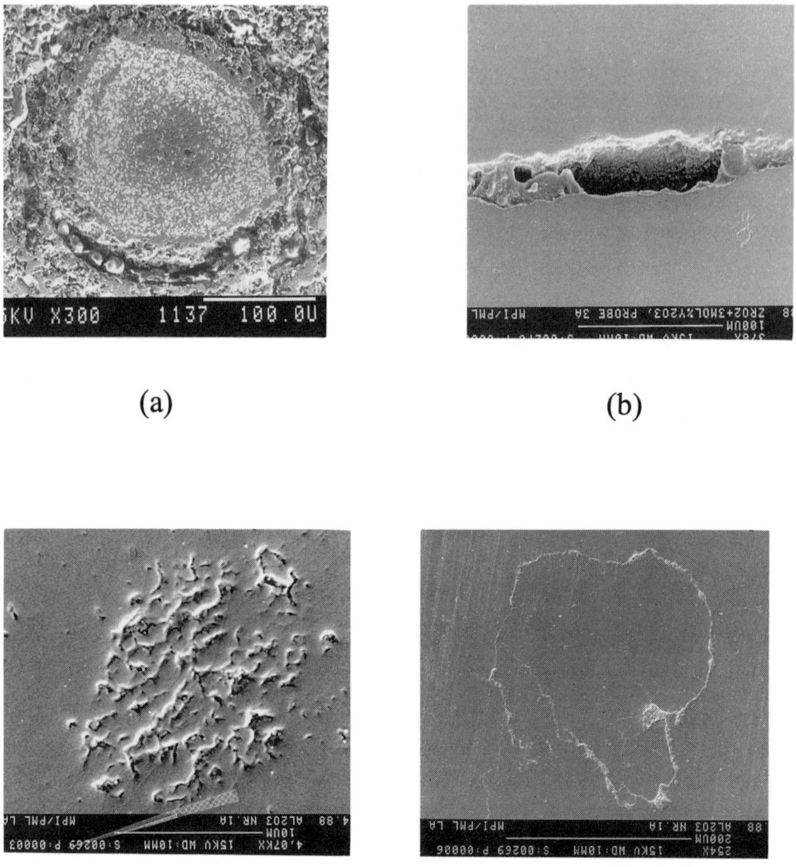

Figure 1.32: Processing-derived Strength-controlling Defects

a) Wc/tib$_2$ Agglomerate in Fine Grained Matrix with a Circumferential Crack
b) flat Pore in 3y-tzp, Burnt Organic Compound
c) low-density Agglomerate in Alumina
d) agglomerate in Alumina with Contaminated Surface

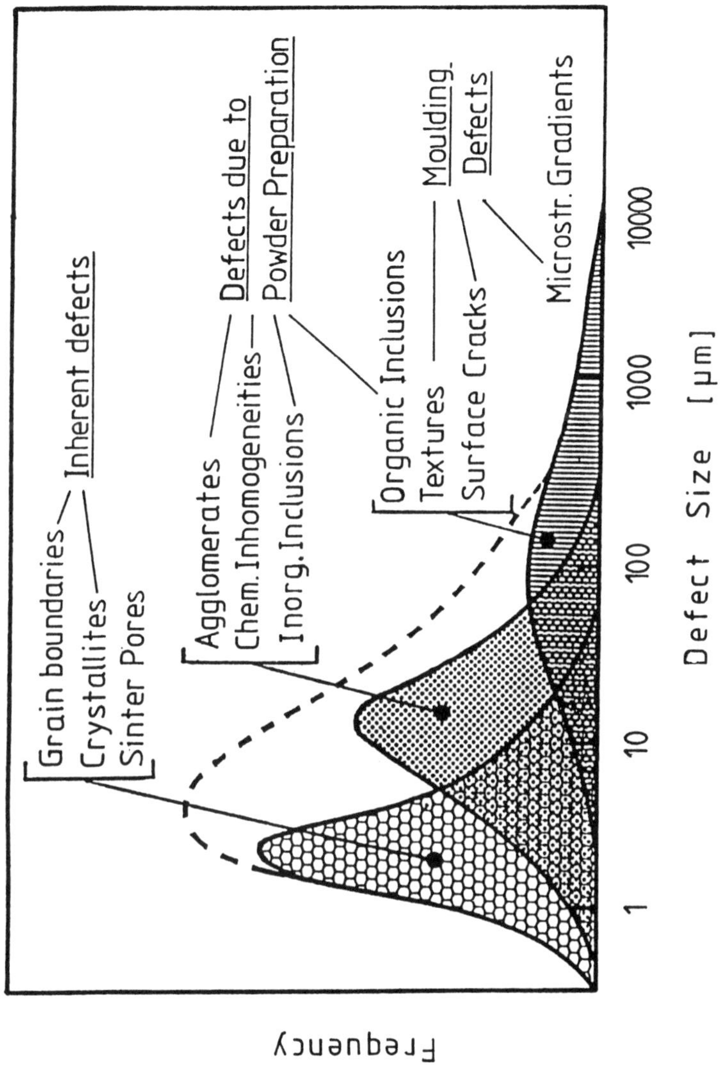

Figure 1.33: Frequency/size Distribution of Microstructure-immanent and Process-derived Flaws

chemical impurities, and inorganic impurities resulting from cross-contamination by other powder species. Agglomerates, in particular, may extend into the critical range of 200-500 μm and thereby represent a significant cause of failure.

The least frequent but dimensionally most hazardous defects are organic particles, pressing defects, and machining defects that may become several millimeters large and result in a probability of fracture approaching certainty. One must therefore avoid the relatively small number of large defects or detect them early by materials testing. In that sense, the bend test and tension test become a metallographic method of characterizing the microstructure.

As a result of the even tighter requirements, the development of improved production processes has aimed at the powder purity (total metallic impurities <0.1 wt.%), on a reduction in particle size (<1 μm, preferably <100nm), and well-defined alloy compositions in powders of high chemical homogeneity. Furthermore, controlled surface conditions of the powders, such as surface purity, agglomeration behavior, and rheological behavior are important factors in the shaping processes. Novel powder production methods that provide such well-defined powders are colloidal processes as well as reaction processing in which some critical steps of powder production and processing can be avoided. For conventional powder metallurgical methods, the development of high purity ultrafine powders with good dispersion and flow properties have a high priority, and also the development of novel mixing, doping, and shaping processes that exclude, as far as possible, contamination by foreign particles. Further areas of interest are the development of novel densification processes, such as sinter-HIP, sintering under gas pressure and nonlinear sintering (shrinkage-controlled sintering). The goal is always maximum densification combined with minimum grain growth.

5.0 CONCLUSION AND OUTLOOK

Concerning the future development of high-performance ceramics, it can be stated that, of the known strengthening mechanisms, transformation strengthening and platelet reinforcement are the most promising at the present time. Nevertheless, the potential for possible improvements is far from being exhausted, and the available spectrum of materials is not being fully utilized. A pool of unexplored materials is found in the transition metal borides, carbides, nitrides, and even some silicates that in turn may also be introduced into suitable matrices as reinforcement. It must be kept in mind that the strategies for materials optimization discussed here are essentially designed for static loads at room temperature.

The complex interactions mentioned in connection with cyclic tension -compression loading are far from being understood at either room temp-erasure or high temperatures. Here especially, there is an enormous need for innovative materials development that may have to abandon the paths that proved successful for static loading, thus blazing new trails for the future.

These developments present new challenges in the field of materials characterization. The quantitative characterization of microstructures has to be rethought to encompass grain boundary properties and microstructural defects. Methods of analysis and testing must be developed that permit the rapid acquisition of data that provide valid predictions for large structural components. Present solutions of such problems are available in the quality control methods of metal working industries but ceramics require a much higher spatial resolution of such procedures. Because of the requirement of high-volume specimen throughput, this means that fully automated high-resolution methods of analysis will have to be developed, comparable to those in microchip production, with the difference that the microstructural analysis must be three-dimensional for ceramics. Another key word is *interface atomistics.* At present, this area of ceramic research is the domain of a few specialists; however, some day it will become a routine part of ceramic quality assurance. Thus, high-performance ceramics, as a key engineering material, promise to revolutionize both materials applications and materials charact-erization in a major way.

REFERENCES

1. R. Danzer, Lebensdauerprognose hochfester metallischer Werkstoffe im Bereich hoher Temperaturen, Metallkundlich-Technische Reihe 8 (G. Petzow and F. Jeglitsch, eds.), Borntrager Berlin, Stuttgart (1988).
2. R. W. Rice, Mechanisms of toughening in ceramic matrix composites, in Proc. 9th Annual Conf. Composites and Advanced Ceramic Materials, Cocoa Beach, Florida, 1985; and in Ceram. Eng. Sci. Proc 7, The American Ceramic Society Inc., Columbus, OH (1985), 589 ff.
3. K. T. Faber and A. G. Evans, Crack deflection processes-I. Theory, II. Experiments, Acta Met. 31:565-584 (1983).
4. M. J. Hoffmann, H. Liu, and G. Petzow, Whisker reinforcement of ceramics—opportunities and problems, in Proc. Int. Symp. on Fine Ceramics, Arita/Japan (1988), pp. 169-182.
5. A. G. Evans and K. T. Faber, Crack-growth resistance of microcracking brittle materials, J. Am. Ceram. Soc. 68:255-260 (1984).
6. N. Claussen, Microstructural Design of Zirconia Toughened Ceramics (ZTC), Advances in Ceramics, Vol. 12, Science and Technology of Zirconia II, The American Society Inc. Columbus, OH (1984), pp. 325-351.
7. N. Claussen, Strengthening strategies for ZrO_2-toughened ceramics at high temperatures, Mat. Sci. Engin. 71, (1985), pp23-28.
8. P. Greil and G. Petzow, Strengthening of silicon nitride ceramics, in Proc. Int. Symp. on Fine Ceramics, Arita 86, Arita, Saga Pref. (1986), pp. 5160.
9. G. Petzow, Hochleistungskeramiken - eine neue Werkstoffgeneration, Max-Planck-Gesellschaft, Jahrbuch 1986, Verlag Vandenhoeck und Ruprecht, Gottingen (1986), pp. 62-77.

2

Deformation and Fracture of Ceramic Materials

Ioan Marinescu, Mariana Pruteanu

1.0 DEFORMATION

When a load is applied to a material, deformation occurs because of a slight change in the atomic spacing. The load is defined as *stress* (σ), and it is typically measured in units of pounds per square inch (psi) or megapascals (Mpa). The deformation is defined as *strain* (ϵ): measured in inches (or centimeters) of deformation per inches (or centimeters) of the initial length or in percent.

Figure 2.1 shows the typical ceramic materials fracture in a brittle mode with only elastic deformation prior to fracture. Typical metals fracture in a ductile mode with initial elastic deformation followed by plastic deformation.

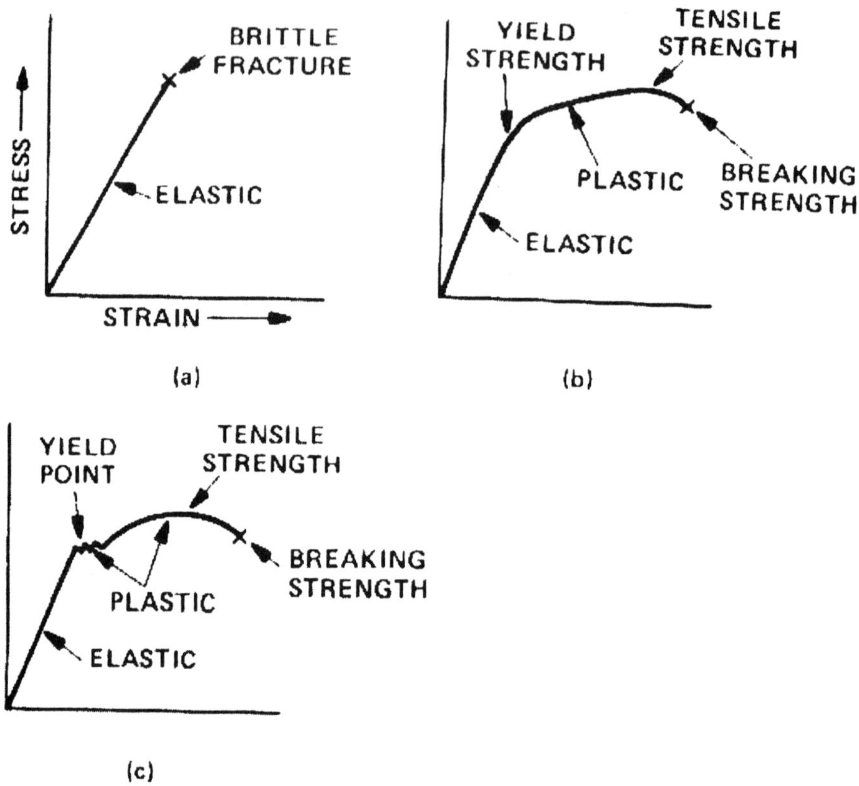

Figure 2.1: Types of Stress-strain Behavior:
(a) brittle fracture typical of ceramics.
(b) plastic deformation with no distinct yield point.
(c) plastic deformation with yield point.[1]

2.0 DISLOCATION

The mechanism of plastic deformation involves movement of dislocations. A dislocation is a defect in the way planes of atoms are stacked in a crystal structure. There are two types of dislocation: Edge Dislocation and Screw Dislocation.

Edge Dislocation

Edge Dislocation: a partial plane of atoms terminated within the crystal structure.

Screw Dislocation

Screw dislocation: produces a line of discontinuity in the crystal structure.

Under an applied load, these types of dislocations can form and multiply leading eventually to the fracture of the ceramic material.

The criteria for plastic deformation are the same as for metals, as follows:

- The presence of dislocations;
- The mechanism of generation of new dislocations under an applied load; and
- A path along which the dislocations can move.

Ceramics are known as very brittle materials. The phenomena of plastic deformation due to the dislocation activity are limited to very high temperatures especially for polycrystalline ceramics. At room temperatures, the random orientation of the grains severely inhibits dislocation motion (which terminates at the grain boundaries). The two mechanisms for plastic deformation are slip and twinning.

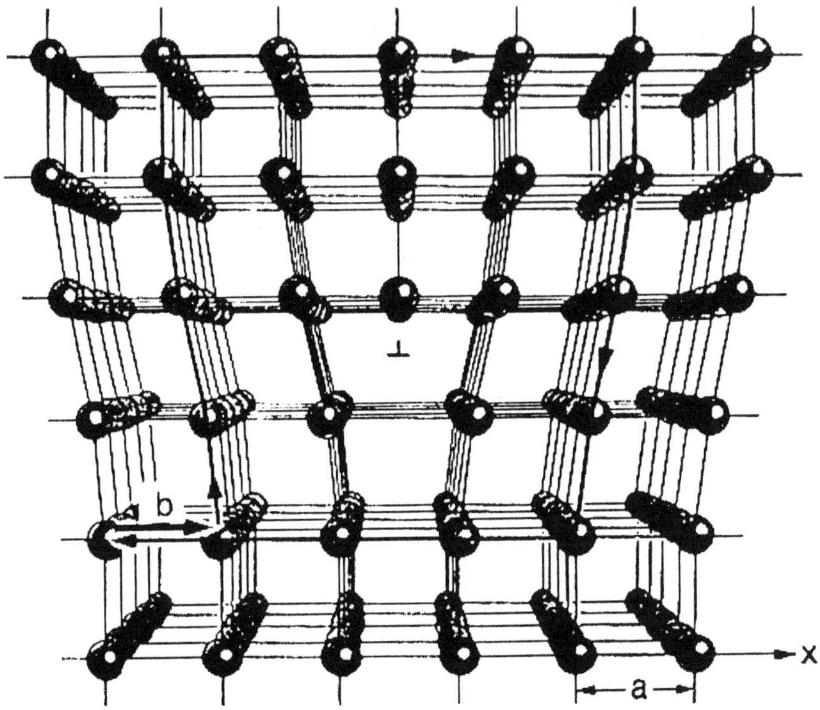

Figure 2.2: Simple Schematic Illustrating an Edge Dislocation and Showing that the Displacement b (Burgers Vector) is Equal to One Unit Cell Edge.[1]

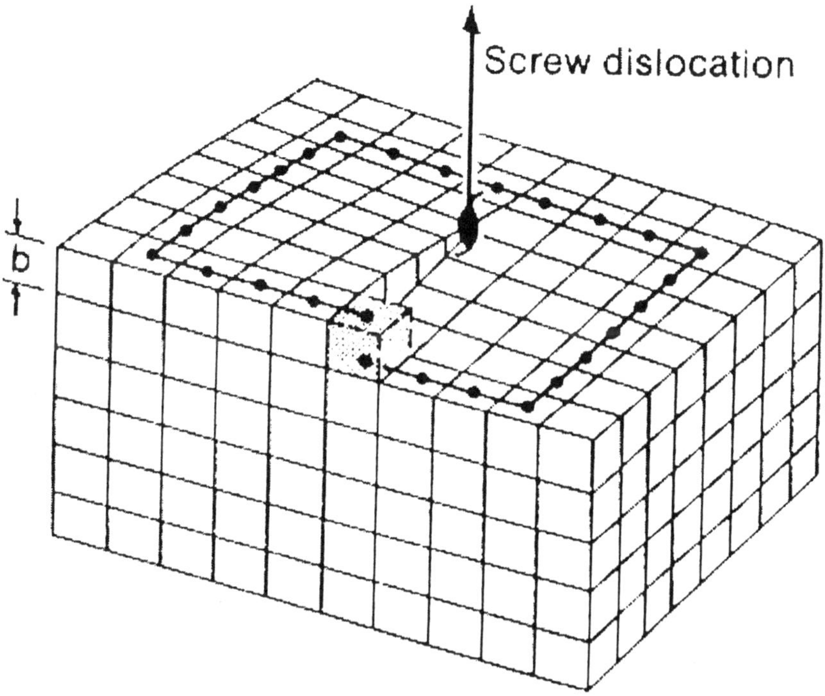

Figure 2.3: Simple Schematic Illustrating a Screw Dislocation.[1]

3.0 SLIP MECHANISM

Usually slip in ceramic materials occurs on two different slip systems. In most cubic metals, the crystallographically equivalent slip systems are sufficiently numerous to permit complete flexibility. This means that glide of dislocations on slip planes can produce all the strain components and thus produce any change of shape. In ceramics, a minimum of five independent systems is needed to permit an arbitrary change of shape.

In ceramics with ionic bonding, the slip systems depend not only on the crystal structure but also on the ionic positions in the dislocation core. Ceramic materials do not exhibit dislocation mobility at room temperatures. But at high temperatures, limited plasticity is possible, and this may have several origins including dislocation motions, grain boundary sliding, or softening of minor faces. Kronberg (1957) has discussed dislocation motion for alumina in detail. The unit cell for the alumina lattice is made up of six layers of ions parallel to the basal planes that are comprised alternately of aluminum and oxygen ions. All sites are filled on the oxygen planes; in the aluminum planes only two thirds of the sites are occupied. The unoccupied sites (or holes) are arranged regularly but differently on each of the three aluminum planes in the unit cell.

Figure 2.4 shows two of the adjacent basal planes, a completely filled oxygen plane, and partially filled aluminum ions laying in the interstices between the oxygen ions. Deformation of single alumina crystals, at temperatures of 1300°C or higher, shows that slip can occur on the {0001} basal plane. This is the closest packed plane of alumina, and this represents the expected slip plane. The detected slip direction is <1120>, and it is at 30° to the close-packed direction of the oxygen ions (see Figure 2.4). These geometrical patterns of both aluminum atoms and holes must be restored after shear. Reference to the holes in Figure 2.4 shows that for the <1010> direction, the repeat distance for holes is $\sqrt{3}$ times greater than that in the <1120> direction.

Grooves and Kelly (1963) developed a procedure for determining the number of independent slip systems.

Slip occurs most easily on the basal slip system when the temperature of operation is the lowest and the tensile flow stress the smallest of the three possible slip systems. To activate slip on the second and third (prismatic and

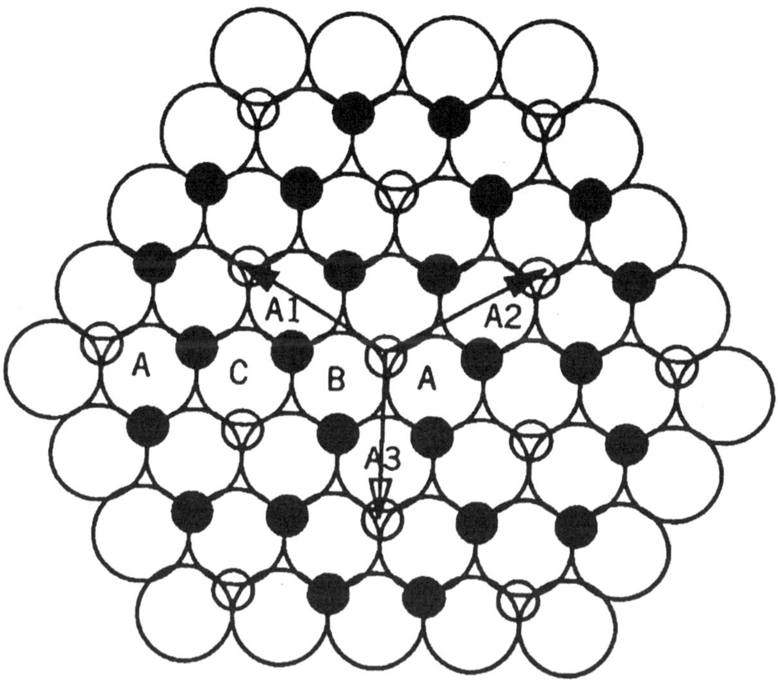

Figure 2.4: Oxygen Ions (large open circles), Aluminum Ions (small solid circles), and Unoccupied Octahedral Interstitial Sites = Holes (small open circles).[11]

Table 2.1: Slip Systems in α-Alumina[3]

System Name	Slip System	Remarks
Basal	(0001)1/3 <2110>	Dominant system under shear super-imposed on 1-atm pressure
Prismatic	{1210} <1010>	Occurs above 1600°C under shear superimpo-sed on 1-atm pressure
Pyramidal	{1011}1/3 <1101> or possibly {1012}1/3 <1101> or possibly {2113}1/3 <1101>	Occurs above 1600°C under shear superimpo-sed on 1-atm pressure

pyramidal) slip system, the temperature must rise significantly. The basal slip system can be activated most easily, followed by the prismatic and pyramidal ones. For instance, at 1500°C the stresses needed to activate the prismatic and pyramidal slip systems exceed the stress to activate the basal slip system by a factor of 8 and 16 respectively. Thus, it can be concluded that alumina oxide is a ceramic material characterized by a strong deformation anisotropy. To achieve manoscopic deformation of polycrystalline body by slip, five independent slip systems must operate.

4.0 TWINNING MECHANISM

Twinning is an important mode of plastic deformation, and it has been observed both in single crystal and polycrystalline ceramics. As compared to plastic deformation by dislocation glide, twinning occurs at low temperatures and high strain rates. The experiments on polycrystalline alumina between room temperature and about 500°C have shown twinning to be the predom-

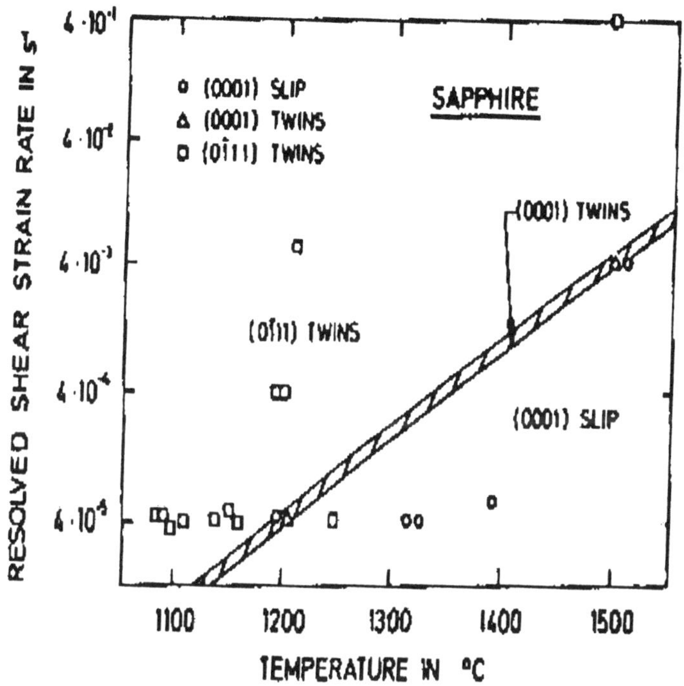

Figure 2.5: Occurrence of Twinning and Slip as a Function of Strain Rate and Temperature[2]

inant mechanism.[2]

Figure 2.5 illustrates a strain-rate temperature map with three important deformation modes:

- At elevated temperatures and small strain rates (0001) slip prevails;
- At low temperatures and high strain rates, rhombohedral twinning predominates; and
- For a very narrow band of experimental parameters there is a range of (0001) twinning.

5.0 FRACTURE OF CERAMIC MATERIALS

It is very important to understand the strength of real polycrystalline ceramics. For this, it is necessary to emphasize the behavior of microcrack under stress. A lot of scientists have studied the linear elastic fracture mechanism (LEFM); this refers to a crack in a continuous body without taking into consideration what happens on an atomic scale. Among them are Kelly and MacMillan (1986), B. Lawn (1993), Hasselman, Munz, Sakai and Shercshenko (1992).

Inglish (1993) says that, considering a sample under a tensile stress σ, if an elliptical flaw is introduced, the tensile stress will change especially near the end of the flaw.

The flaw has the length 2c in the X direction and 2b in the Y direction. The equation of the ellipse is:

$$\frac{x^2}{c^2} + \frac{y^2}{b^2} = 1 \tag{2.1}$$

and ρ - the radius of the curvature is:

$$\rho = \frac{b^2}{c} \tag{2.2}$$

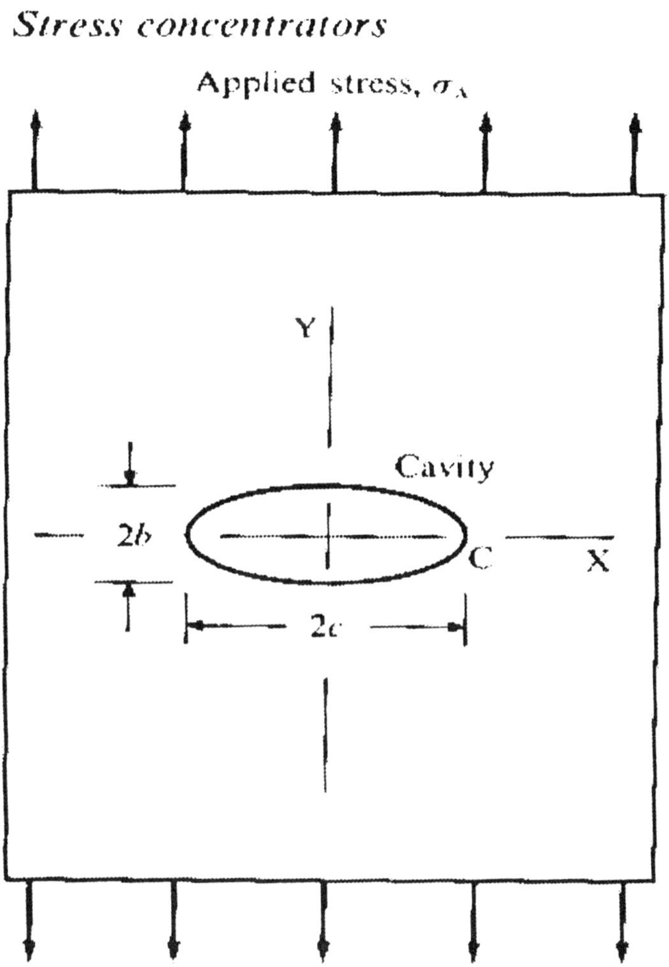

Figure 2.6: Plate Containing Elliptical Cavity , Semi-axes b, c, Subjected to Uniform Applied Tension σ. C denotes "notch-tip".[3]

The stress distribution around the flaw is very complex:

$$\sigma_{yy} = \sigma(1 + \frac{2c}{b}) = \sigma[1 + 2(\frac{c}{p})^{1/2}] \qquad (2.3)$$

In this equation, there are two characteristics of the elliptical crack: the total length of the crack and the radius of the curvature at the end of the crack perpendicular to the initial stress.

Griffith (1920) presents a theory of brittle fracture based on energy considerations. This theory is still considered the basis of fracture mechanics. He based his treatment on the idea that the brittle body contains small flaws that act as stress concentrators when an external tensile stress is applied. Stresses cannot relax by local plasticity, and the local stress in the vicinity of the most severe flaw may reach the theoretical strength, thus causing the failure of the body.

The Griffith equation of stress is:

$$S = \frac{1}{y}(\frac{2\Upsilon_1 E}{c})^{\frac{1}{2}} \qquad (2.4)$$

where: S is the strength, γ_1 is the fracture energy per unit area necessary to initiate failure, E is Young's modulus, c is the flaw size, and y is a dimensionless constant that depends on the flow size: sample size ratio.

The above equation shows that the strength of ceramics is mainly controlled by two variables: fracture energy and flaw size ratio.

Before discussing the two variables, we present two types of flaw: extrinsic flaws and intrinsic flaws. Extrinsic flaws are defects generated by a particular process of fabrication. The presence of this type of flaw doesn't depend on the type of ceramic material. Intrinsic flaws are originated at pores, large grains, or second phase particles. Also, they may be generated due to the anisotropy of elasticity or thermal expansion.

Fracture Energy

If a material fractures, energy is required to create the new surface. In polycrystalline ceramics the fracture mode can be transgranular or intergranular, and it is a complicated process compared with the cleavage of a single-crystal or the separation of a bi-crystal. The experimental values of the fracture energy of polycrystals γ_{pc} are bigger than the experimental values of the single crystal γ_{sc} by a factor of 3-10. According to some scientists (Rice and Davidge), this happened because the true fracture surface of a polycrystal is bigger than the projected surface used for calculating the fracture energy per unit area. They also say that the tortuosity of the fracture surface determines a lot of microscopic deviation of the local crack front from the path of the microcrack. These processes depend on the microstructure: pores, grain size, and the presence of second phases.

When porosity increases, the fracture energy decreases because an increase in porosity makes the cross-sectional area to be fractured smaller.

$$\Upsilon_1 = \Upsilon_0 \exp(-bP) \qquad (2.5)$$

where: γ_0 is the fracture energy per unit area (necessary to initiate failure), b is an empirical constant, and P is the volume fraction of pores.

The influence of the grain size on fracture energy is contradictory depending on the testing method and on the type of specimen used to determine γ. Some authors showed that when the grain size is smaller than 10 μm the fracture energy increases with decreasing the grain size. The grain size dependence on the fracture energy may be due to the microscopic details of crack propagation. For example, in a NB test (notched bend specimen) fracture starts from a single flaw, the dimensions of which are small compared to the specimen dimensions.

Fracture energy increases when microcracking occurs. In ceramic materials, the microcracking can be due to high thermal expansion anisotropy (TEA) or to porosity. The thermal expansion anisotropy provokes stresses across the grain boundaries that become bigger when the grain size is bigger. The dependence of fracture energy on temperature can be seen in Figure 2.8. The fracture energy decreases when temperature increases above 1000°C.

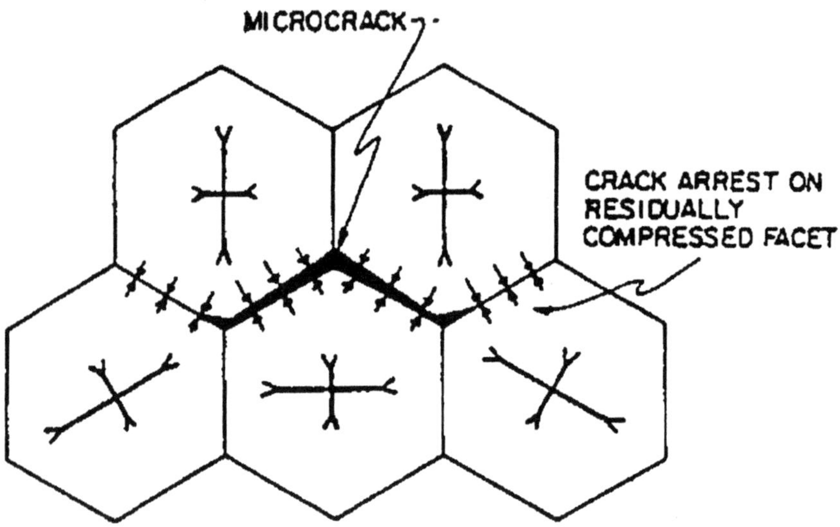

Figure 2.7: The Location of Microcracks with Respect to the Residual Stress on the Grain Facets.[11]

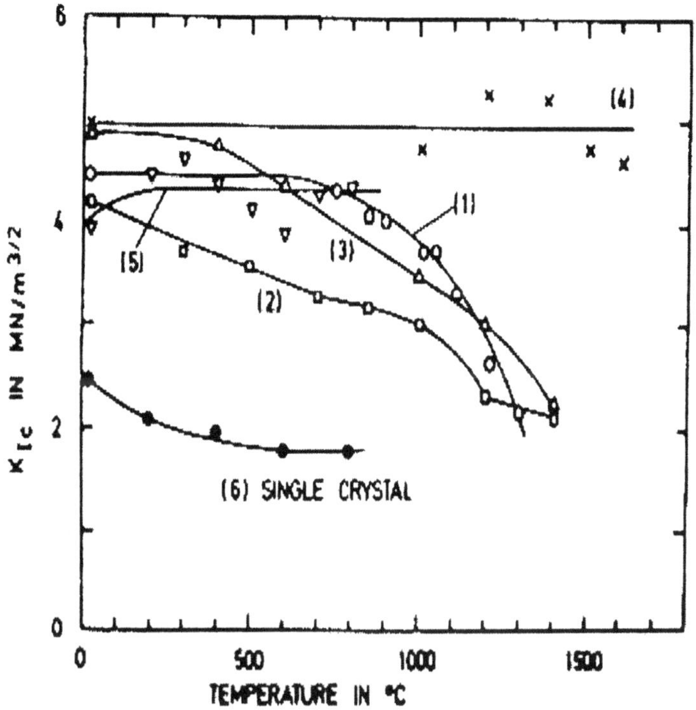

Figure 2.8: Temperature Dependence of Fracture Toughness. Curves (1) to (5): Polycrystalline Alumina with different grain sizes. Curve (6): Single-crystal Alumina.[2]

Types of Flaws

The flaw size c is an important variable that controls the strength of ceramic materials. Flaws are considered to be pre-existing or stress-induced. The first category has two types of flaws: extrinsic flaws (due to external conditions, like machining and sintering), and intrinsic flaws (which depend on material properties and microstructure).

One type of extrinsic flaw is the stepped flaw. They form by the linking up of the flaws in two or more different planes parallel to the fracture surface. When the ceramic material has a small grain size ($c>G$, which means the flaw size is bigger than the grain size) the flaw depends only on the machining conditions. In this case, the strength of material is independent of the microstructure.

The residual pores determine another type of extrinsic flaw after sintering. Rice observed that the uniform porosity consisting of many small pores is less dangerous than a small number of large pores. When $c \approx G$ (the flaw size equals the grain size) the strength does not depend anymore on matching processes because the size of the largest grains determines the critical flaw size.

Intrinsic stress-induced flaws may be divided into two groups: those generated by dislocation interaction and those generated by mechanical twinning. Some authors showed that the flaws are generated by dislocation interaction. Other authors showed that the flaws generated by dislocation appear because of the formation of Griffith cracks. Fracture of ceramics can also be determined by mechanical twining in both basal and rhombohedral planes. For example, in polycrystalline aluminum oxide, mechanical twins (twin bands) were detected in many samples. When these twin bands intersect grain boundaries, microcracks can occur and failure appears to be caused by the coalescence of such microcracks.

Different types of flaws found in aluminum oxide are shown in Table 2.2.

Table 2-2: Flaw Types Found in Aluminum Oxide[2]

(A) Pre-existing flaws	
Extrinsic flaws	Intrinsic flaws
Edge flaws Stepped flaws Induced flaws	Large surface grains Large subsurface grains Groups of large grains Large grain plus one pore Impurity particles
(B) Stress-induced flaws	
Dislocation-nucleated flaws in single crystal aluminum oxide Twin-nucleated flaws in single-crystal aluminum oxide Twin nucleated flaws in polycrystalline aluminum oxide	

6.0 INDENTATION IN CERAMIC MATERIALS

Many authors (Lawn, Marshall, Evans and Lankford) described another kind of fracture named indentation. Indentation is very important because it enables us to quantify the brittleness of ceramic materials.

Indentation refers to the contact of a hard indenter on a sample surface, the response of the material, and the crack formation and propagation. In describing indentation fracture, we must consider the contact stress field within which the cracks evolve. This field is very dependent on the indenter shape and material properties (hardness, toughness, and elastic modulus).

Lawn and Marshall demonstrated that hardness H (which represents resistance to deformation) and toughness K_c (which represents resistance to fracture) are very important characteristics to describe the brittleness of ceramic materials. In fact, they proposed that the ratio H/K_c be an index of brittleness.

They studied the evolution of the deformation/fracture pattern for

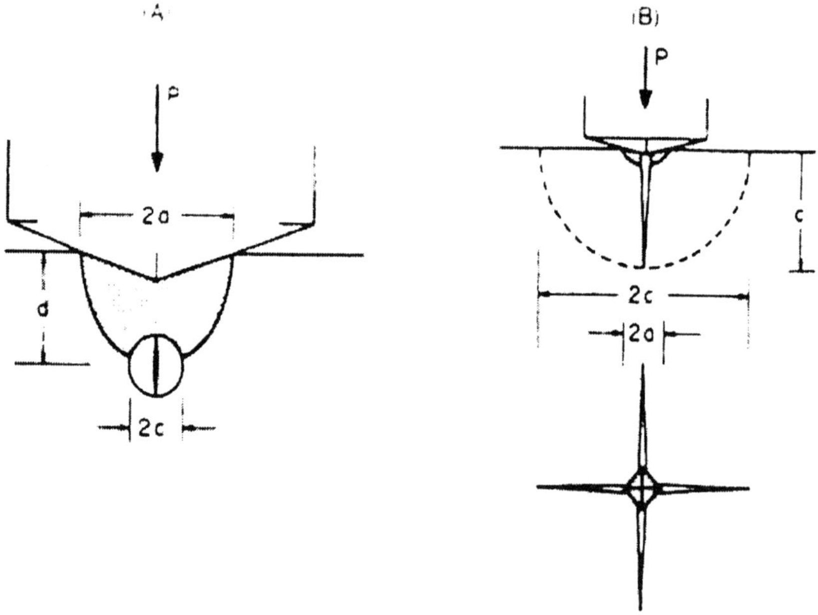

Figure 2.9: Idealized Deformation/Fracture Pattern for Vickers Pyramid Indentation: showing (A) initiation and (B) propagation stages of median crack development.[7]

Vickers pyramid indentation, and they also obtained a universal deformation/fracture diagram. Figure 2.9 shows two distinct stages during indentation: initiation and propagation. The initiation cracks begin as full pennies at the boundary between the deformation zone and the surrounding elastic matrix (Figure 2.9a) where the tensile stress is maximum. At a critical load, the subsurface cracks become unstable and expand toward the indented surface. Figure 2.9b shows that finally the stable halfpenny configuration is attained. The equation for the deformation zone is:

$$\frac{P}{a^2} = \alpha_0 H \tag{2.6}$$

where P is the load, a is a characteristic dimension of the impression, α_0 is an indenter constant, and H is the hardness.

For the propagation stage, the median cracks develop under center loading conditions and satisfy the equation:

$$\frac{P}{c^{3/2}} = \beta_0 K_c \tag{2.7}$$

where β_0 is an indenter constant, K_c is the toughness, and c is a characteristic crack dimension.

The *c(P)* function for initiation stage is more complex:

$$P* = \lambda_0 K_c (K_c/H)^3 \tag{2.8}$$

$$C* = \mu_0 (K_c/H)^2 \tag{2.9}$$

where λ_0 and μ_0 are geometrical constants.

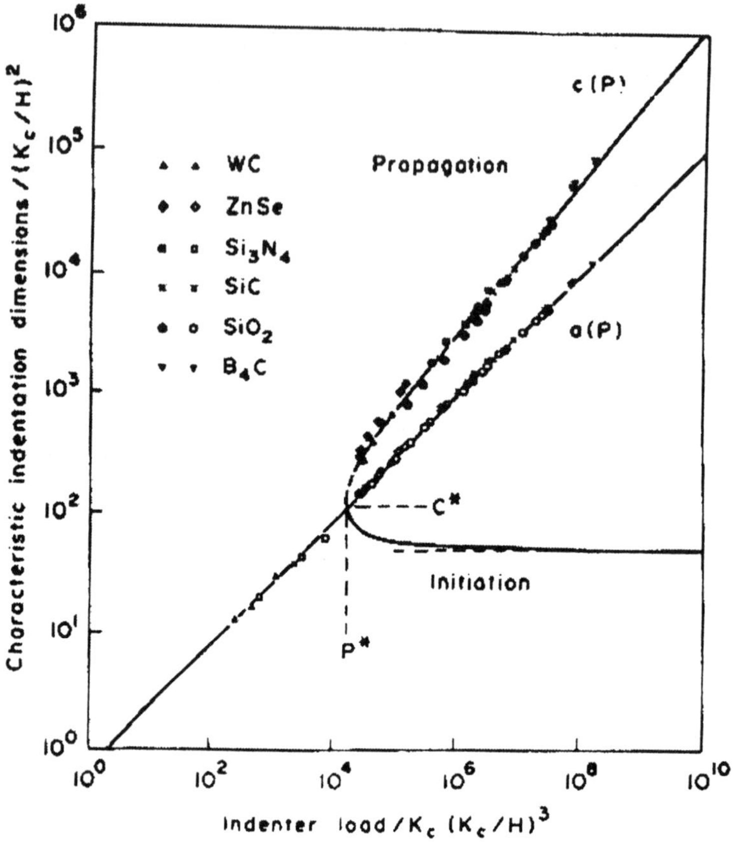

Figure 2.10: Universal Deformation/Fracture Diagram. Deformation Curve is (*P*) and Fracture Curve is *c(P)*.[7]

Figure 2.10 presents three distinct regions of mechanical behavior:

a) $P^*>P$ (small-scale events): the contact is insufficient to develop crack nuclei. This means that the damage is totally deformation controlled (usually metals are in this situation).
b) $P^*<<P$ (large-scale vents): median cracks are well developed, $c>>a$ and, therefore, the damage is fracture controlled (typical covalent solids are in this situation).
c) $P^* \approx P$ (intermediate-scale events): this region represents the threshold where fracture and deformation are comparable.

When the size of crack nuclei (a) becomes comparable with (c), the mechanics response is changed from hardness-controlled to toughness-controlled. So, the ratio K_c/H determines the problem of ductility, and the ratio H/K_c becomes an index of brittleness.

S. Malkin and T.W. Hwang (1996) showed that the indentation fracture mechanics refer to deformation and fracture under a normal contact with a Vickers pyramidal indenter. They observed under the indenter a zone of plastic deformation and two types of cracks that emanate from this zone: median/radial and lateral cracks. Lateral cracks determine the material removal and median/radial cracks determine strength diminution (Figure 2.11).

The authors studied this problem for a normal load (static indenter) and also for a tangential load (moving indenter). In the first case, both median and lateral crack extension occurred not only during loading but also during unloading.

The median crack propagation is divided into two parts: an elastic component (reversible) and a residual component (irreversible). The elastic component initiates the median crack and causes it to extend during loading. The residual component provides continued crack extension, as the indenter is withdrawn. In the second case (moving indenter), there is not only a normal load but also a tangential load in the direction of motion. The tangential force will increase the tensile stress normal to the direction of motion, which means that the median crack propagation will be higher in the plane of motion (for example, test specimens showed more strength degradation when they were ground transversely than longitudinally).

Lateral cracks appear during unloading near the bottom of the elastic zone, and they will develop laterally on a plane almost parallel to the specimen surface. If these kinds of cracks extend toward the specimen surface, this

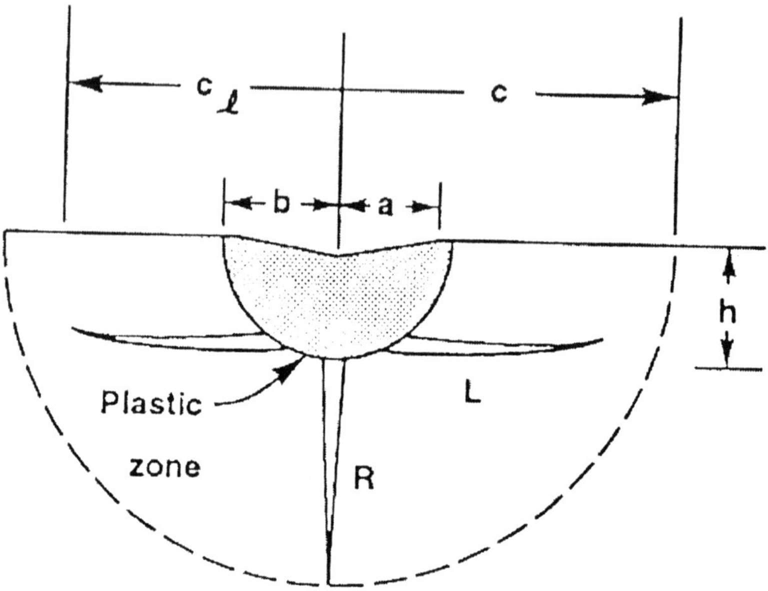

Figure 2.11: Plastic Zone, Median/radial (R), and Lateral Cracks (L) for Vickers Indentation.[8]

means that material is removed by fracture (chipping). The crushed zone boundary beneath the moving indenter occurs by fractures originating at pre-existing flaws where the elastic strain energy density exceeds a critical value.

Hyo-Sok Ahn, Lanhua Wei, and Said Jahanmir (1995)[9] illustrated the subsurface damage in ceramics using an ultrasonic technique (normal-incident compression waves) and a thermal wave measurement technique. They used samples from silicon nitride and soda lime glass; the surface and subsurface cracks were introduced by Vickers indentation. The subsurface lateral cracks were detected by ultrasonic technique while the lateral cracks and the median/radial cracks were detected using the thermal wave measurement techniques (Figure 2.12).

For the experiment, they used normal loading by sharp indenters (Vickers and Knoop diamond pyramidal indenters), and they obtained three types of cracks (radial, median, and lateral cracks) under and around the indenter. Figure 2.12 shows that radial cracks begin from the edge of the contact impression in a direction perpendicular to the specimen surface. Median cracks are situated under a plastic deformation zone (also perpendicular to the surface with a circular geometry in the crack plane; they are named penny-shaped median cracks). These cracks can extend toward the surface and join with the radial cracks resulting in halfpenny cracks. The lateral cracks appear during unloading below the surface, and they extend parallel to the specimen surface.

Figure 2.13 shows the optical micrographs of the Vickers indent for the two different loads (7.84 N and 98.0 N) for silicon nitride specimen. The lower load produced a Vickers impression without visible median/radial cracks, and the larger load produced this type of crack at the indent corner.

The thermal wave measurement technique is very sensitive and can be used to detect surface and near-surface microcracks because microcracks in a material change the local heat flux.

The results of the ultrasonic measurements showed a close correlation between the location of the median/radial cracks and the distortion of the contours.

To conclude, the authors demonstrated that the ultrasonic technique is able to detect subsurface lateral cracks while the thermal wave measurement technique is used to determine both median/radial and lateral cracks in ceramic materials. These types of cracks can be separately identified when using two different deflection components of the probe beam in the thermal wave measurement technique.

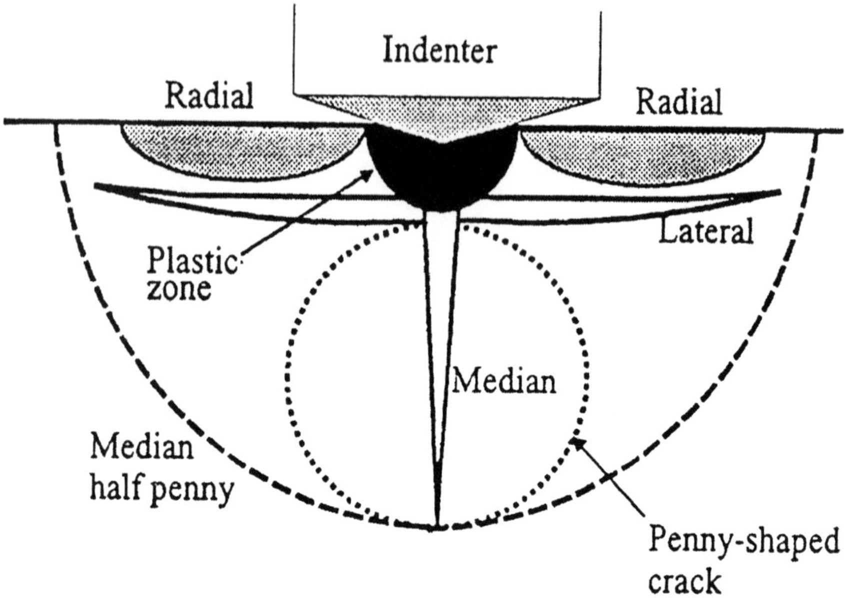

Figure 2.12: Schematic Diagram of Cracks Generated by Indentation of Brittle Materials with a Sharp Indenter.[9]

(a)

(b)

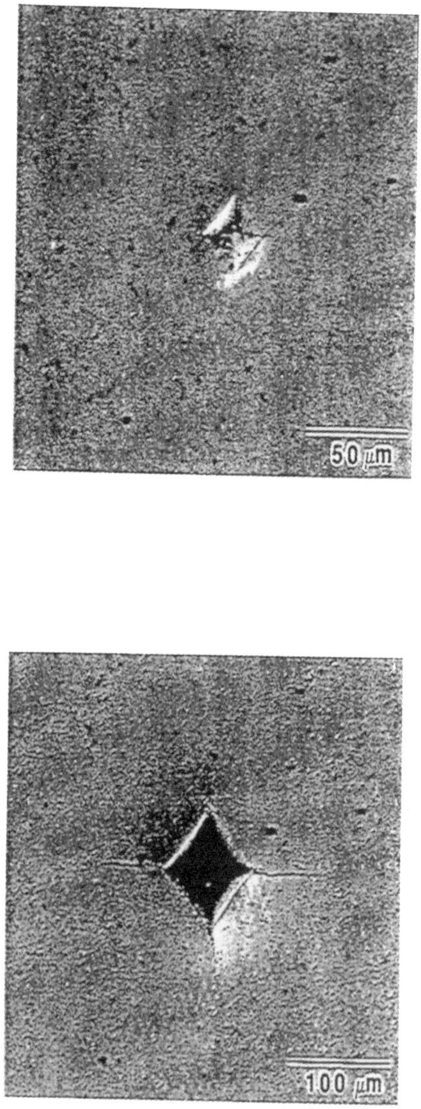

Figure 2.13: Optical Micrograph Indents in Silicon Nitride[9]:
(a) 7.84 N load
(b) 98.0 N load

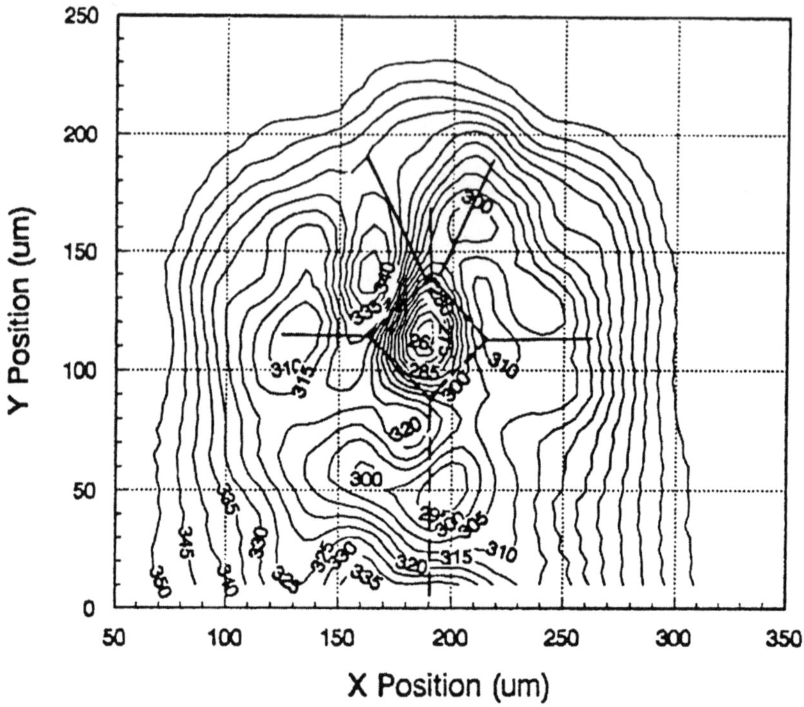

Figure 2.14: Ultrasonic Echo-Amplitude Contour Plot of the Upper Indent, using a 50 MHZ Transducer.[10]

Modes of Crack Tip Deformation

Usually, the load plane is symmetrical with respect to the crack plane. This situation is named "opening mode" (mode 1).

Figure 2-15 illustrates three possibilities of deformation:

- opening mode (mode I): has symmetry on the (x, y) and (x, z) planes;
- sliding or shear mode (mode II): has anti-symmetry on the (x, z) plane and symmetry on the (x, y) plane; and
- tearing mode (mode III): has anti-symmetry on the (x, y) and (x, z) planes.

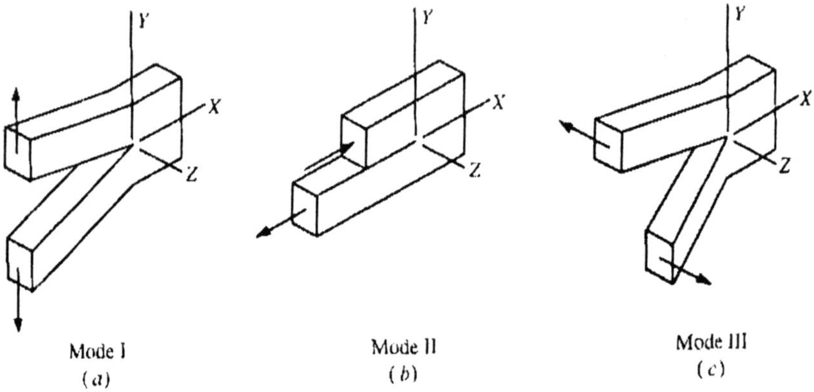

Figure 2.15: Three Modes of Failure[6]:
(a) Opening Mode, or Mode I,
(b) Sliding Mode, or Mode II, and
(c) Tearing Mode, or Mode III.

REFERENCES

1. David W. Richerson - "Modern Ceramic Engineering; Properties, Processing and Use in Design", Second Edition, 1992 by Marcel Dekker Inc.
2. E. Dbrre, H. Hijbner - "Alumina in a", Springer - Verlag, 1984
3. John B. Wachtrnan - "Mechanical Properties of' Ceramics", A Wiley-Interscience Publication by John Wiley & Sons Inc., 1996
4. Brian Lawn - "Fracture of Brittle Solids", Second Edition, Cambridge University Press, 1993
5. Shigeyuki Sornlya - "Advanced Technical Ceramics", Academic Press Inc., 1984
6. P.P. Benham, R.J. Crawford, C.G. Armstrong - "Mechanics of Engineering Materials", Second Edition, Longman Group Limited, 1996
7. B.R. Lawn, D.B. Marshall - "Hardness, Toughness, and Brittleness : An Indentation Analysis", J. of the American Ceramic Society, vol 62, no.7-8, pg.347-350, 1979
8. S. Malkin, T.W. Hwang - "Grinding Mechanisms for Ceramics", Annals of the CIRP, vol. 45/2/1996, pg. 569-580
9. Hyo-Sok Ahn, Lanhua Wei, Said Jahanmir - "Non-Destructive Detection of Damage Produced by a Sharp Indenter in Ceramics", J. of American Ceramic Society, Jan. 1995
10. R.F. Cook, E.G. Lineger, R.W. Steinbrech, F. Deuerler - "Sigmoidal Indentation Strength Characteristics of Polycrystalline Alumina", J. of American Ceramic Society, vol.77, no.2, pg.303-314, 1994
11. G. Evans - "Fracture in Ceramic Materials Toughening Mechanisms, Machining Damage, Shock", Noyes Publications, 1984
12. R.W. Davidge - "Mechanical Behavior of Ceramics", Cambridge University Press, 1979

3

Abrasive Processes

Ioan Marinescu, Brian Rowe, Yin Ling, and Hans G. Wobker

1.0 TYPOLOGY OF ABRASIVE PROCESSES[1][2]

The term *abrasive machining* usually describes processes in which the individual cutting edges are not only distributed but also more randomly oriented.

There are several processes and types with various bases for their typology. The type of tool is accepted as the first classification criterion.

Bonded Abrasive Processes

Grinding

The grinding process is most easily understood with reference to the abrasive grits (hard particles with sharp edges) bonded into a wheel which rotates with a high speed. The orientation of individual grains is random; therefore, a grain may encounter the workpiece surface with a positive, zero, or, as is most likely, with a negative rake angle (Figure 3.1). The consequences of this process depend on the depth of engagement.

The geometry of grinding is as multivariable as other machining processes (Figure 3.2). Some common grinding operations include surface grinding with a horizontal spindle (Figure 3.2(a)), surface grinding with a vertical spindle (Figure 3.2(b)), cylindrical grinding (Figure 3.2(c)), internal grinding (Figure 3.2(d)), centerless grinding (Figure 3.2(e)), and form (plunge) grinding in (Figure 3.2(f)).

Coated Abrasive Machining

The abrasive grit may be bonded to a paper, fabric, or metal surface with a water-soluble or water-resistant adhesive. The grit is spaced in a controlled manner and is often deposited electrostatically to align its sharp edges perpendicular to the backing surface.

Honing

Honing is the finishing process that renders the smoothest surface texture and the greatest accuracy. The abrasive is made into a slab (stone, stick), and is moved in a rapid oscillating motion over the surface (usually a hole) to be finished (Figure 3.3a). Superfinishing is a variant in which oscillating motion is imparted to a fairly large stone, and the surface pressure is kept very low. Thus, as the surface becomes flatter, it builds up its own hydrodynamic lubricant film which terminates the action of the abrasives (Figure 3.3b).

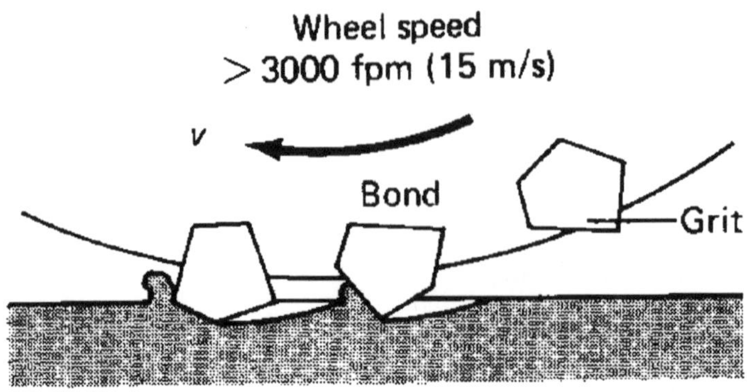

Figure 3.1: The Process of Grinding.[1]

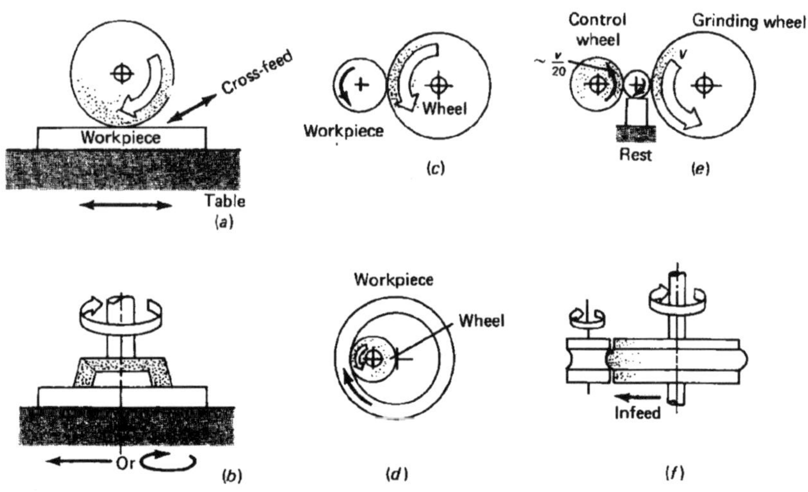

Figure 3.2: Various Grinding Processes: (a) Surface Grinding, Horizontal Spindle; (b) Surface Grinding, Vertical Spindle; (c) Cylindrical Grinding; (d) Internal Grinding; (e) Centerless Grinding; and (f) Form (Plunge) Grinding.[1]

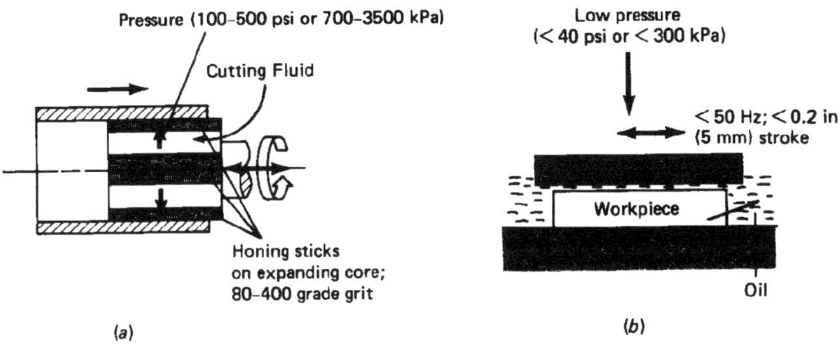

Figure 3.3: Schematic Illustration of Random-motion Abrasive Machining with Bonded Abrasives: (a) Honing and (b) Superfinishing.[1]

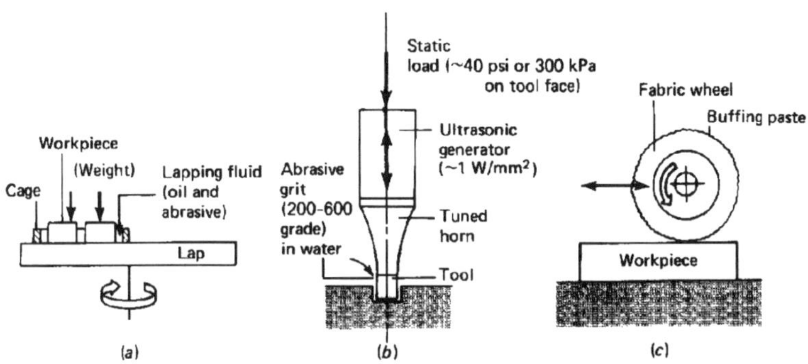

Figure 3.4: Machining with Loose (Unbonded) Abrasives: (a) Lapping, (b) Ultrasonic Machining, and (c) Buffing.[1]

Machining with Loose Abrasives

Lapping

Lapping is a process that employs a fine, loose abrasive in a fluid suspension to produce very fine surface texture and a high degree of flatness (Figure 3.4a). The workpiece describes a planetary movement and acquires a uniformly-machined, random, flat finish.

Ultrasonic Machining

The specialized process of ultrasonic machining utilizes a small-amplitude (about 0.04 mm) and high-frequency vibration (for example, 20,000 Hz). The ultrasonic vibrations, produced by an ultrasonic wave generator, drive the tuned form tool that is made of malleable metal. Abrasive grit is supplied as a slurry to the interface, and the workpiece surface is gradually eroded (Figure 3.4b).

Buffing, Polishing, and Burnishing

In most instances, the abrasive is applied to a soft surface such as the cylindrical surface of a wheel composed of felt or other fabric (Figure 3.4c). For *buffing*, the abrasive is located in a semisoft binder. The abrasive may be used dry or submerged in oil, water, or other carrier/lubricant for polishing purposes. Both buffing and polishing are capable of producing surfaces of high reflectivity due to a smearing of surface layers but not with greater smoothness. Burnishing more quickly provides the same result, especially for holes. Furthermore, when the burnishing roll penetrates to some depth, it produces a more favorable residual stress pattern. Burnishing is therefore preferred for applications where fatigue resistance is important.

Barrel Finishing

A completely random process, barrel finishing or tumbling is of great value in removing burrs and fins from workpieces and, generally, in improving their surface appearance. In principle, the workpieces are placed into a barrel (often with a many-sided cross section to prevent bunching up of the workpieces). When a barrel is rotated, mutual impact removes surface protuberances. Much-improved surface texture occurs when a tumbling medium is added, either in a liquid carrier (wet) or by itself (dry). The medium is chosen to suit the purpose and may range from metallic or nonmetallic balls to chips, stones, or conventional abrasives.

Grit and Shot Blasting

Abrasive particles are projected with high velocity at the workpiece surface to remove surface films, such as oxides, and to impart to the surface a uniformly mat surface texture. In grit blasting, the particles are irregularly shaped. In shot blasting, the particles are approximately spherical. Controlled shot-blasting promotes plastic surface deformation and residual compressive stresses. The required velocities may be produced by compressed air or high speed wheel rotation with or without paddles. Contact with the wheel accelerates the grit particles.

A closely controlled jet of dry abrasives is sometimes used to cut slots or holes into very hard materials.

A special form of grit blasting is hydrohoning in which the abrasive medium is suspended in a liquid which is then directed onto the surface in the form of a high-pressure jet.

2.0 TRIBOLOGY OF ABRASIVE PROCESSES [3][4][5][6]

Abrasive finishing processes are manufacturing methods used to generate surfaces of desired characteristics with an economically acceptable rate of surface generation. Invariably, these processes utilize wear-resistant abrasives in the surface generation process. Sometimes, these processes are described as "finishing methods using random cutting edges" to distinguish them from methods which primarily rely on machining with defined cutting edges.

To obtain good surface quality and durability of a workpiece made of a brittle material, it is necessary to identify and use an appropriate process. This area of abrasive technology has developed from a long period of convergence of research in the fields of material science, tribology, and machining technology. The abrasive finishing process can be investigated with experimental and analytical methods, if one focuses on the interactions occurring in the "grinding zone". The "grinding zone" constitutes a complex friction pair,[6] which is suitable for study by using a tribology approach. The objective is to increase the workpiece wear rate (stock removal rate) while reducing the abrasive tool wear rate to a very low level to increase the tool life. The grindability of brittle materials is inversely proportional to the resistance to abrasive wear of the material to be ground. An increase of material removal rate is limited by the reduction in the life of the abrasive tool as well as by reduced quality of the machined surface.

A description of the interactions is followed by their grouping into "machining" and "tribological" interactions at the grinding zone. The understanding, control, and management of these interactions can lead to significant advancements in abrasive finishing processes.

In all abrasive finishing processes, two requirements are observed. One requirement is the generation of the surface and the effectiveness or efficiency with which it is achieved. The second is the characteristics or quality of the generated surface. The two requirements are related to each other, as shown in Figure 3.5.

A Systems Approach for the Abrasive Finishing Processes

Abrasive finishing processes may be broadly described in terms of the abrasive tool, the machine tool, the work material, and the operational variables. As illustrated in Figure 3.6, these four input groups interact with each other, culminating in the output or abrasive finishing process results. Fur-

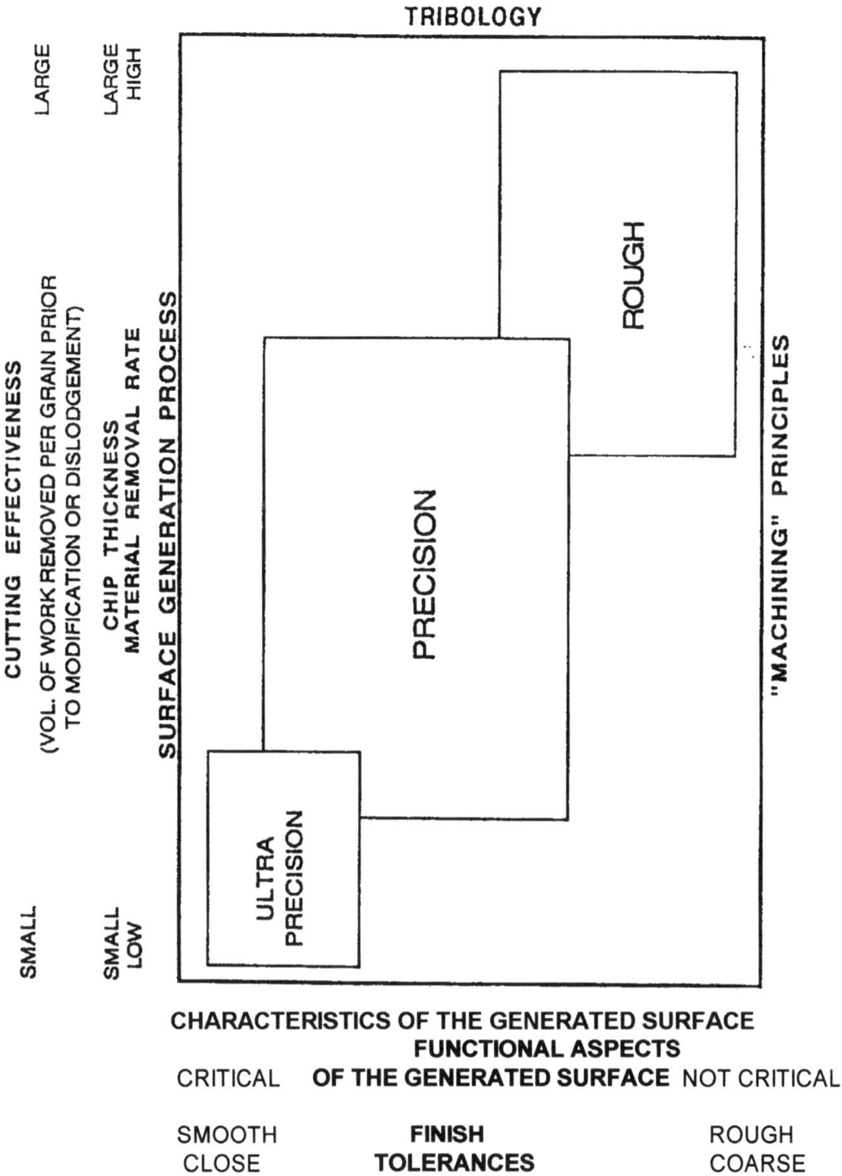

Figure 3.5: Overview of Abrasive Finishing Processes.[5]

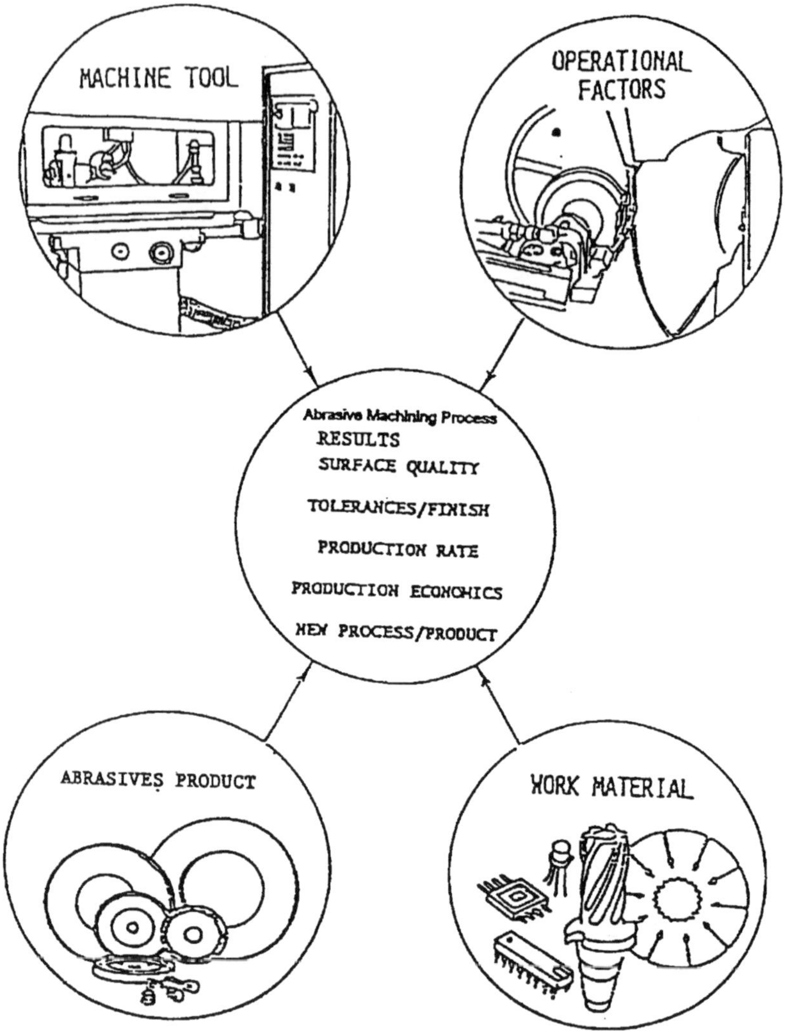

Figure 3.6: Schematic Representation of the Production Grinding System.[5]

thermore, each of these input groupings is multivariable, as described in Table 3.1. Indeed, there exist a few hundred variables for each abrasive finishing process which could cause one to believe that the abrasive finishing process is complex and can only be managed as an art. However, this need not be the case.

Irrespective of the choice of variables in the four input categories, for every abrasive finishing process it is possible to visualize four interactions between the abrasive product and the work material (Figure 3.7). Of these, the abrasive/workpiece interaction is the most critical, which in many respects is analogous to machining processes with cutting tools.

In ceramics grinding, from a tribological viewpoint, to consider the grinding wheel as a microcutting tool is an oversimplification.[7] The random distribution and geometry of abrasive grains on the active surface of the wheel are modified during the grinding process. The modifications are due to the simultaneous actions in the contact zone involving three different mechanisms: microcutting, plowing, and rubbing.[8] In addition to the effects on distribution and micro-geometry, the abrasive process is also affected by the type of bond, the nature of the workpiece material and coolant, and by the mechanical working parameters.[9][10]

Tribometers with monograins[11][12] can model very simplified abrasion machining processes without indicating the type of wear of abrasion materials from tools for long life. For this purpose, a modern grinding machine with variable speed and high accuracy of depth of cut should be used.[13][14] Table 3.2 proposes the correlation between the main tribological parameters and technological parameters specific to the grinding process.

Tribological Interactions in Abrasive Finishing Processes

Grinding processes for brittle material include cutting (plastic deformation), plowing and rubbing/sliding. The last two mechanisms are characteristic of tribological processes. In order to understand the tribological interactions in brittle material grinding, it is necessary to investigate the interaction between the following:

Table 3.1: Selected Variables Influencing Abrasive Finishing Processes.[5]

MACHINE TOOL FACTORS

DESIGN
- Rigidity
- Precision
- Dynamic Stability

FEATURES
- Controls
- Horse Power
- Spindle Speed
- Slide Movements/Axes
- Positioning Accuracy
- Repeatability
- Thermal Stability
- Truing and Dressing Equipment

COOLANT SYSTEM
- Type, Pressure, Flow
- Nozzle Arrangements
- Filtration System

WORK MATERIAL FACTORS

PROPERTIES
- Mechanical
- Thermal
- Chemical
- Abrasion Resistance
- Microstructure

GEOMETRY
- Wheel-Part Conformity
- Access for Coolant
- Shape/Form Required

PART QUALITY
- Geometry
- Tolerances
- Consistency
- Surface Characteristics

GRINDING RESULTS
- SURFACE QUALITY
- RETAINED STRENGTH
- PRODUCTION RATE
- COST PER PART
- COMPONENT PERFORMANCE

ABRASIVE PRODUCT

ABRASIVE
- Type
- Properties
- Particle Size and Shape
- Size Distribution
- Content/Concentration

BOND
- Type
- Hardness/Grade
- Stiffness
- Porosity
- Thermal Conduction

WHEEL DESIGN
- Size/Shape
- Core Material
- Form or Profile

OPERATIONAL FACTORS

FIXTURES AND WORK HOLDING

WHEEL BALANCING

TRUING, DRESSING AND AND CONDITIONING
- Devices
- Techniques
- Parameters

GRINDING CYCLE DESIGN

WHEEL/WORK CONFORMITY

CHIP THICKNESS

INSPECTION METHODS

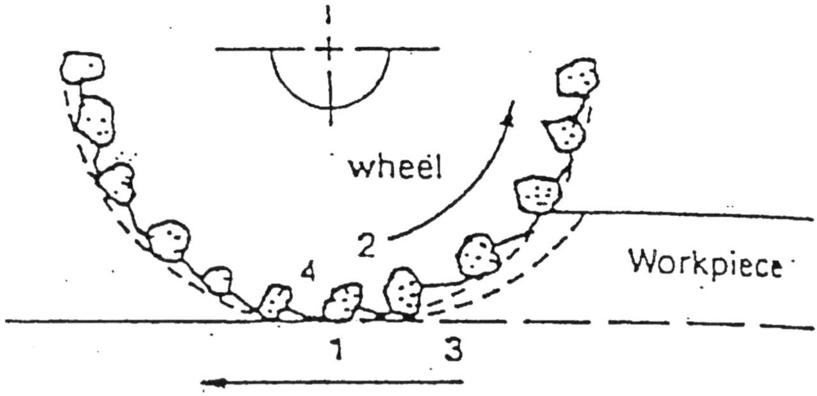

Figure 3.7: Interactions in the Grinding Zones.[5]
(1) Abrasive/workpiece Interface
(2) Chip/bond Interface
(3) Chip/workpiece Interface
(4) Bond/work Interface

Table 3.2: Correlation Between Tribological and Technological Parameters.[6]

Tribological Parameters	Technological Parameters
Wear particle width b_w *[μm]*	Equivalent grinding thickness h_{eq} *[μm]*
Linear wear of workpiece ***D**h_w [mm]*	Ground material thickness ***D**h_w [mm]*
Linear wear of wheel **D** h_s *[mm]*	Radial wear ***D**r_s [mm]*
Workpiece wear V_w *[mm³]*	Material removal V_w *[mm³]*
Wheel wear V_s *[mm³]*	Wheel wear V_s *[mm³]*
Wear time t_s *[s]*	Grinding time t_s *[s]*
Workpiece wear rate Q_w *[mm³/s]*	Stock removal rate Q_w *[mm³/s]*
Wheel wear rate Q_s *[mm³/s]*	Tool time *T [h]*
Wear ratio ***Φ** [-]*	Grinding ratio *G [-]*
Load F_n	Normal force F_n
Frictional force F_f	Tangential force F_f
Friction coefficient *μ [-]*	Force ratio μ_A *[-]*
Friction energy W_f *[J]*	Specific grinding energy ***u** [J/ mm³]*

- Superabrasives-workpiece,
- Workpiece-bond,
- Chip-bond, and
- Chip-workpiece.

The mechanism is complex, and it is necessary to consider the nature of the active layer of the wheel in order to understand the tribological aspects.[15]

The relative motion between the wheel and workpiece constitutes a 'friction-pair'.[16] If grinding is to be considered as a tribological process the wear of the wheel and of the workpiece must be considered. The wheel wear is defined by V_s, and workpiece wear by V_w, which is the material removal. The grinding ratio G is a convenient indicator of the tool life and tool cost and also gives an indication of correct wheel selection for the performance required.

The dominant wear mechanism of the active abrasive layer appears to be by grain attrition, grain micro-fracture, and abrasion of the bond, rather than poor grit retention caused by weakening of the bond matrix. Figures 3.8a, 3.8b, 3.8c, 3.9a, 3.9c show the wear flats and fatigue wear induced grain fracture caused by the large compressive stresses.

Figures 3.8a, 3.8b and 3.8c also show a relatively uniform distribution of the grains on the active layer surface with a comet-tail effect during the grinding process. In this effect, the bond remaining behind each grain provides a buttressing which has a positive effect upon the retention of the grains in the bond material.

Fatigue wear evidenced by grain and bond fractures provides automatic self-sharpening of the wheel which helps prevent burn due to the development of wear flats. The effect is evident in Figures 3.8d, 3.8e, 3.9b and 3.9d. The wear flats can be seen in Figures 3.8g and 3.8h. These figures also indicate a degree of bond erosion evidenced by loss of grains.

In Figures 3.8h, 3.9d and 3.9e, we observe signs of adhesive wear, also known as transfer wear. Small particles are transferred from the workpiece surface to the top of the super-abrasive grain surface. This process is similar to the three body abrasive process but is not a predominant wear mechanism for brittle material grinding.

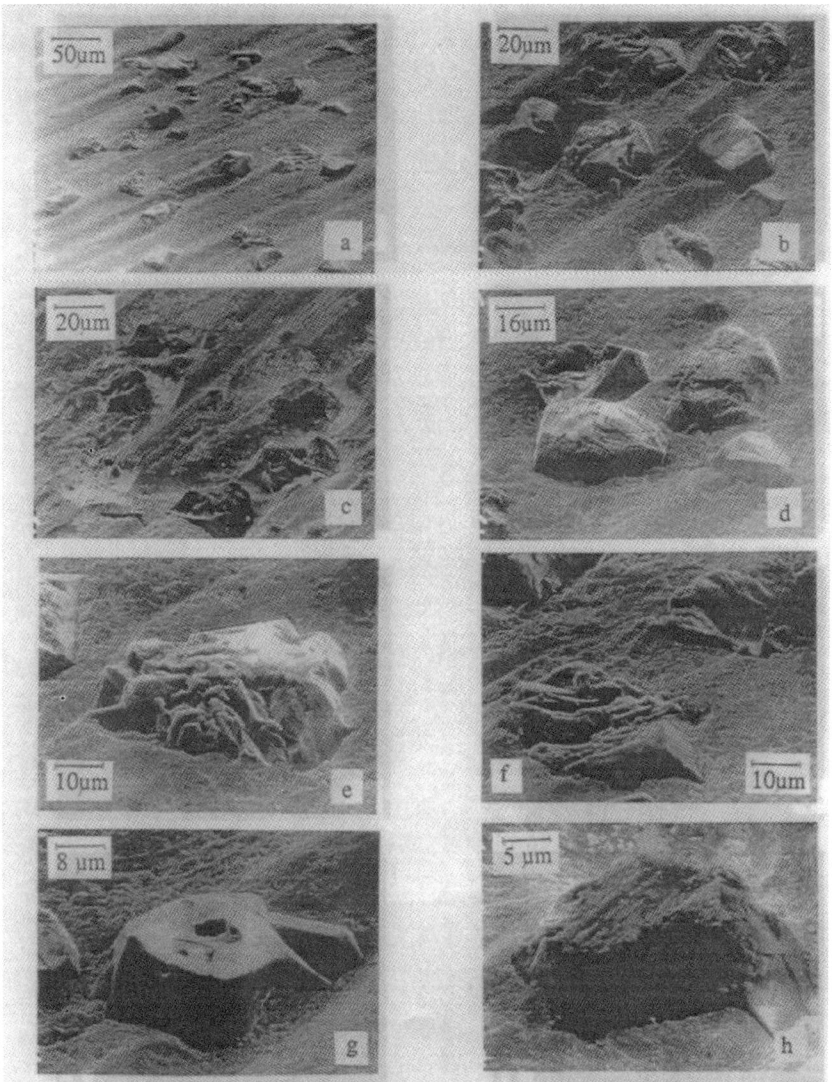

Figure 3.8: The Distribution of the Grains on the Active Layer Surface.[1][6]

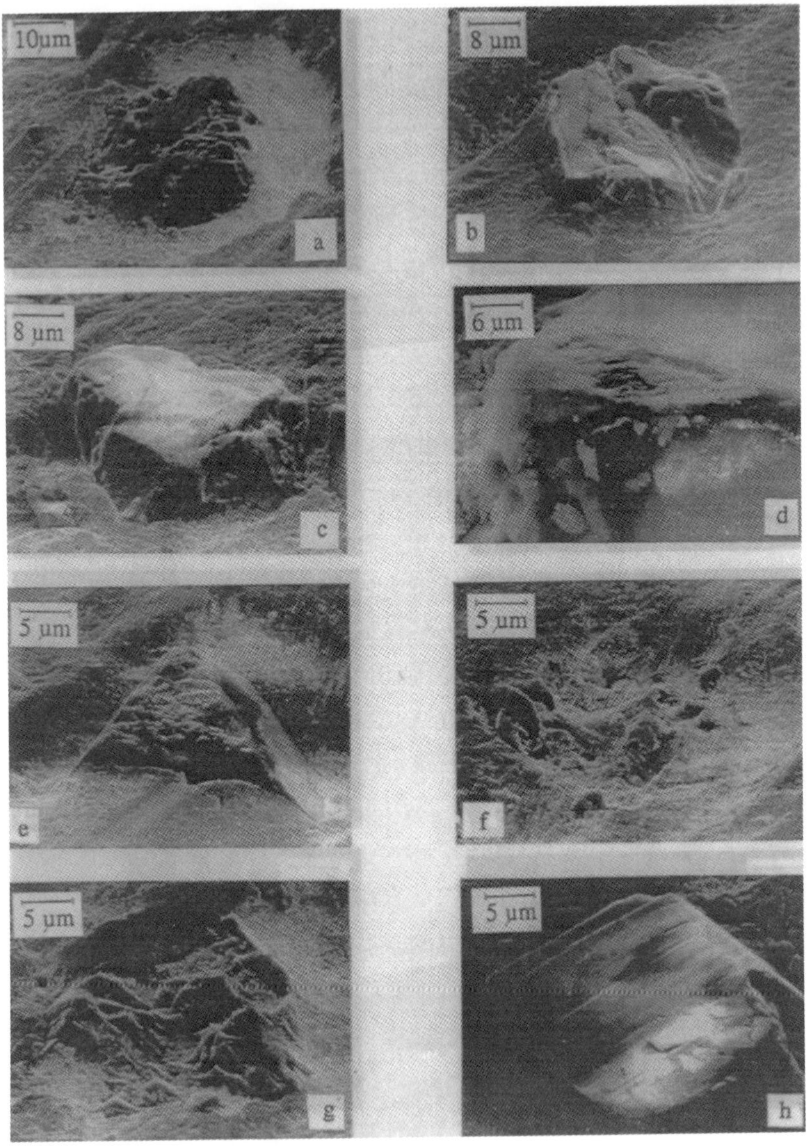

Figure 3.9: The Distribution of the Grains on the Active Layer Surface.[11][6]

There are only a few previous studies that attempted to link the wear of diamond wheels with the quality of the obtained surfaces for both brittle materials and ductile material grinding.

3.0 SINGLE POINT SCRATCH TESTS

Single point scratch tests have been performed by pin-on-disc sliding, fly-milling, scratching and ploughing, etc. in order to understand the removal mechanisms of ceramics machining and the frictional behavior of diamond abrasives during the machining processes.

Pin-on-disc-sliding [17]

The friction measurement, as well as the abrasive grain wear test, is shown in Figure 3.10. The friction force is measured by a strain gauge mounted on the thin part of the steel holder. Thus, both the variation of the friction force with sliding distance and the coefficient of friction under steady state conditions can be measured. Diamond abrasive grains with a grit size of 2.0 mm are used in the test. The disc (100 to 200 mm in diameter) materials are different ceramics. The vertical load is either 2.5 or 3.5N and the sliding speed can be varied between 5.0 and 15 m/s. A scanning electron microscope is used to examine the worn surface of the diamond abrasive grains. An abrasive grain slides in its own particular track during the measurement of friction, and each separate grain has its own track. Scatter in the coefficient of friction is small and, therefore, is determined as the average for three individual grains. The wear volume of the diamond abrasive is determined from its loss of height and the worn flat area. Wear tests on the diamond abrasives are carried out at the same speeds and normal loads as in the friction tests, where the sliding distance is almost 4000m, but the volumetric wear is based on the average of about six experiments due to the increased scatter of the data. The groove volume of wear for the disc is also measured, using a surface profilometer.

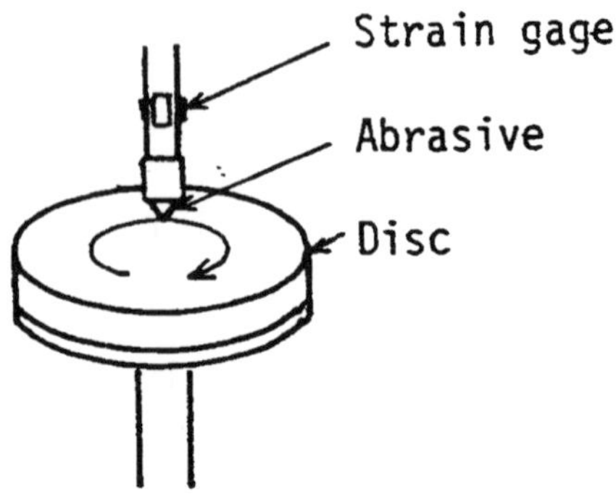

Figure 3.10: Measurement of Frictional Force.[17]

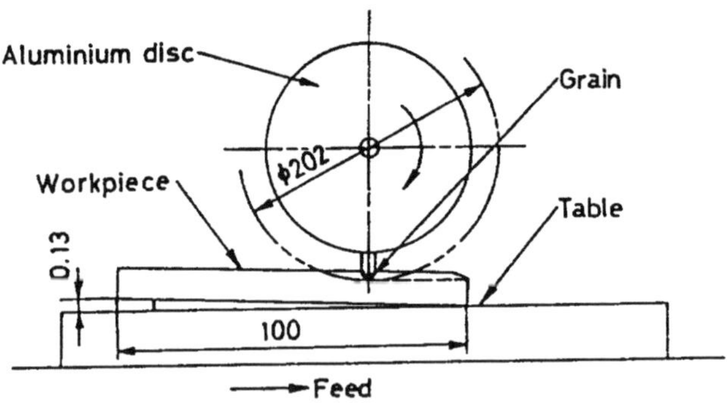

Figure 3.11: Set-up for Fly-milling Test with an Abrasive Grain.[18]

Fly-milling [18]

Overcut fly-milling tests on diamond abrasive grains are carried out with a surface grinder and with the ceramic workpiece slightly inclined, as shown in Figure 3.11. The method is similar to that of Stanislao (1969) and Brecker (1973). When making a measurement, the metal wheel is lowered until it just touches the workpiece surface at its lowest end. As the workpiece is fed past the wheel, the depth of the groove is automatically increased, due to the slope of the work surface (but it may cycled up and down if the abrasive grain chips). Each subsequent cut increases in depth, until the grain fractures, while attrition wear of the tip occurs. The measurement of attritious wear is made by tracing across the grooves with a stylus instrument to determine the decrease in depth of cut between successive grooves. The conditions for the fly-milling tests can be chosen in accordance with surface grinding parameters. The scratched grooves and worn diamond grains can be examined by a scanning electron microscope.

Scratching and Ploughing [19]

Figure 3.12 diagrams an apparatus used for scratching experiments. The sample *S* is moved horizontally underneath the diamond *D* by the micromanipulator *Mi*, which is driven by the motor *Mo*. The diamond is attached to the arm *A* of a balance. This arm is connected to the frame by a leaf spring *LS*, so that it can be swung up and down but remain stiff horizontally. The balance is brought to equilibrium and displaced vertically until the diamond just touches the ceramic sample; the desired load *L* is then applied. The tangential force on the sample is measured by the strain gauges *SG* on the bending element *BE* (Figure 3.12b). In experiments, diamonds were ground to the shape of a square pyramid; the scratches were made in the one plane as in Figure 3.13. The scratching speed was 1 μm/s.

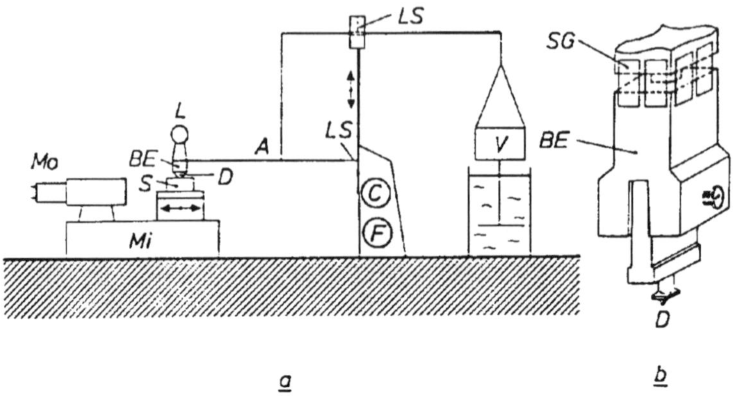

Figure 3.12: (a) Diagram of the apparatus. The sample *S* is displaced underneath the diamond *D* by the motor *Mo* and the micromanipulator *Mi*. The diamond is attached by the bending element *BE* to the balance arm *A*, which can be swung up and down but is stiff horizontally. After the balance has been brought to equilibrium and adjusted vertically the load *L* is applied. *C, F* coarse and fine vertical adjustment. *V* counterweight and damper. *LS* leaf springs. (b) The bending element *BE* with the diamond *D* and strain gauges *SG* for measuring the tangential force.[19]

Scratching and Indenting [20]

A specially designed nano-indentation and scratching system used in ceramic damage tests is illustrated in Figure 3.14; it includes several improvements over an earlier system. It includes an electromagnetic driver with attached diamond tool (indenter) and a position-determining eddy current sensor. The sensor measures the relative displacement between the diamond tool and the test specimen. Some elastic deflection of the system is included in the measured displacement but constitutes a small proportion of the signal, e.g. 40 nm at a load of 300 mN. A computer-controlled digital waveform is fed

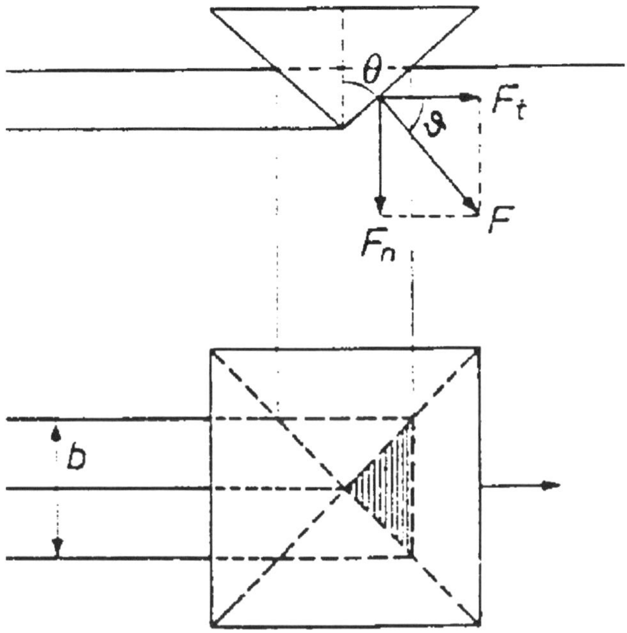

Figure 3.13: Geometry of diamond and groove; vertical longitudinal section and plan view. The diamond is a square pyramid (half apex angle θ) that moves with one plane leading. *F* is the force the diamond exerts on the sample, F_t is the tangential component and F_n the normal component, *b* is the groove width. ('Tangential' and 'normal' relate to the surface of the sample, not to the leading plane of the diamond). In the ploughing model in its simplest form is perpendicular to the leading plane.[19]

to the driver to apply the load to the tool. Vertical position measurement sensitivity is about 1 nm and loads are directly measured by a load cell to about 0.1 mN. Thermal drift of the system in the vertical direction is monitored and

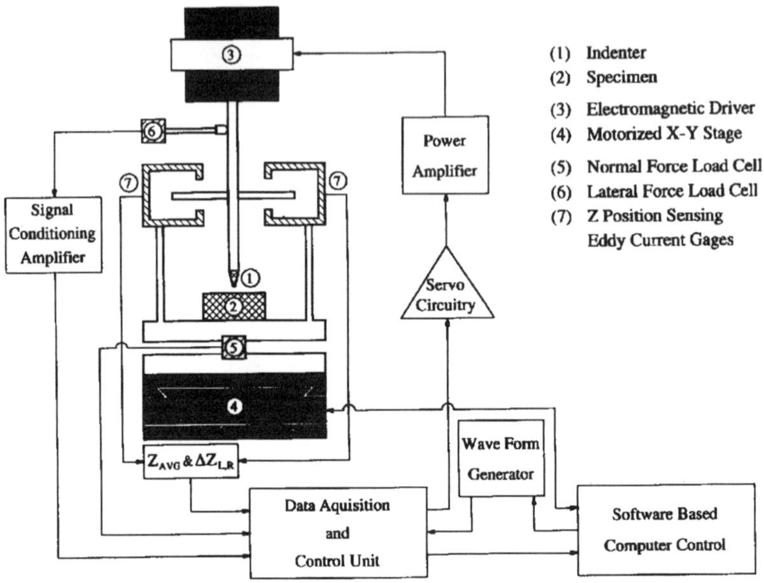

Figure 3.14: Schematic Illustration of the Indentation-scratching System.[20]

automatically corrected by the system software. A digitally-controlled *X-Y* position stage allows generation of multiple indentation patterns of any form and also permits scratching by translation of the specimen during indenter loading.

The load-time waveform used for indentation in these studies was basically triangular, typically ramping linearly from zero to maximum set load in 60 s, then holding for 5 s, before ramping to zero load in 60 s. Surface contact (zero depth) was determined by the change in the indenter tip vibration amplitude as it moved into actual contact with the specimen surface. Scratching was conducted in this work at selected loads over the range 5-500 mN. Either stepped (constant) or linear increasing load waveforms were used for the individual scratches. For example, in one study, sets of five scratches with 80 μm length were programmed with different linear ramping loads of 10-40, 40-80, 80-120, 120-160 and 160-200 mN. The ramping load scratches made it possible to detect threshold load values for cracking. The stepped-load scratches were used to examine uniformity of material response, e.g. from grain to grain, at each load increment. In a typical test, motion was not started until the load had been increased to the desired value. At the conclusion of the scratch, stage motion was stopped and the load was ramped down to zero. Scratching speed was in the range 2-10 μm s^{-1} for all tests. A load cell measured the lateral force resisting the scratching of the material, and an eddy current sensor provided tool penetration data. Analysis of data recorded during testing along with post-test scanning electron microscope (SEM) observations of scratch damage morphology provide information about the scratching and indenting process.

Two diamond tool shapes are used, one of which is a four-sided diamond pyramid. Vickers indenter with a tip angle of 136°, face-to-face. The shape of the tip apex (determined from SEM photographs) consists of a short wedge tip segment of length 300 nm. Another tool is wedge-shaped with a curved edge (0.5 mm radius), a -45° rake angle and an 8° clearance angle. The edge is oriented perpendicular to the sliding direction. This type of tool is used in single-point diamond turning of crystalline materials. Illustrations of both tools are found in Figure 3.15. The curved edge scratching tool is expected to produce a more elastically stressed volume because of a wider contact width and area, higher compressive stresses ahead of the tool (due to the flat face-on orientation), and a lower average contact pressure for a given load. The Vickers diamond pyramid, moved along an edge, is expected to produce more shearing and lateral displacement of the material ahead of the tool. Wear of the pyramid tool occurs and necessitates occasional tool replacement while wear of the curved edge tool is barely noticeable.

The environments in which indenting and controlled scratching are carried out include air (RH=25-50%), deionized water, purified mineral oil, mineral oil with 0.25 wt.% stearic acid and two water-based commercial grinding fluids containing chemical additives. The liquid environments are freshly applied prior to each test, and some make-up liquid is added as requ-

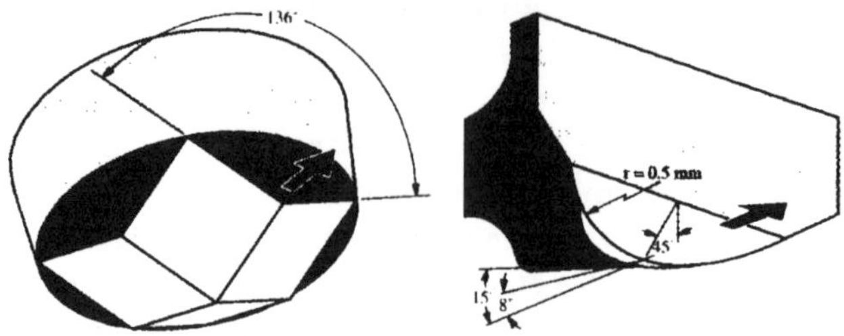

Figure 3.15: Schematic Illustration of the Two Diamond Tool Shapes used, (a) Pyramidal (Vickers) and (b) Wedge-shaped, Curved Edge.[20]

ired during the test.

The single point scratch tests provide a better understanding of the microprocesses. The experimental conditions can be clearly defined, e.g. for the geometry of the diamond and force on the diamond. However, grinding is a multi-point process. Grinding experiments with diamond wheels are necessary for the investigation of mechanisms of machining ceramics.

4.0 MULTI POINT SCRATCH TESTS

Grinding is a complex process. As the abrasive particles scratch the surface of the material several different microprocesses come into play (flow, cracking, fracture). The result depends on a number of factors, including the shape, force and the velocity of the particles, the humidity of the environment, the hardness and the crack resistance of the material and the distribution of microcracks, cavities, and other flaws in the materials. In view of cost and the importance of grinding for workpiece quality, it is useful to closely analyze this complicated process. Many researchers have been working on ceramics grinding. In general terms, the objectives of the investigation are to find out how the specific energy, i.e. the energy required for the removal of unit volume of the material, and the damage to the surface of the workpiece depend on the setting of the grinding machine and the properties of the materials. Unfortunately, ceramics are much more susceptible to surface damage during grinding than are metals, and their subsequent behavior under load is much less forgiving of induced grinding damage. Furthermore, of all the conventional machining processes, grinding is usually the most expensive per unit volume of stock removal. In the manufacture of precision components, grinding alone may constitute a significant portion of the total cost.[21] Therefore, the economic feasibility and competitiveness of high performance ceramics depends significantly on how efficiently they can be machined by grinding. Reduction of grinding costs by operating at faster stock removal rates is constrained mainly by surface damage to the workpiece, and this factor is especially critical for grinding of ceramic materials. Efficient grinding of high performance ceramic materials requires identification and control of the grinding conditions to maximize the removal rate while maintaining the required surface integrity.

The machining approach to investigating grinding mechanisms typically involves measurement of grinding forces coupled with microscopic observations of grinding debris and surface morphology. A fundamental parameter derived from force measurements is the specific grinding energy. Any proposed mechanisms of abrasive workpiece interactions must be consistent with the magnitude of the specific energy and its dependence on the operating parameters.

Grinding Debris

Examination of grinding debris for various brittle materials suggests material removal mainly by fracture. For example, particles from grinding of hot-pressed silicon nitride in Figure 3.16a appear to have been fractured from the workpiece. At higher magnification in Figure 3.16b, the particle on the right shows an irregular fractured surface and the polycrystalline microstructure. However, the orientation of the particle on the left reveals a striated

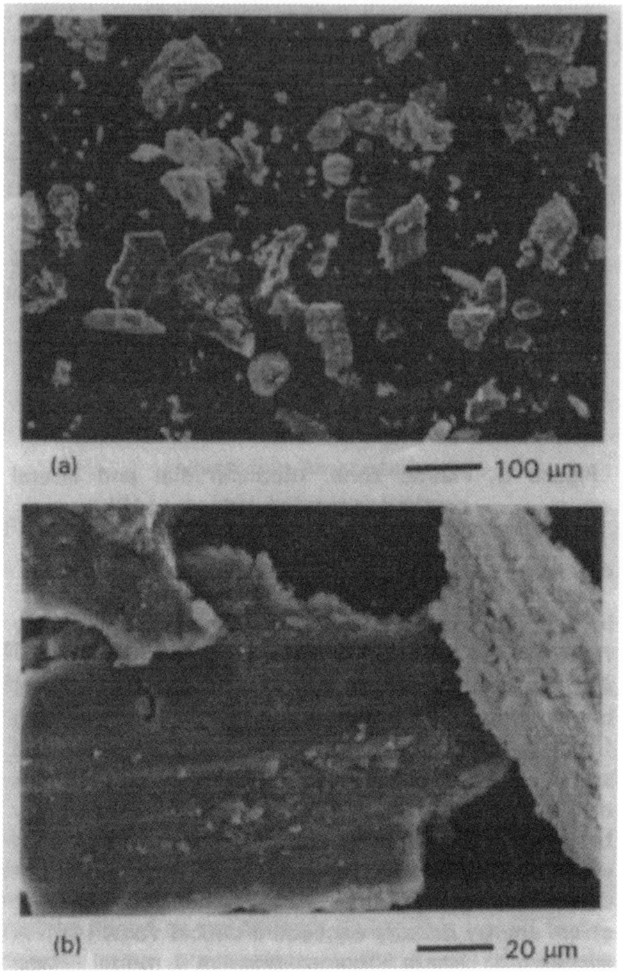

Figure 3.16: SEM Micrographs of Grinding Debris: Si_3N_4 (hot pressed).[25]

ground surface on one side appearing to illustrate plastic deformation, which may have been generated during the previous grinding pass.

Microscopy of Ground Surfaces

Grinding involves a multitude of scratches and interactions between adjacent scratches. The effect of interactions between adjacent scratches on stock removal has been investigated for hot-pressed alumina by making a series of parallel scratches with separation distances from 0 to 250 μm.[22] The results in Figure 3.17 show a strong dependence on both the separation distance and the load. When the separation is so large that the scratches do not interact or so small that scratches overlap, the material removal is relatively small. At intermediate distances, the material removal becomes much larger. Figure 3.18 shows a ground surface before and after etching (reaction bonded Si_3N_4, a=38 μm, v_w=200 mm/s, v_s=10 m/s) and exhibits ductile flow. After etching (49% HF, 100°C, 1 hour), most of the deformed striations are removed. This would indicate that much of the "deformed" surface layer consisted of glassy (amorphous) material. Some grooves generated by abrasive grains can still be seen, although other observations show almost no evidence of ductile flow after etching.[23]

Grinding Energy

Apart from the microscopic observation, it is also important to consider the specific energy in grinding of ceramics. The specific grinding energy is defined as the energy per unit volume of material removed. For metallic workpiece materials, the total specific grinding energy can be modeled as the sum of chip formation, plowing, and sliding components.[24] For ceramics and other brittle materials, the relationship between the grinding mechanisms and specific energy are more complicated.

Force and grinding energy data have been obtained for various ceramics over a wide range of conditions.[25] For straight surface grinding of hot-pressed silicon nitride (HPSN) with diamond wheels, specific energies have

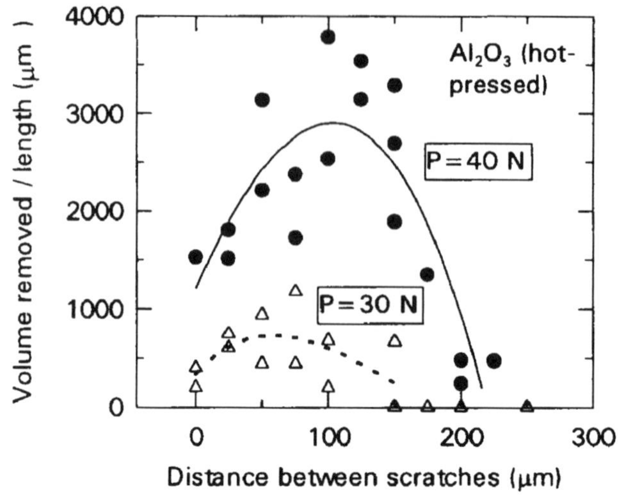

Figure 3.17: Material Removal per Unit Length Versus Distance Between Scratches.[22]

been reported ranging from about 15 J/mm^3 at extremely high removal rates to as much as 800 J/mm^3 at very low removal rates. A simple explanation for this type of behavior is that a decrease in removal rate results in a smaller undeformed (uncut) chip size, causing more flow and less brittle fracture. Results for specific grinding energy versus average uncut chip cross-sectional area in Figure 3.19 for aluminum oxide and silicon nitride show two distinct regimes.[26] Below an average uncut chip area of about 0.05 μm ,^{2}which corresponds to a maximum uncut chip thickness of about 0.35 μm, the specific energy increases steeply with smaller chip size. Above the transition, specific energy decreases more slowly as the chip size increases. The uncut chip thickness at the transition may correspond to the critical threshold depth of cut for ductile regime grinding. This would suggest material removal mainly by ductile flow below the transition and brittle fracture above the transition.

SEM observations of grinding debris indicate material removal mostly by brittle fracture, except under extremely fine grinding conditions. In an

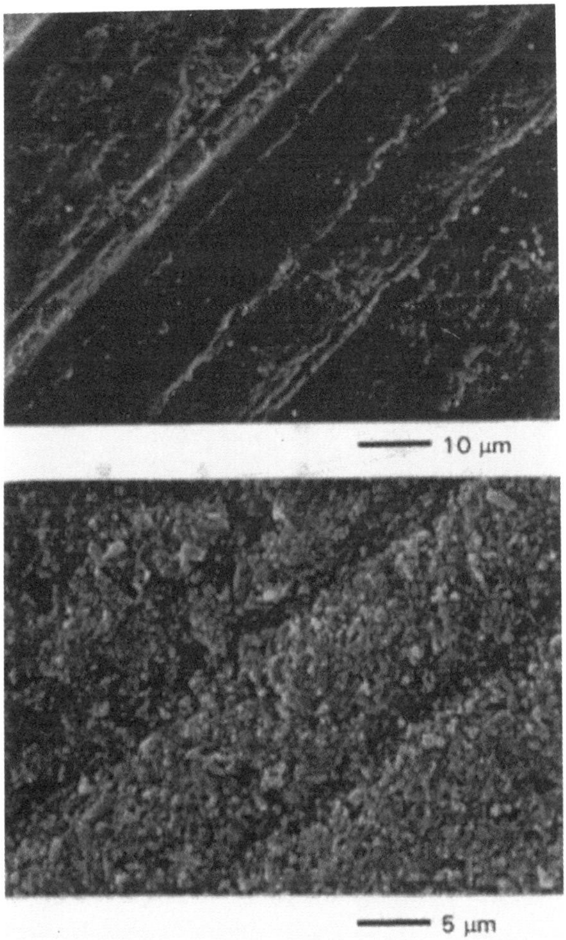

Figure 3.18: Ground Surface Before and after Etching: Si3N4 (reaction bonded), a=38 μm, V_w=200 mm/s, V_s=10 m/s.[25]

attempt to account quantitatively for the magnitude of the specific grinding energy, the portion of energy associated with brittle fracture can be estimated as the product of the surface area generated by fracture and the fracture energy

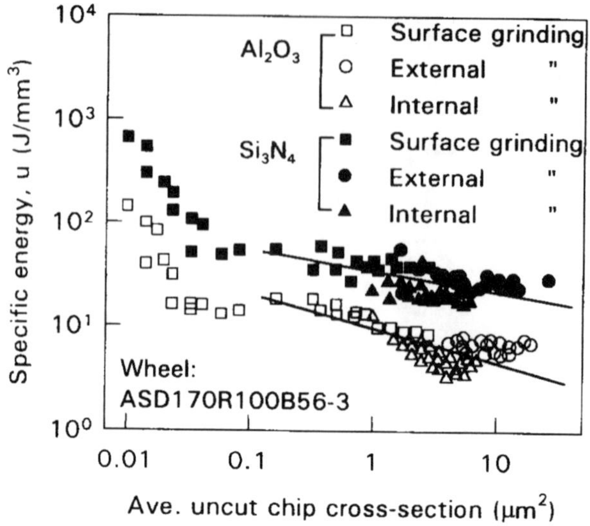

Figure 3.19: Specific Energy Versus Average Uncut Chip Cross-sectional Area.[26]

per unit area of surface generated.[25] For simplicity it is assumed that particles removed by grinding are cubes of dimension b. Approximating the fracture surface energy by half the critical energy release rate G_c ($G_c=K_c^2/E$, where E is the elastic modulus and K_c the fracture toughness) for crack formation (two surfaces), the specific fracture energy for grinding is:

$$u_f = 3G_c/b \quad \textbf{(3.1)}$$

The smallest particles removed by grinding are approximately 1 μm in size, although many are much larger. For hot-pressed silicon nitride, $G_c \approx$ 80 J/m^2. Using these values for b and G_c, $u_f \approx 0.24$ J/mm^3, which is only about 1% of the specific grinding energy.

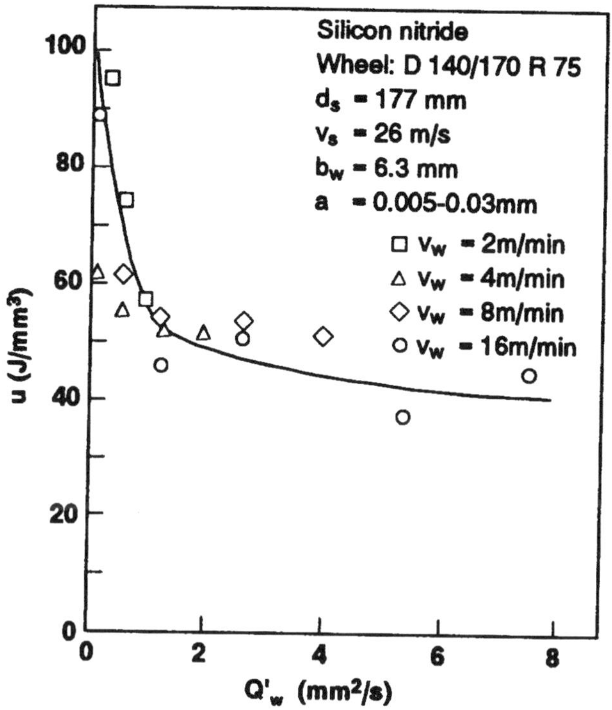

Figure 3.20: Specific Energy Versus Removal Rate per Unit Width for Straight Surface Grinding of a Hot Pressed Silicon Nitride.[27]

Some specific energy measurements are shown in Figure 3.20, which range from about 100 J/mm^3 at low removal rates down to about 40 J/mm^3 at high removal rates. Grinding of reaction bonded silicon nitride (RBSN) tends to give slightly lower specific energies. Specific energies for high density aluminum oxide are about half as large.[27]

It is interesting to note that the specific energy obtained for grinding of silicon nitride is similar to values found for grinding of steels. Paragraph analysis therefore suggests that most of the grinding energy is expended by flow-type mechanisms of abrasive-workpiece interaction rather than by a brittle fracture mechanism. Brittle fracture can account for only 1% of the measured specific energy. This suggests that most of the energy is not expended by brittle fracture but by ductile flow, even though most of the material removal is by brittle fracture.

The material removal rates in grinding of ceramics vary widely depending on the application. With recent advances in our understanding of the grinding of ceramics, it is possible to achieve material removal rates comparable to that in metal grinding. However, current practice achieves only about a tenth of these material removal rates. There is significant room for improvement in these production applications by systematically implementing the results published in the literature. Summarizing the research on ceramics grinding, here are some important observations:[28]

- Grinding is carried out with diamond wheels
- Process is operator dependent
- Process is expensive
- Approximately 75% of total part costs can be attributed to grinding damage frequently imparted to the workpiece by grinding
- The chip-formation process is predominantly characterized by brittle fractures; although plastic deformation also occurs, chips are formed in front of and behind the chip-removing abrasive grain

5.0 GENERAL MODEL OF ABRASIVE PROCESSES

Brittle Fracture Grinding Mode

Ceramic materials are generally much more brittle than metallic materials and show very little plastic deformation under load up to fracture. For this reason, it might be expected that the mechanisms of abrasive-workpiece interactions for ceramic materials would mainly involve brittle fracture. Microscopic examination of high density polycrystalline alumina surfaces ground with diamond typically show fractured areas, which are consistent with a brittle fracture mechanism. However, evidence of plastic flow with striations along the grinding direction is also observed. Transmission electron microscopy reveals dense dislocation arrays in the ground subsurface, which is unequivocal evidence of plastic flow. SEM examination of ground silicon nitride surfaces also reveals striations indicative of ductile sideflow, analogous to the plowing regime in grinding of steels, interspersed with fractured areas. This would suggest that both flow and fracture play an important role in the grinding process for ceramics. On the other hand, debris produced by grinding of silicon nitride appears to consist mainly of fragments produced by brittle fracture.

Ductile-Regime Grinding Mode[29]

Ductile-regime has been used to describe the material-removal mechanisms in grinding of ceramics. The transition from brittle to ductile material-removal at smaller cutting depths can be argued purely from considerations of material-removal energy. Specifically, for lower machining depths of cut, it can be shown that plastic flow is a more energetically favorable material-removal process than fracture. The material property characterizing resistance to plastic flow is the yield stress, σ_y. The energy (E_p) required to plastically deform a specific volume of material (V_p) can be written as:

$$E_p = \sigma_y V_p \tag{3.2}$$

The material property characterizing resistance to fracture is the Griffith crack propagation parameter, G. The energy (E_f) required for fracture is a function of the area (A_f) of new surface created by crack propagation. Thus:

$$E_f = G A_f \tag{3.3}$$

For a machining depth d, it is reasonable to assume that the order of magnitude of both V_p and a_f are determined by d. That is:

$$V_p \sim d^3 \tag{3.4}$$
$$A_f \sim d^2 \tag{3.5}$$

The ratio of material-removal energies, then, is given by:

$$\frac{Plastic\ \ Flow\ \ Energy}{Fracture\ \ Energy} \sim \frac{E_p}{E_f} \propto d \tag{3.6}$$

Consequently, as the scale of machining decreases, plastic flow becomes an energetically more favorable material-removal mechanism. The depth at which a brittle-ductile transition occurs is a function of the intrinsic material properties governing plastic deformation and fracture.

Consideration of this energy argument leads to a generalization termed the "Ductile-Regime Grinding Hypothesis." This hypothesis states that, for any material, if the dimensional scale of material-removal is made small enough, material-removal will proceed by a mechanism of plastic flow and not fracture. In grinding, the "scale" of material-removal is characterized by the depth of cut imposed on a particular abrasive grit. Figure 3.21 depicts the development of a grinding apparatus capable of ductile-regime grinding of brittle materials. The machine has a piezoelectric infeed system, plunge-grinding configuration, high machine stiffness and provides controlled cutting depths as small as 2 nm.

Specimen ground in the ductile regime exhibit enhanced strength and surface texture. Using scanning electron microscopy as a post-grinding analysis tool, the ground material surfaces were examined for evidence of surface fracture, which would be indicative of the grinding ductility. Figure 3.22 illustrates this brittle-ductile transition as a function of grinding infeed rate for fused silica. An expression for the critical (threshold) depth of cut d_c,

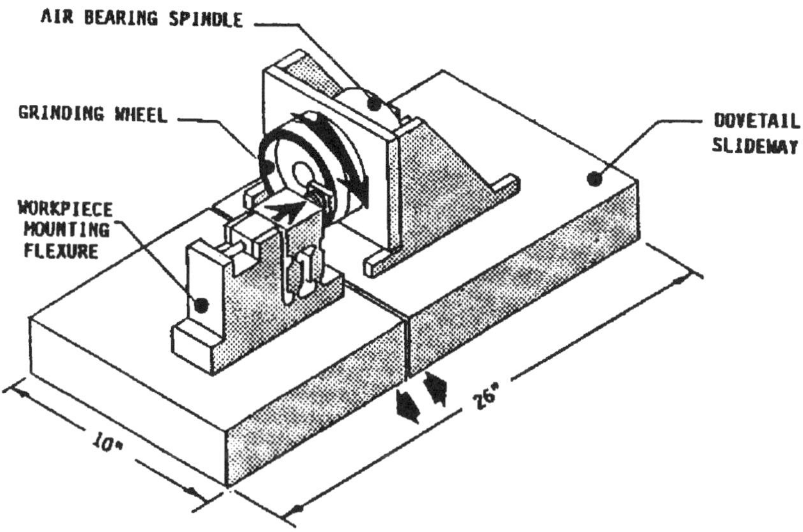

Figure 3.21: Schematic of the Pegasus Machine.[29]

corresponding to the threshold load per grain for fracture was derived as:

$$d_c = \beta(E/H)(K_c/H)^2 \tag{3.7}$$

where β is a constant, E is Young's modulus, H is the hardness, and K_c is the fracture toughness. Experimental results for 10 percent fracture (90 % ductile flow) on the ground surface, which was designated as the ductile regime limit, gave critical threshold infeeds ranging from about 3 nm to 200 nm. These results followed Equation (3.7) fairly well with $\beta = 0.15$, assuming that the depth of cut is equal to the machine infeed.

Ductile mode grinding enhances surface quality, but is very slow and costly. One possible method of promoting ductile flow and achieving removal rates is by using high wheel speeds. Higher wheel speeds reduce the underformed (uncut) chip thickness, and result in a smaller force per grit, increased ductile flow, and decreased strength degradation. Straight surface grind-

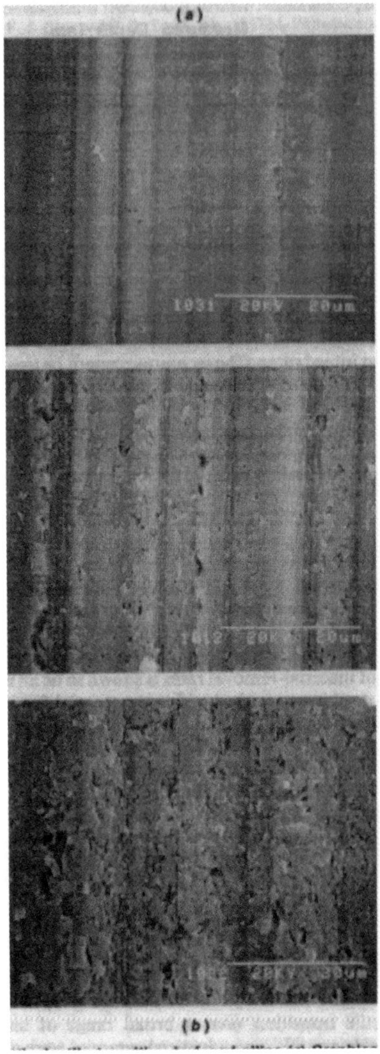

Figure 3.22: Brittle-ductile Transition in Fused Silica (a) Graphical Representation of Surface Fracture Versus Infeed Rate (b) Microphotographs Corresponding to Three Different Grinding Infeed Rates. From Top to Bottom: 2, 7.5 and 37.5 Nm/rev.[29]

ing results in Figure 3.23a show reduced normal force per grit as the wheel speed is increased up to 178 m/s. The corresponding results for percent are surface fracture (pull-out). Figure 3.23b indicate more flow at faster speeds for a given removal rate. However, the highest removal rate (10.8 mm^2/s) gave the least surface fracture, but not the lowest force per grit. This anomalous behavior may be related to the grinding temperature. While elevated temperatures are generally considered to have a deleterious effect on workpiece quality, exposure of hot pressed silicon nitride to a temperature of 800°C has been shown to increase strength due to glassy phases which blunt the flaws. Therefore, the reduction in surface fracture and apparent increase in flow may be associated with glassy phase formation at elevated grinding temperatures. It remains to be seen whether the reduction in percentage fracture at higher grinding temperatures also results in reduced strength degradation.

Powder Regime Grinding Mode [30]

The powder regime grinding mode hypothesis proposes that the material removal mechanism of ceramics grinding is mainly due to material pulverization rather than ductile deformation.

Material pulverization has been observed in both single-point grinding and diamond wheel grinding processes as shown in Figures 3.24 and 3.25. Figure 3.24 shows SEM observations of surface and subsurface regions of alumina and silicon nitride after single-point grinding tests. No cracking or chipping has been found at a depth of cut less than the respective critical value (2.5 μm for silicon nitride, 2.0 μm for alumina, 0.5 μm for silicon carbide). However, when the depth of cut was larger than the critical values, cracks were observed in the subsurface layers. These cracks were covered by a powder-regime (a pulverized layer) with a smooth surface.

Micron sized ceramic grains in the grinding region are pulverized into submicron sized grains, thereby forming a powder regime on the ground surface. The grinding-induced powder regime layer on silicon nitride was observed at a higher magnification by TEM, as shown in Figure 3.26. The TEM photo-graph indicates a variation of grain size distribution with small grains in the top and large grains in the bottom. In the middle part of the pho-

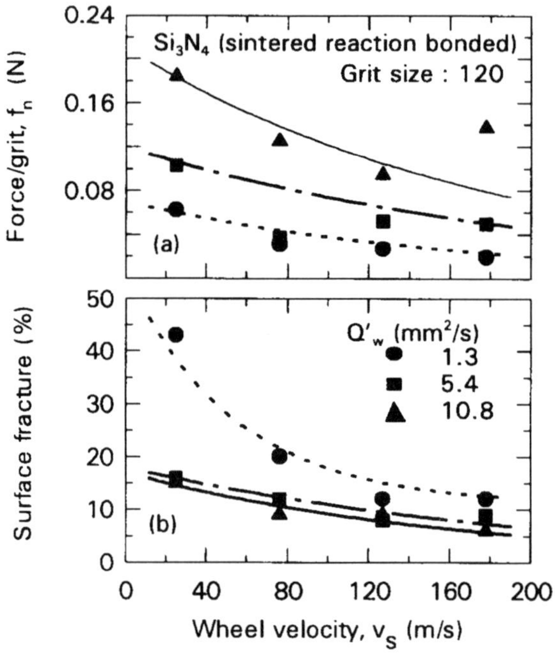

Figure 3.23: Effect of Velocity and Removal Rate.[25]

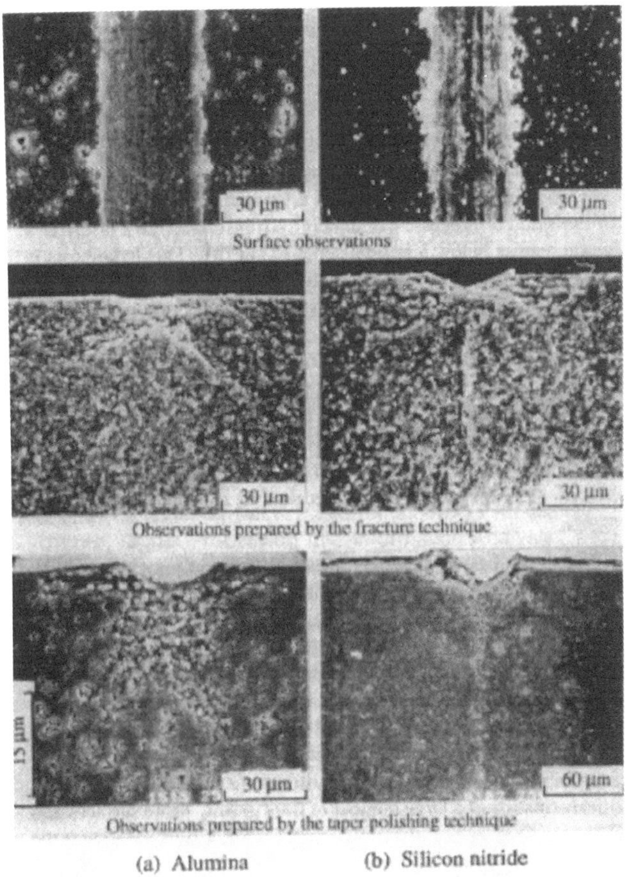

Figure 3.24: SEM observations of surface and subsurface regions of alumina and silicon nitride ground by single-point diamonds. The top row shows the surface observations; The "powder regime" is to be observed in the bottom row.[30]

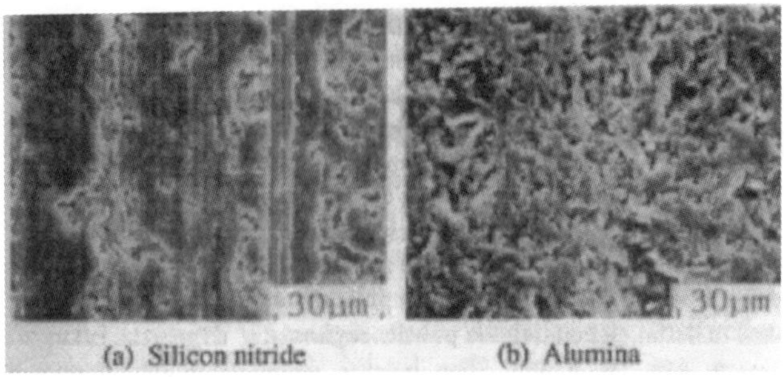

Figure 3.25: SEM Surface Observations of Silicon Nitride and Alumina Ground by Sd140-n75b Wheel.[30]

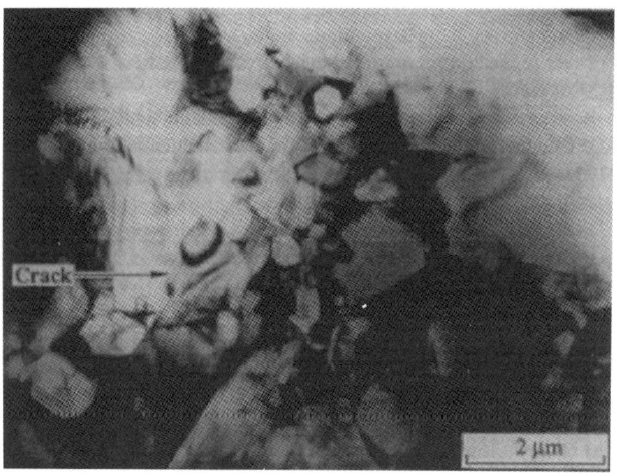

Figure 3.26: Observation of grinding-disturbed layer of silicon nitride by TEM. A cluster of small grains is approximately 0.3 μm in grain size. A crack can be seen on a one micron sized grain.[30]

tograph, there is a cluster of small grains with an average grain size of 0.3 μm, the remaining are grains of different sizes. The grains, mainly hexagonal crystals, have appreciable gaps between them, indicating a loose bonding condition. These submicron-sized grains are randomly oriented, and some are marked by dislocation movements. A crack is present on a one-micron-sized grain which would tend to cleave the grain into several submicron-sized grains. This crack arrests at the grain boundary which hinders the further propagation of cracks into the surrounding grains. The crack arrests by grain boundaries localize the damage induced by the grinding process. The submicron-sized grains can be formed by splitting micro-sized grains along the cleavage planes.

The powder regime in single-point grinding was quantitatively measured on silicon nitride and alumina. Figure 3.27 shows results on both silicon

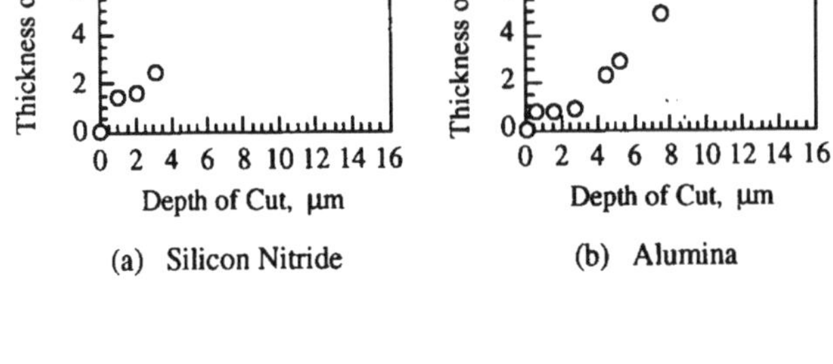

Figure 3.27: Correlation of Thickness of Powder Regime with Depth of Cut in Single-point Grinding.[30]

nitride and alumina ground by a single-point diamond with a tip radius of 1 μm. The thickness of the powder regime layer increases as the depth of the cut increases.

Figure 3.28 schematically shows the powder regime generated by a single-point grinding process. Side flows of the pulverized material, on the other hand, are caused by the special stress state that forms material pile-up on both sides of the ground groove as shown by the SEM photographs of Figure 3.24. Furthermore, due to the elevated temperatures in the contact region, the pulverized material could be resintered, which would also contribute to smoothing the surface of the ground ceramics.

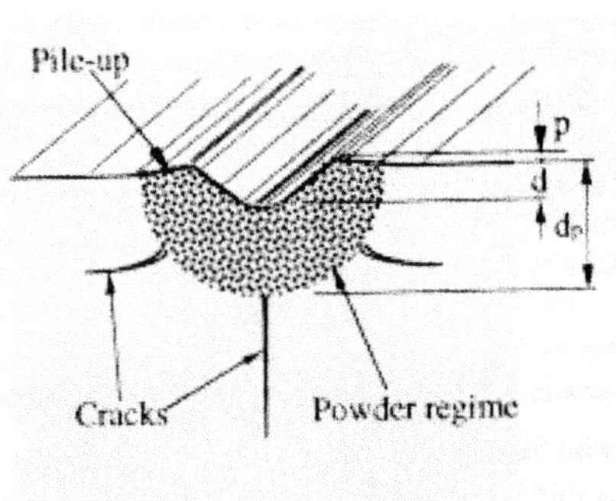

Figure 3.28: Schematic Showing the Powder Regime Layer Induced by the Single-point Grinding Process.[30]

As observed in Figure 3.24, the powder regime generated by sideways material pulverization flows forms material pile-up on both sides of a groove. In order to evaluate the degree of the material side flows, a measure has been taken of the material pile-up. The pile-up coefficient *e* is defined as $e=p/d$ where p is the height of a pile-up, as measured from SEM photos, and from the surface profile of a groove. The value *d* is the grain depth of cut.

Figure 3.29 shows coefficients of material pile-up in single-point grinding of silicon nitride and alumina. As the depth of cut decreases, the coefficients of pile-up are slight larger for silicon nitride than for alumina. This can be attributed to the fact that the thickness of the powder regime layer generated on silicon nitride is larger than on alumina at the same depth of cut. The larger the powder regime generated in a grinding process, the greater is the material sideways movement. Powder regime grinding as envisioned results in a surface consisting of small crystalline particles, compacted and semi-sintered.

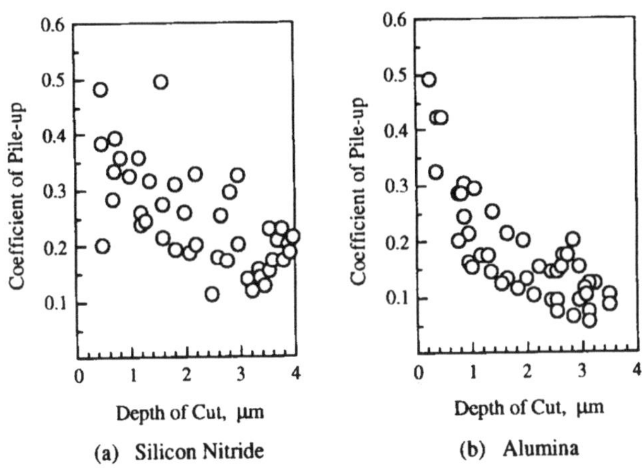

Figure 3.29: Coefficient of Pile-up Vs. Depth of Cut.[30]

The Double Fracture Model[31]

Double fracture model means that in micro-grinding of ceramics (lapping, for example), it is possible to have two types of stock removal mechanism, one of which is the double fracture mechanism. The fracture is a macro-fracture and is the effect of a grain which acts as an indentor, as shown in Figure 3.30.

The fracture dissipates a quantity of energy as heat because of the friction and deformation. Because of a small shock during the unloading phase of the grain-workpiece interaction, micro-fractures will appear on the particles which were just separated by the micro-fracture as shown in Figure 3.31.

This mechanism provides a possible explanation of what occurs in grinding where the phenomenon is more intense because of the dynamics of the process. Even in ductile grinding where the material removal is considered to be in a plastically-deformed mode because of the thermal shock, the plastic chips will suffer a thermal fracture and some are transformed into a powder.

The SEM picture of chips collected from ductile grinding of alumina oxide (Figure 3.32) shows a spiral deformed shape like a turning chip.

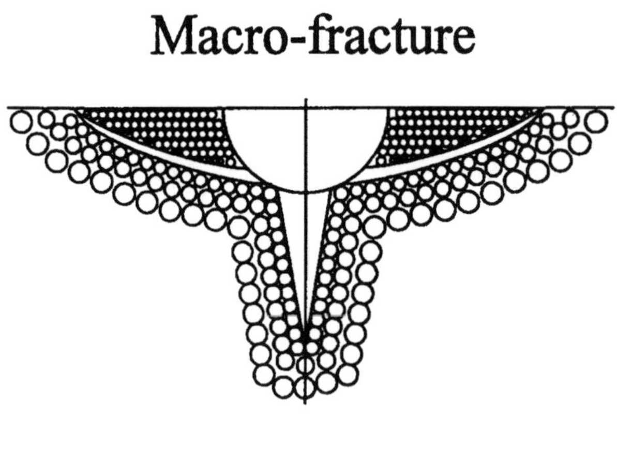

Figure 3.30: Macro-fracture of Ceramics.[31]

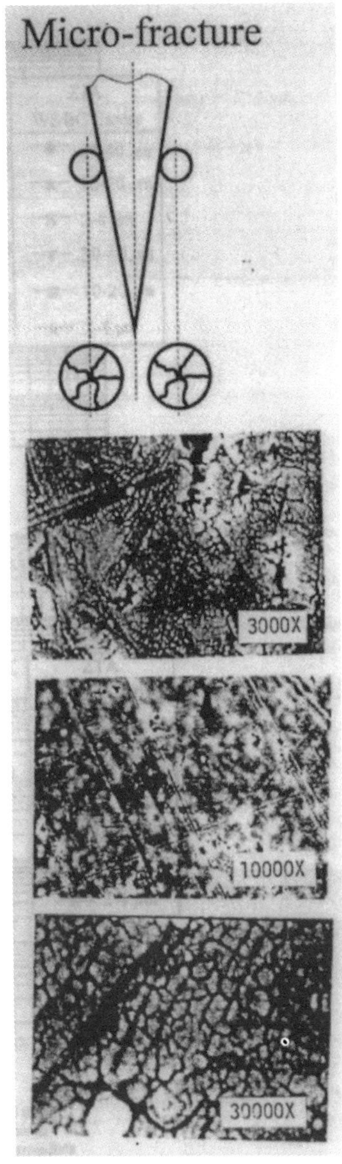

Figure 3.31: Micro-fracture of Ceramics.[31]

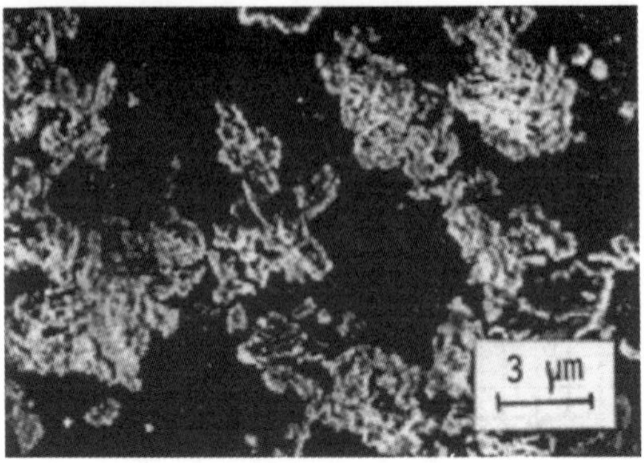

Figure 3.32: Ductile Ceramic Chips.[31]

Because of thermal shock, the ground surface can be affected and some cracks can be observed.

In double fracture micro-grinding (lapping), the results indicate that two different trends occur. It appears that during lapping with the 30-40 μm and 10-20 μm abrasive size, material removal is through the brittle fracture mode. The process is very stable from the viewpoint of surface finish, stock removal rates (Q_w), and specific energy (u), as seen in Figure 3.33a-b. Specific energy values are higher for the 10-20 μm abrasive size than the 30-40 μm. This is due to the greater number of diamond grains associated with the smaller size. There are approximately eight times as many grains with the 10-20 μm abrasive than in the same volume of 30-40 μm abrasive. Corresponding with this trend, the abrasive coefficient, μ_a, or force ratio, has the same relationship. The stock removal rates Q_w are practically equal for both sizes.

From Figure 3.33a-b, it is noted that the behavior of the 2-4 μm abrasive differs from the other two abrasive used. The behavior of this grit size is primarily because these grains are much smaller, thus the number of grains is much larger than in the other cases. For the same concentration of diamond powder, the 2-4 μm powder has 1000 times more grains than the

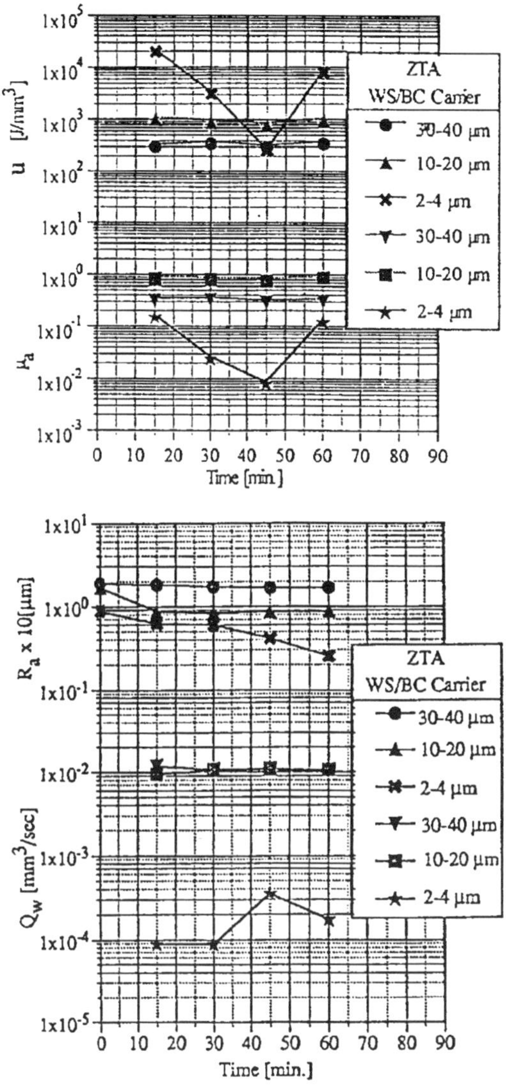

Figure 3.33: Results for Tests Using Three Abrasives. Each Symbol Represents an Average of Three Parts. (A) Specific Energy E_c and Force Ratio μ_A; (B) Average Roughness R_a and Stock Removal Rate Q_w. [31]

30-40 μm powder. From the graphs, we can see that the specific energy was at a minimum after 45 minutes of lapping. Specific energy then increased with a longer lap time. Based on ceramic grinding experience, the increase in the specific energy is usually an indication of the transition of the material removal mechanism from brittle mode to ductile mode. Further research focusing on brittle fracture/plastic deformation in micro-grinding needs to be performed.

Semi-Ductile Grinding Mode[32][33]

The term semi-ductile grinding is used to mean the brittle-ductile transition process in grinding of ceramics: one is due to plastic deformation on the characteristic slip plane and the other is due to brittle fracture on the characteristic cleavage plane. In machining of homo-geneous materials, the slip or cleavage planes coincide with the plane of maxi-mum shear or tensile stresses. When the resolved shear stress τ_{slip} in the easy slip direction exceeds a certain critical value τ_c inherent to the work material before a cleavage takes place, a plastic deformation occurs in a small, stressed field in the work material with a specified scale which can correspond to the depth of cut. Alternatively, a cleavage occurs when the resolved tensile stress normal to the cleavage plane σ_{cleave} exceeds a certain critical value σ_c preceding a plastic deformation. The mode of material removal depends on which criterion dominates or precedes $\sigma_{cleave} > \sigma_c$ or $\tau_{slip} > \tau_c$ for the stress state under a particular machining condition.

For increased understanding of the transition mechanism, the scale of machining can be classified into three regions based upon the defect density of the work material.

Larger than Micron Scale

In this region, microcracks and dislocations are usually included in the stress field. σ_c sensitivity decreases as the scale of machining increases because the number of microcracks in the stressed field increases. Both macroscopic defects such as pre-existing crack, grain boundary and precipitates and atomistic defects such as stacking faults and platelets in the work material function as a microcrack. On the other hand, τ_c is not sensitive to the defects. Therefore, brittle mode material removal is the predominant criterion in this region.

Micron to Sub-micron Scale

σ_c does not decrease in this region as the size of the stress field decreases because the microcracks seldom occur in the stressed field. As dislocations are included in the stressed field, τ_c does not show remarkable change. Therefore, material removal mechanisms tend to change from brittle to ductile.

Sub-micron to Nanometer Scale

In this region, machining result in a defect free surface. Both τ_c and σ_c increase to the same level as the intrinsic strength of perfect (defect-free) work material. These intrinsic values, σ_c and τ_c, are occasionally affected by environmental temperature. The analysis based on continuum mechanics shows that the plastic deformation occurs before cleavage in indentation on defect free surfaces.

An explanation of the formation of ductile bands, along with fractured bands is shown in a modified form of König & Sinhoff's model in Figure 3.34. Experimental results on machining of mono-crystalline Si and $LiNbO_3$ support the applicability of the criteria.

6.0 SURFACE TOPOGRAPHY AND SURFACE INTEGRITY

Hans G. Wobker

During the sintering process, ceramics undergo a volumetric shrinkage and are thereby far from meeting high demands on their form and size

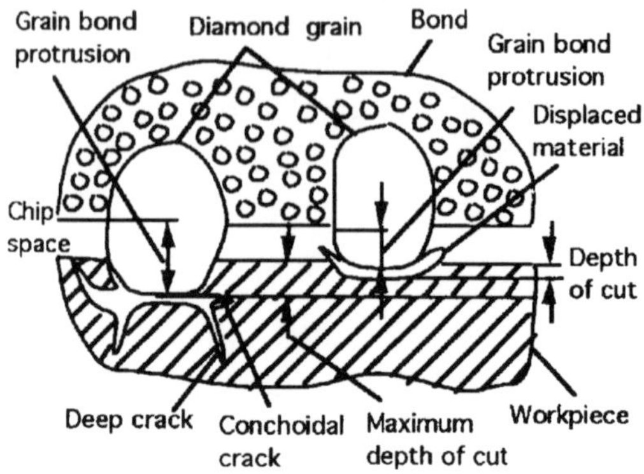

Figure 3.34: A Modified Form of Konig's Model to Explain Semi-ductile Grinding. A Typical Grain (On the Left Side) Produces Fracture While a Slightly Protruding Grain (On the Right Side) Produces Ductile Streaks. [33]

accuracy. To meet these demands, machining processes such as grinding, lapping, honing or polishing with diamonds are necessary finishing processes to obtain functional surfaces. The properties of these machined surfaces are mostly decisive for the functionality of components. Until now, the connection between the machining process and the component's functionality was not completely known. However, the distinction between surface topography and surface roughness has been established through recent research and industrial developments.[34][35][36]

Definition of Surface Topography

Machined surfaces are not geometrically exact; they include form errors, waviness, and surface roughness.[37] Directional machining processes cause directional, grooved surface topographies. Usually, their roughness is perpendicular to the machining direction but is higher than parallel to the machining direction. However, ceramics show this behavior in addition to other less brittle or ductile materials. The diamond grits of the machining processes cause plastic deformations as well as brittle spall-offs on surfaces.

Definition of Surface Integrity

In addition to the surface topography and the material microstructure, the machining process influences the properties of components such as strength and tribological behavior. Concerning the machined surface, there is more to observe than the geometric properties, i.e. evenness, measure, or roughness. The mechanical and/or tribological properties of machined components are essentially influenced by machining processes because the surface layer of the workpiece will be altered due to the interaction of the tool and the material. The surface properties of components, as a result of the machining process, have become a very important factor required to understand the connection of process and component operability. To abstract the expression 'surface integrity' and its demarcation to other properties of a workpiece, it is necessary to define it.[38]

Subsurface

The subsurface of a workpiece ought to be defined as the volumetric part of a workpiece that is influenced by the machining process. The depth of this layer depends on the process that produced the influenced zone. Von Weingraber subdivided the surface into an interior and exterior boundary layer as in Figure 3.35.[37] On the outside, the surface is covered by a very thin layer in which the atoms and molecules are exposed to forces besides the interior forces. Although the mass of this layer is very small compared to the whole body, the surface energy of this layer determines many physical processes. Its thickness is atomic and can be detected with surface analytics and chemical characterization. The interior layer of the machined workpiece, called the subsurface, determines the functional properties. Compared to the body structure, this layer is alternated.

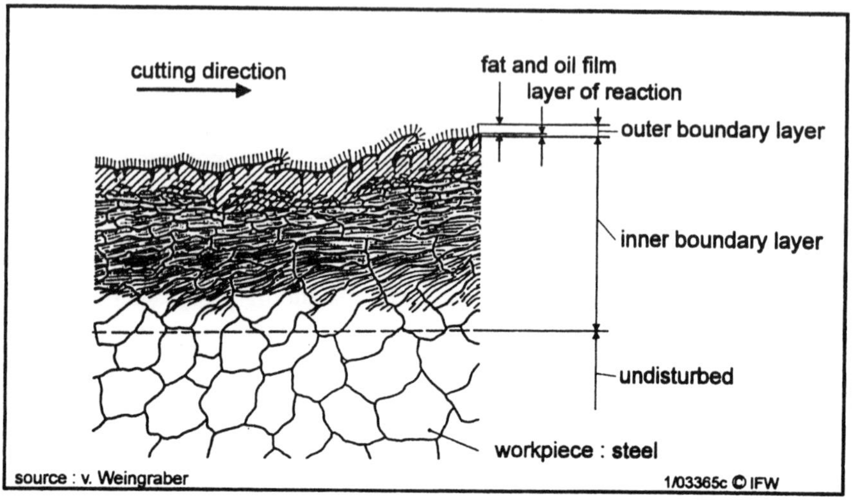

Figure 3.35: Interior and Exterior Layers of a Component

Surface integrity and Classification

Surface integrity covers the interior effects of a machining process. This includes the outermost layers whose properties differ measurably from those of the base material. Figure 3.36 shows a principal machined surface and subsurface, and their schematic physical and technological properties: mainly the properties of machined surfaces.

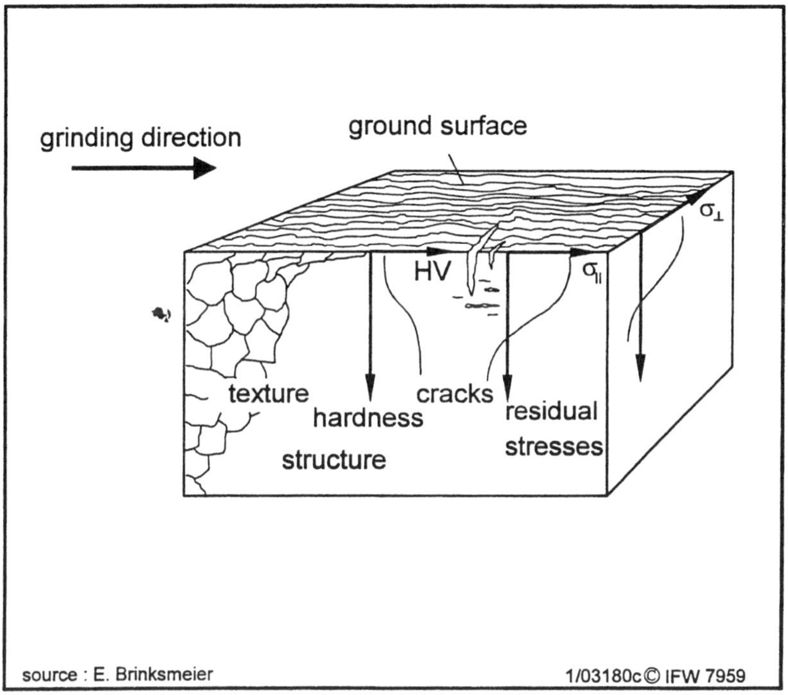

Figure 3.36: Properties of Surface Layers in a Machined Workpiece

The American Standard for Surface Integrity classifies the surface alterations by the principal modes that produce them:

- mechanical,
- metallurgical,
- chemical,
- thermal, and
- electrical.

This functional classification after the topics cannot differentiate between cause and effect of the surface integrity.

'Surface integrity' is defined as the properties of a part that are influenced by physical and chemical effects of the machining process.[39] Regardless of whether they are physically or chemically effected, geometrical properties are also included. Material deficits such as pores, cracks, material excesses, or residual stress-induced deflections could be such geometric properties. The surface material properties grouped by geometrical, physical, and chemical properties are shown in Figure 3.37.[40] The physical properties of the surface layer can be divided into fundamental groups of physics. The function of a component is influenced geometrically by cracks and flaws, physically by mechanical stresses, and by the structure of the material. The subsurface structure is based on chemical effects which cause the physical

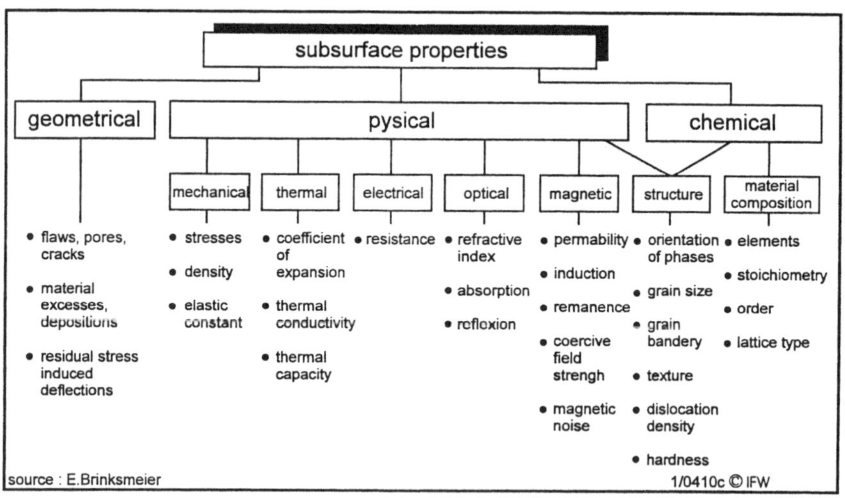

Figure 3.37: Table of Subsurface Properties

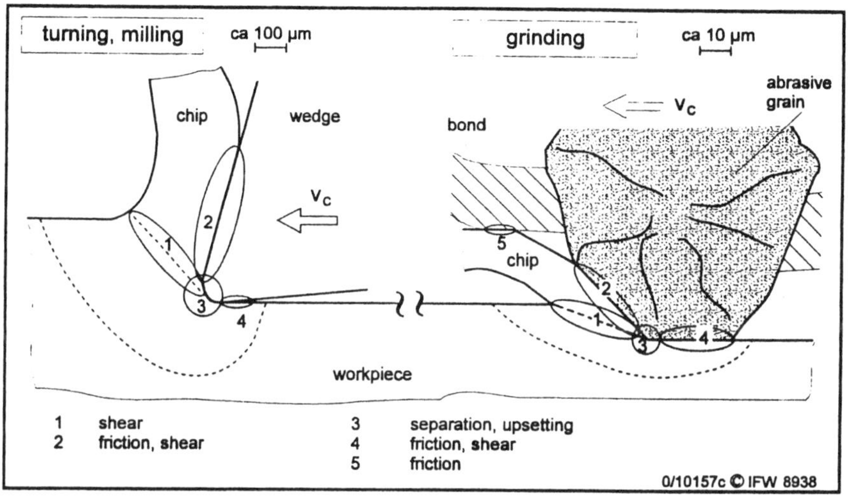

Figure 3.38: Mechanisms of Energy Transformation in Machining Processes.

properties of a material and is characterized by grain size, grain boundary, texture, dislocations, or hardness. Normally, this can be analyzed by using metallurgy or ceramography. Strictly, the hardness depends on the measuring method and is more a technological than a physical property.

Influences on Surface Integrity

Starting from an initial state of the workpiece, cutting processes influence the surface, subsurface, and the properties of the machined component. Beside the process itself, the initial state of the workpiece and the surroundings are important for surface integrity.

The concentrated energy consumption at the contact region of a machining process can be divided into mechanical, thermal, and chemical effects which lead to differing intensities of surface property alterations. Figure 3.38 depicts the areas of differing energy consumption mechanisms in cutting with geometrically defined and undefined tools. The generation of mechanical and thermal effects are comparable in both, but the geometric dimensions differ in at least one order.

The mechanical, thermal, and chemical effects overlay during the process and decide on the output quantity which is the result of the process. The thermal and mechanical effects are the most important consequences of the process. Figure 3.39 shows, for most cutting processes, how these effects influence the material removal, and therefore the surface integrity in grinding. Both influences are rendered inseparable by the process and lead to fundamentally different effects on material removal mechanisms, and therefore in surface integrity.

The input quantities of a process are composed of the disturbances machine settings, and system parameters: wheel and workpiece speed, depth of cut, or wheel specifications. These factors lead to process quantities which mainly influence the surface integrity. The process quantities are physical quantities such as: cutting forces and moments, temperatures, thermal power, and energy. Figure 3.40 shows different machine settings and system parameters of a grinding process, which influence the temperature in the contact region.

In the contact region of a grinding, lapping, honing, or polishing process, the cutting grits lead to elastic-plastic deformations of the surface. In machining steels, dislocations or hyinnings lead to hardness increase or decrease. In machining brittle materials, cracks are induced by mechanical or thermal effects. After the abrasive passes, the material rebounds and either cracks or spalls-off occur due to the acting tensile stresses.

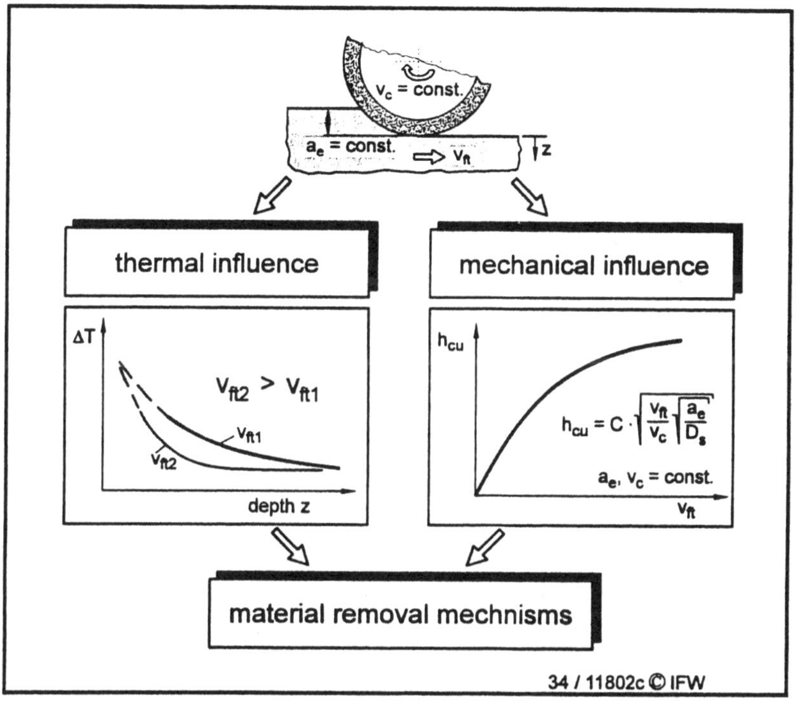

Figure 3.39: Thermal and Mechanical Effects Influence the Material Removal Mechanisms.

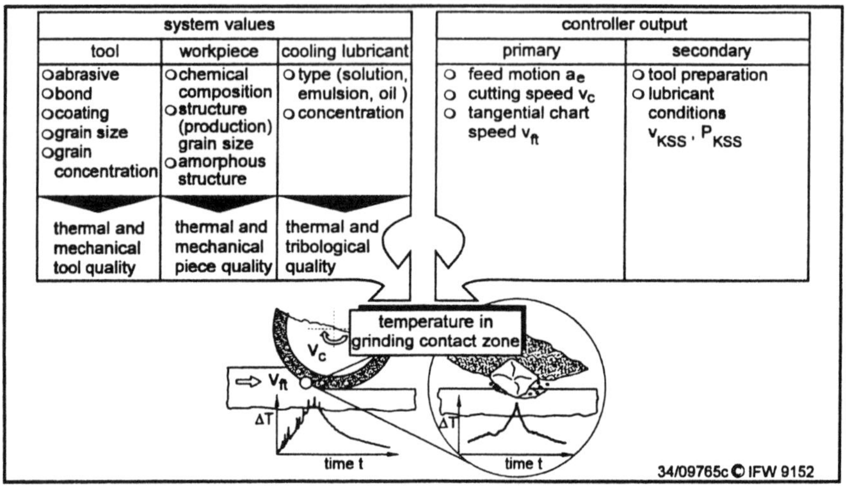

Figure 3.40: Influence of System and Setting Quantities on Temperature in Grinding.

A very high sensitive indicator for thermal or mechanical influences on surface integrity are residual stresses, especially in machining metals. The mechanical effects of the grits lead to plastic deformations and cracks. The strong negative effective cutting angle in grinding induces plastic deformations on the surface. Therefore, changes in lattice lead to micro- and macro-residual stresses and strains. The material compression and deformation cause surface compressive residual stresses. In grinding steels, temperature effects lead to tensile residual stresses.

Roughness

The quality of the surface topography - the surface roughness - can be defined as the deviation from the geometrical ideal surface. The most common measuring system of surface roughness are the contact stylus analyzer or optical measuring systems. Figure 3.41 shows the most common roughness data: R_z, R_p, R_q, R_{max}, and R_a. R_z is the average value of 5 single measurements for each evaluation length. The arithmetical mean deviation of the profile R_a, the root-mean-square deviation of the profile R_q, and the maximum profile peak height R_p are integrated roughness quantities.

Evaluation Technique for Surface Integrity

Generally, one can distinguish between mechanical, thermal, and chemical impacts that influence the state of surface and subsurface conditions. Chemical reactions in grinding of ceramics have not yet been discovered and may be negligible because most ceramics are chemically inert. Mechanical impacts may result in plastic deformations, compressive residual stresses, and cracks; however, thermal impacts may result in tensile stresses, cracks, phase transformations, and soft annealing.

Some of the structural, chemical, and physical surface and subsurface properties which are of interest in precision and ultra-precision machining are outlined in Figure 3.42. The dimensions range from 1 nm to more than 1 mm. None of the available measuring tech-niques shown in Figure 3.43 are effective in identifying structural parameters over the full range of dimensions. This can only be reached by using different techniques with different types of physical interaction with the material.

The most important specific techniques to yield surface integrity data are shown below. Many of these techniques are established in industry and research. Information on special facilities for new evaluation techniques, are relatively unknown.

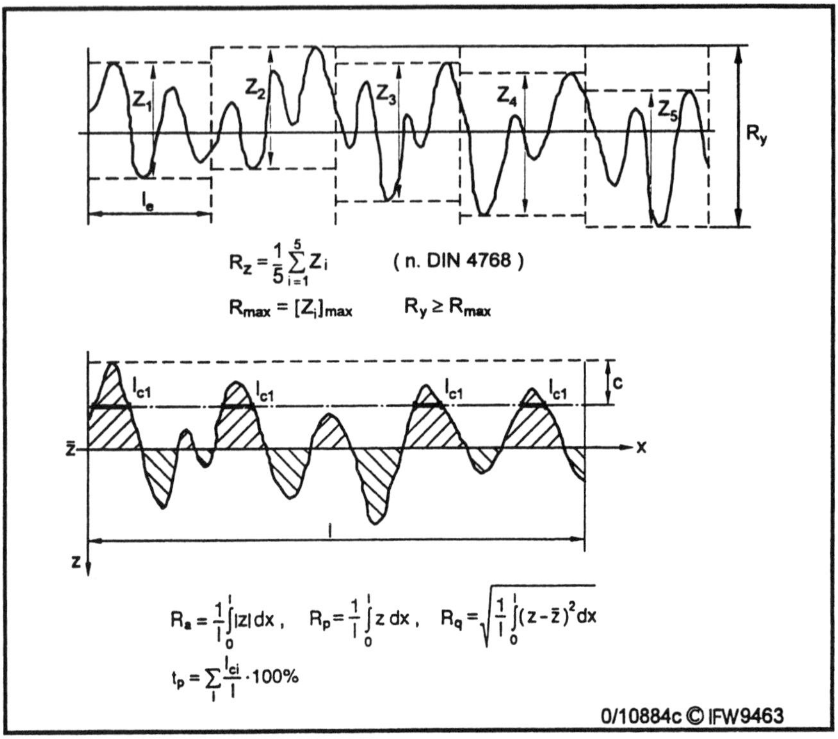

Figure 3.41: Surface Characterization.

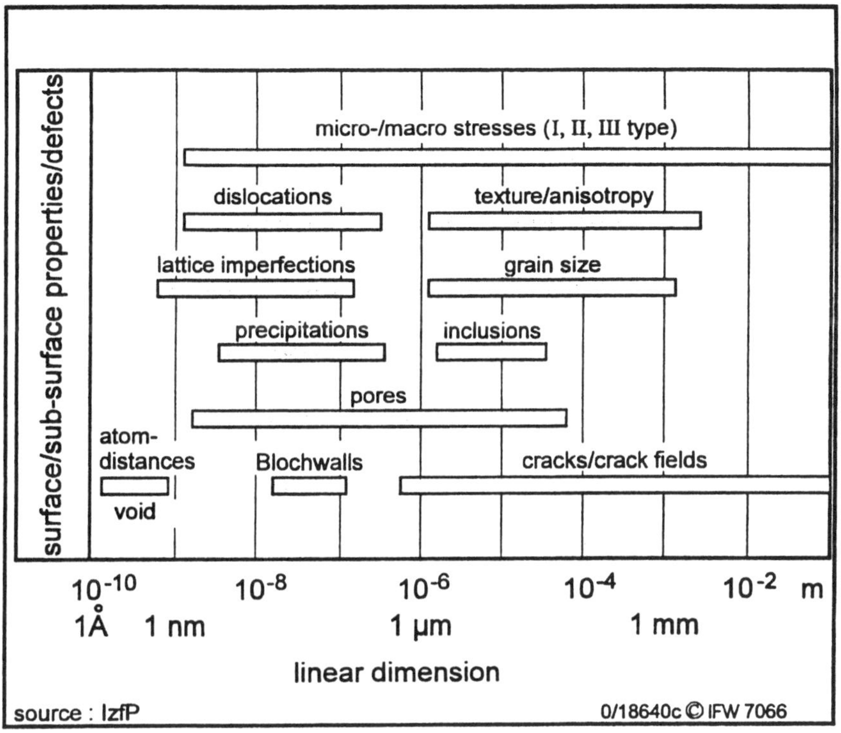

Figure 3.42: Linear Dimensions of Structure and Defects

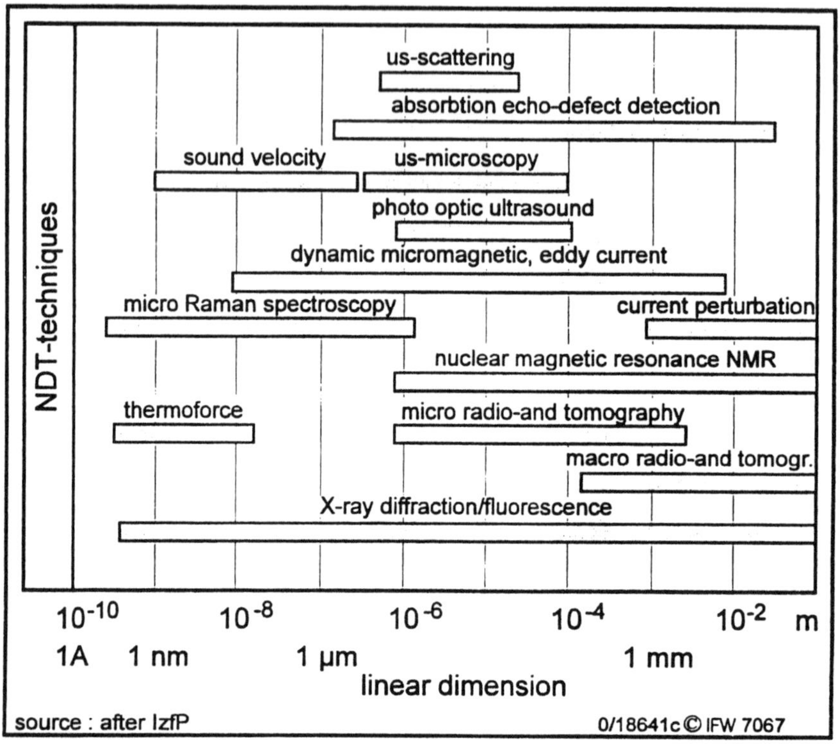

Figure 3.43: Non-destructive Testing Methods and the Linear Dimensions of their Interaction.

Visual Examinations

Light-optical microscopy is very often used in ceramography. The magnification is limited to a factor of 1000. Because of the wavelength of light, the resolving power is limited to 0.3 pm. For characterization of machined surfaces, the most important disadvantage is the small depth of sharpness. At a magnification of 1000, the depth of sharpness is 0.01 pm. Therefore, light-optical microscopy requires optically-even surfaces.

For characterization of fracture surfaces, scanning electron microscopy (SEM) is most commonly used to investigate the surface structure and to detect flaws and open pores. A very thin electron-beam (diameter app. 0.01 µm) linearly scans the surface. The impact of the primary electrons of the beam causes secondary electrons to emerge from the specimen; these electrons are then detected. These secondary electrons give the picture deeper regions radiate less, higher regions radiate more electrons. The resolving power is limited to 0.01 µm. At a magnification of 1000, the depth of sharpness is 35 µm. Depending on the primary electron current, the maximum magnification can reach up to 200,000.

Microstructure Examinations

To estimate the influenced depth of the surface structure and the state of the material below the surface, different techniques are applied in the following manner.

A very simple method to determine the penetration depth of the influenced surface layer is by polishing the cross-section. Normally, the region of the influenced layers are very small. During fine-grinding and polishing of the cross-section, the specimen is rounded off. To enlarge the influenced region, the taper polishing technique is suitable. Figure 3.44[22][41][42] presents another simple technique for directly observing subsurface damage in the machined ceramics. The technique requires two polished specimen whose polished surfaces are tightly bonded together. After grinding perpendicularly to the interface, the polished surfaces are taken apart to observe the influenced subsurface without following the polishing procedure.

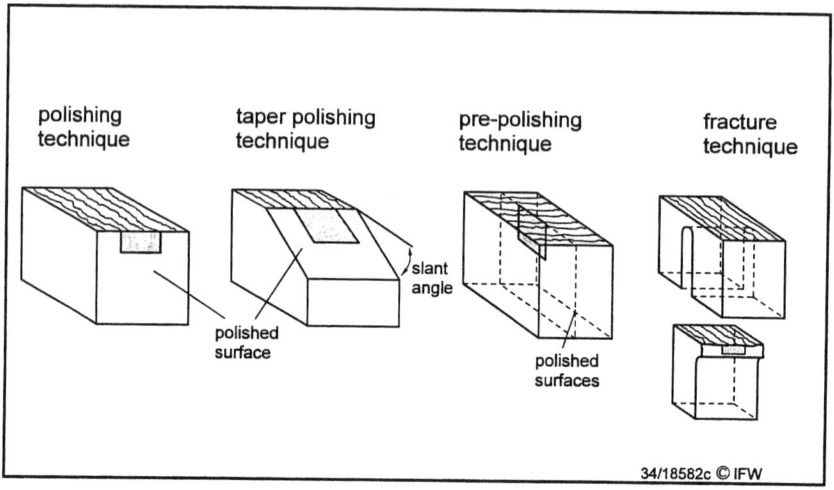

Figure 3.44: Examination Technique of Subsurfaces Micostructure

To observe the subsurface structure after machining, a fracture technique can be used,[30] Figure 3.44. A slit is introduced to a ceramic sample, with a ground surface, from the bottom side. The uncut margin of 4 mm is fractured by a bending moment. The cross-section of the ceramic sample is then observed by SEM.

To characterize structural alterations as dislocations, stacking faults, or finest dispersions, transmission electron microscopy (TEM) is one of the most commonly used analyzing methods. Electron beams transit thin layers. TEM uses the deflection of the electron beam on lattice faults. Depending on the structural state, the beam transition causes transit time differences and leads to interferences of the beam. The beam potential reaches up to 1 million Volts which leads to magnifications up to 10^6 (resolving power less than 1 nm).

Analysis by TEM, necessitates preparing the specimens by thin ground technique. By using electrolytic etching, the thin specimen will be etched until the thin plate develops a hole. The taper to the boundary of the hole will be analyzed by the transmission electron. The interpretation of the observed micrographs is very difficult.

Hardness Measurements

Most hardness tests yield numerical values that correlate with the materials' resistance to indentation by another body. Principally, hardness can be determined by scratching, indentation under static or dynamic load, or by rebounding of the tester from the specimen because of elasticly. Of technical importance is the static indentation thus formed; one then analyzes the indentation area (Brinell, Vickers, and Knoop) or the depth of the indentation (Rockwell), Figure 3.45.

The Brinell indentor is a sphere; the hardness is calculated by dividing the maximum applied load by the area of the indentation. Vickers and Knoop indenters are blunt pyramids. Vickers hardness is calculated by dividing the force by the contacting surface area; Knoop hardness is calculated by dividing the load by the projected area. Because the Knoop indenter penetrates only about half as deeply as the Vickers indenter for the same load, it is frequently preferred for studies of superficial hardness.

The indentation depth of the Rockwell tests correlates inversely with the penetration depth of the indentation. Indenters of different geometries are used for materials of differing range of hardness. For all hardness tests, the indentation is influenced by the form and hardness of the indenter, so hardness values have to be named by the method.

The brittleness and the presence of microcracks in the surface region of ceramics lead to spallings by using indenters. It can be difficult to measure the impression because of the spallings and induced cracks. To avoid this, smaller impressions than the mean crack spacing should be used.

Depending on the load of the indentation hardness, measurement is distinguished between macro-hardness measurement (forces > 50 N) and micro-hardness measurement (forces < 2 N). To analyze the hardness of machined surfaces and subsurface regions, Vickers and Knoop microhardness

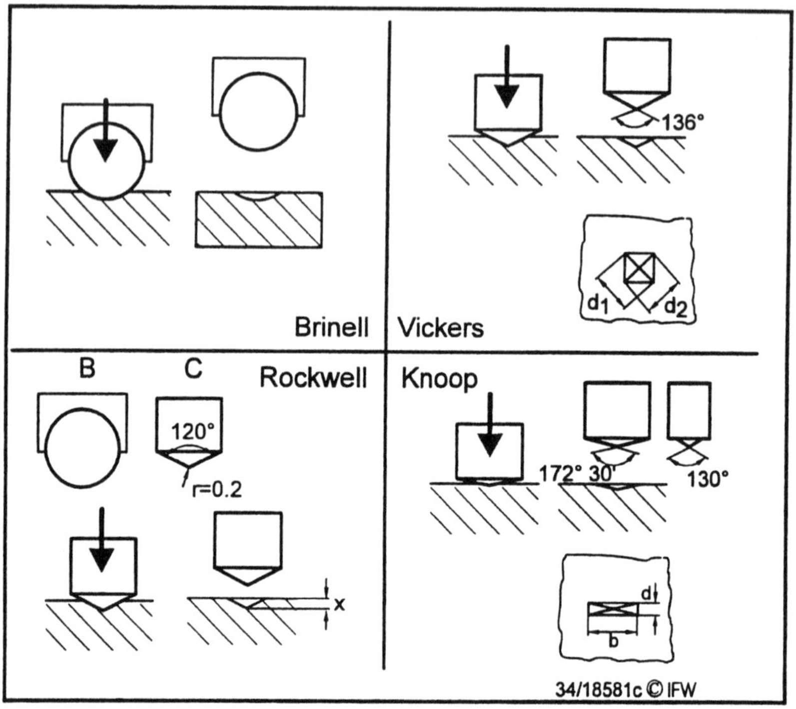

Figure 3.45: Hardness Indenter

measurement methods are the most suitable testing methods. However, the small forces lead to high measuring uncertainties by the optical evaluation of the impression. Therefore, hardness measuring systems were developed which use the depth of the indentation for hardness data. Figure 3.46 shows a microhardness testing system schematically. Principally, force or displacement actuated systems can be used. The instrumented microindentation tech-

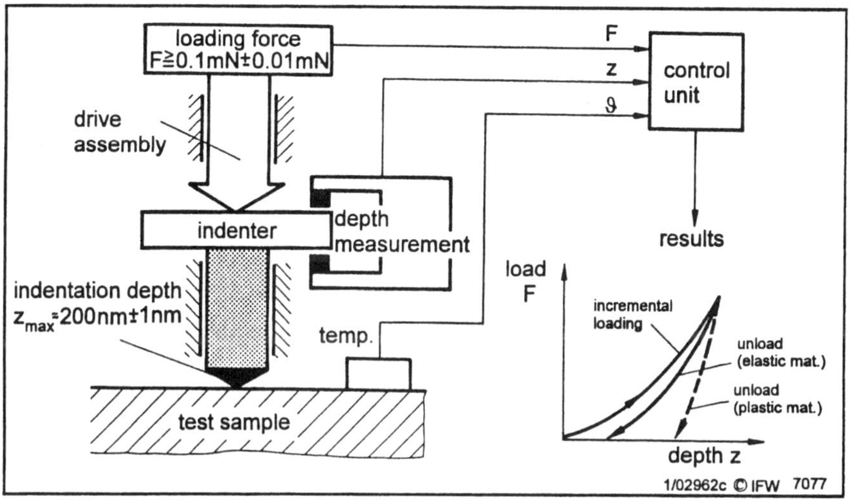

Figure 3.46: Micro-Hardness System

nique is basically a hardness testing procedure at very low force levels in the range between 0.1 mN and about 250 mN. The indentation depth is typically in the sub-micrometer range which classifies the method as quasi-non-destructive one.

One of the most difficult problems in measuring the hardness alteration of machined surfaces is the expansion of the influenced region in a range of microns. The brittleness of ceramics leads to spallings off the surface by measuring the hardness in a cross-sectional arrangement.

Evaluation of Cracks and Flaws

Fracture of ceramics originate from machining flaws as well as from material intrinsic failures. The brittleness of ceramics and their small grain sizes causes critical crack lengths in the range of microns. Therefore, crack inspection methods have to have a high resolution. However, non-destructive testing methods can be grouped into capillary, magnetic, inductive, ultrasonic, and x-ray testing methods.

The dye penetration inspection is very common in non-destructive testing of metals. Capillary effects cause the penetrant to be sucked deep into cracks, pores, and hollows of the surface because of low viscosity and surface tension of the moisture. After cleaning, the workpiece will be prepared and the outcoming dye will show the failures qualitatively.

This method is limited to cracks open to the surface, furthermore, the width of failure also limits its applicability. Often, cracks in ceramics are closed or smeared by plastically deformed material; therefore, the dye penetration test is not suitable. Penetrant inspection is impractical to use on excessively porous test materials.

Because of non magnetic and non electrical conductivity of most ceramics, magnetic and inductive testing methods are impractical for detection of failures in ceramics. Radiographic tests are also not successfully applicable because of the size of failures and ceramic densities.

Conventional ultrasonic techniques, as pulse echo method, were further developed to detect near-surface damage by sharply-focused transducers which generates very short-duration pulses. In order to detect defects by ultrasound efficiency, it is necessary to use wavelengths in the defect size. By using wide-open angels of the incoming ultrasonic beam, Raleigh waves are stimulated. These waves are surface waves with a penetration depth of the wavelength. Surface non-homogencitics, such as flaws, cracks, pores, and the surface roughness, reflect these waves and lead to interferences. This technique is able to determine flaws up to a depth of $d/\lambda = 1/10$. The resolving power is 15 μm at 100 MHz and 0.75 μm typically at 2 GHz.

Photothermal analysis of machined surfaces is a new method of non-destructive and non-contact testing. The technique yields local differences in thermal conductivity. Figure 3.47 shows the principle of the photothermal analysis method.

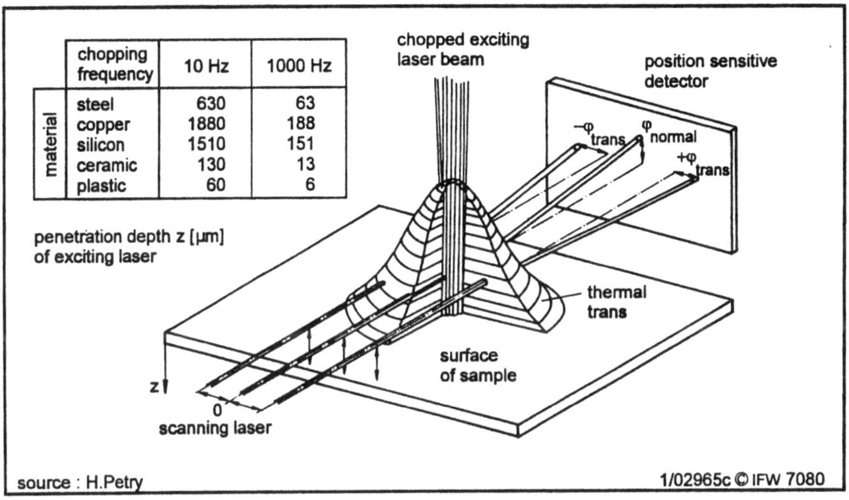

	chopping frequency	10 Hz	1000 Hz
material	steel	630	63
	copper	1880	188
	silicon	1510	151
	ceramic	130	13
	plastic	60	6

Figure 3.47: Photothermal Inspection Method.

The sample is subjected to an intensity-modulated laser beam. The heat is transferred into the subsurface area by heat transfer. Because of the short pulse duration and intensity of the excitation laser, only a limited depth of the subsurface area is influenced by this excitation. Therefore, the depth of the inspected subsurface area can be varied by changing the chopping frequency and the excitation power. Changes in the local thermal properties will be influenced by local defects and impurities including cracks and pores. The thermal wave analysis measures the local temperature at the samples surface, by a second laser beam, to detect changes in the local thermal diffusively which are related to subsurface defects. The size of the detectable defects depends on the focus diameter of the excitation laser. The thermal wave analysis is very suitable for the detection of closed defects underneath the sample surface down to defect sizes of 10 μm.

Residual Stress Measurements

The machining process, which generates the functionally relevant surfaces of a component, has great importance for the development of the physical state of the surface and the residual stress distribution in it. Residual stresses act within a body without external forces or moments. The internal forces form a system of equilibrium.

In many cases, the measuring methods of residual stresses can be classified as destructive or non-destructive. Another more effective classification is to distinguish between direct and indirect methods.

For indirect methods, the equilibrium of internal forces and moments has to be disturbed. The resulting deformations, of the analyzed body or the removed part, deliver the residual stresses of the body. Because of the hardness and chemically resistant indirect methods, destructive methods are less important for residual stress analyses of ceramics.

For direct residual analyzing methods, physical properties of the body are used which can be influenced by stresses. Principally, magnetic, ultrasonic, and x-ray methods are used to determine stress influenced alterations of physical properties. Because of the high state of development and the wide range of experience, x-ray methods are the most common residual stress analyzing methods.

X-Ray-Diffraction Techniques. X-ray diffraction techniques are capable of measuring interatomic spacings, which are indicative of the macro stresses undergone by the specimen. Stress values are obtained from the elastic strains in the crystals; this can be determined by knowing the elastic constants of the material and assuming that stress is proportional to strain which is reasonable for homogeneous, nearly isotropic materials.

BRAGG described x-ray diffraction as a selective diffraction according to the law:

$$\lambda = 2 \cdot d \cdot \sin\Theta \tag{3.8}$$

with wavelength of the x-rays λ, the lattice spacing d, and the diffraction angle Θ. Figure 3.48 shows the principle of the x-ray diffraction technique at a single crystal with and without loading.

The diffraction angle Θ_0 of the unstressed condition will be increased to $(\Theta_0 + \Delta\Theta)$ in the loaded condition. For technical materials such as polycrystalline, the lattice spacing has to be determined for various grains which have the orientation ψ relative to the specimen surface. The lattice strains perpendicular to the surface (ψ=90°) can be calculated mathematically by extrapolation. By setting the lattice strains equal to those which the theory of elasticity would deliver for a biaxial surface stress condition, the basic equation for X-ray stress measurement results:

$$\varepsilon_{\Phi\psi} = \left(\frac{1+\nu}{E}\right) \cdot \sigma_\Phi \cdot \sin^2\psi - \frac{\upsilon}{E} \cdot (\sigma_1 + \sigma_2) \tag{3.9}$$

where:

v, E	:	Elastic constants
σ_Φ	:	stress at an angle of Φ with a principle stress direction in the specimen surface
ψ	:	angle between the surface normal and the normal of the crystallographic planes from which an X-ray peak is diffracted
$\varepsilon_{\phi\psi}$	:	strain in the direction defined by the angles Φ and ψ
σ_1, σ_2	:	principle stresses in the surface plane

The lattice strains $\varepsilon_{\phi\psi}$ are a linear function of $\sin^2\psi$. To calculate the lattice strains $\varepsilon_{\phi\psi}$ from the Bragg equation, the exact peak position of the diffracted intensities for different ψ angels are necessary.

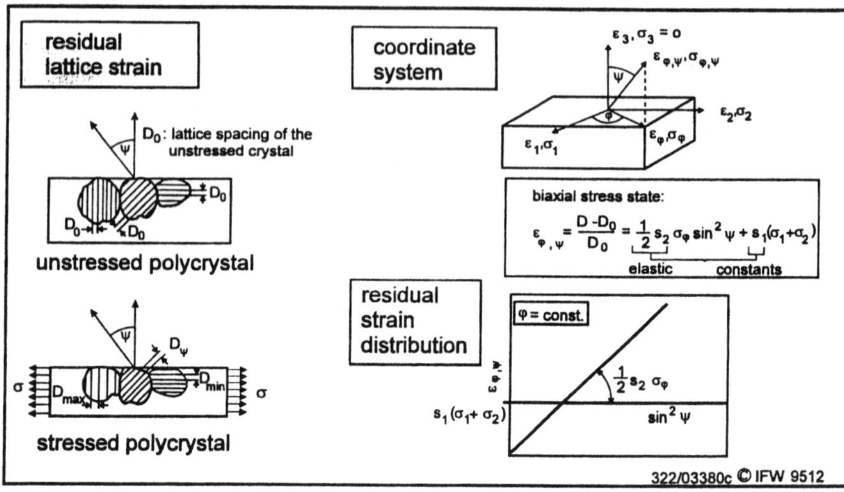

Figure 3.48: Residual Stress Measuring Method.

Figure 3.49 shows the principle of a ψ-diffractometer on the right, and a picture of a computer-controlled x-ray diffractometer on the left side.

The basic equation for X-ray stress measurement is a special case of the general state of stress condition. Figure 3.50 shows the resulting ε-$\sin^2\psi$-distributions for four different subsurface stress conditions schematically. Differences from the linear stress distribution shown in the upper left picture

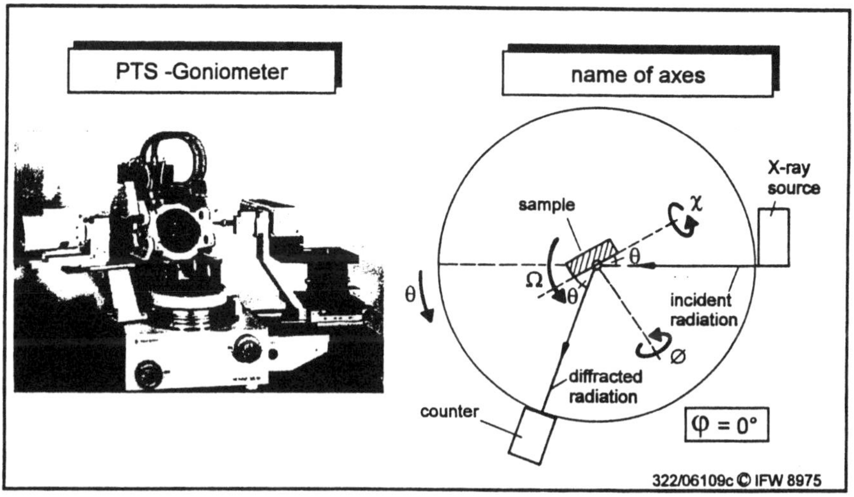

Figure 3.49: X-Ray Diffractometer.

are due to tilted stress conditions or stress gradients in the subsurface. However, the most common method for stress analysis is the linear computation as mentioned above.

Stress gradients. In many cases, machining processes cause stress gradients in the subsurface which lead to ε-$\sin^2\psi$-distribution shown in Figure 3.50. To determine such gradients, different analyzing methods were developed and are used in research and industry.

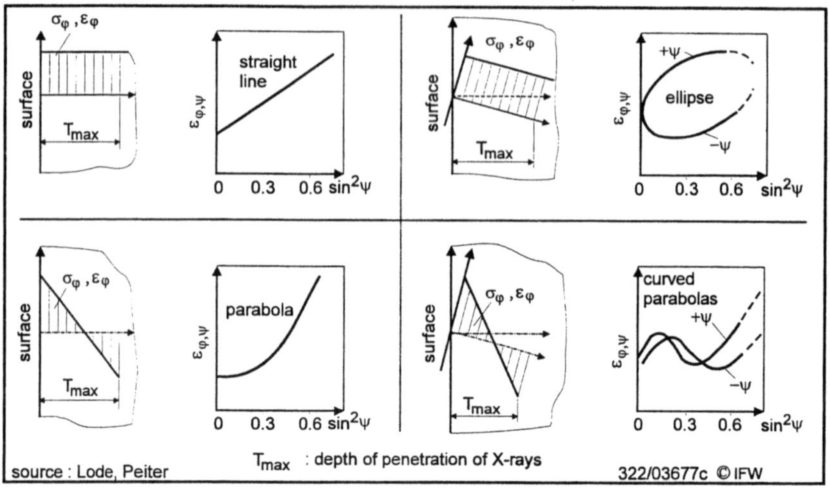

Figure 3.50: ε-sin^{2}ψ-distributions for Different Subsurfaces Stress Conditions.

The penetration depth of the x-rays is a function of the wavelength of the radiation, the analyzed material, and the measuring method. Figure 3.51 shows the penetration depth for different radiations. By using different x-rays, stress gradients can be analyzed. This methodology is very simple but expensive. The diffractometer will be alternated with new x-ray tubes and must be adjusted.

By knowing the reaction of the ε-sin^{2}ψ-distribution on stress gradients, and an increasing curved distribution with increasing sin^{2}ψ based on penetration depth, different computation methods for stress gradients were developed.

An effective technique for the characterization of thin crystalline films and stress gradients uses grazing incidence x-rays. The geometry of the

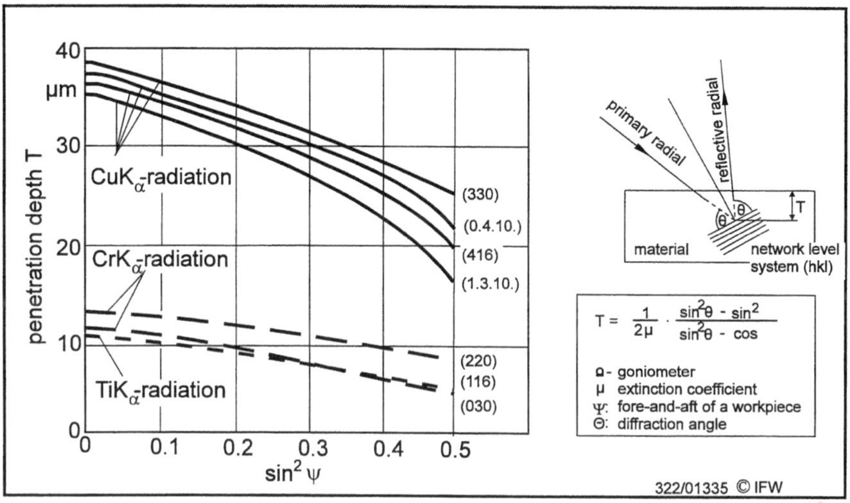

Figure 3.51: Penetration Depth of X-Rays: simple but expensive.

grazing incidence diffraction (GID)-technique and the measuring application is shown in Figure 3.52. The specimen is fixed at a grazing angle with respect to the incident x-ray beam. The detector is scanned along the 2Θ circle in the vertical plane of a diffractometer to record the diffracted x-rays from crystal planes that are inclined to the specimen surface. The penetration depth is a function of the grazing incident angle. The stresses can be calculated from the shifting of the recorded inferences.

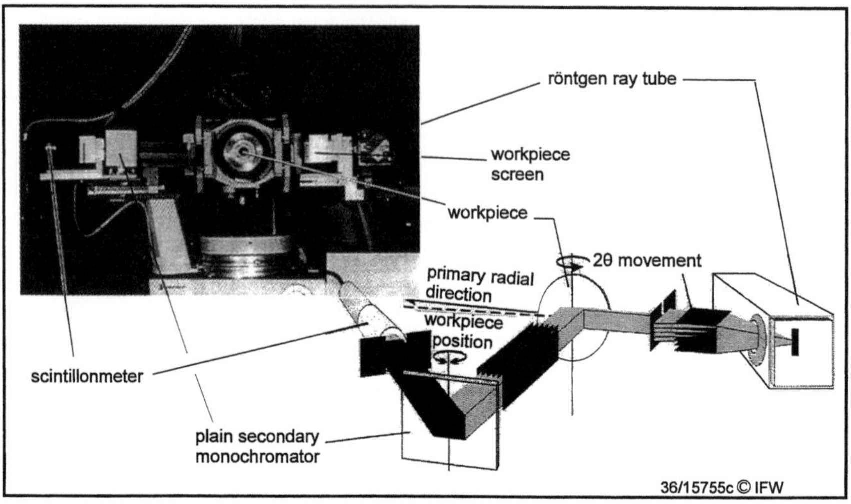

Figure 3.52: Grazing Incidence Diffractometer (GID) Application and Principle

Surface Properties

Surface Structure (Topography)

In general, ceramics are brittle materials and their machining is influenced by this. The surface structure as a part of the surface integrity of a component can be an indication of the functionality. In machining ceramics, the state of the material removal mechanisms is important, because rough surfaces are unacceptable for many applications. Therefore, the post-grinding polishing operation typically serves to improve the surface finish. Unlike grinding, the polishing process precedes predominantly ductile or plastic material removal. Polishing is a more expensive process and not accessible for complex geometries. For the investigations of the surface structure, scanning electron microscopy (SEM) is the most common technique.

Figure 3.53 present micrographs of the surface structure of ground alumina to determine the influence of depth of cut in grinding. To increase

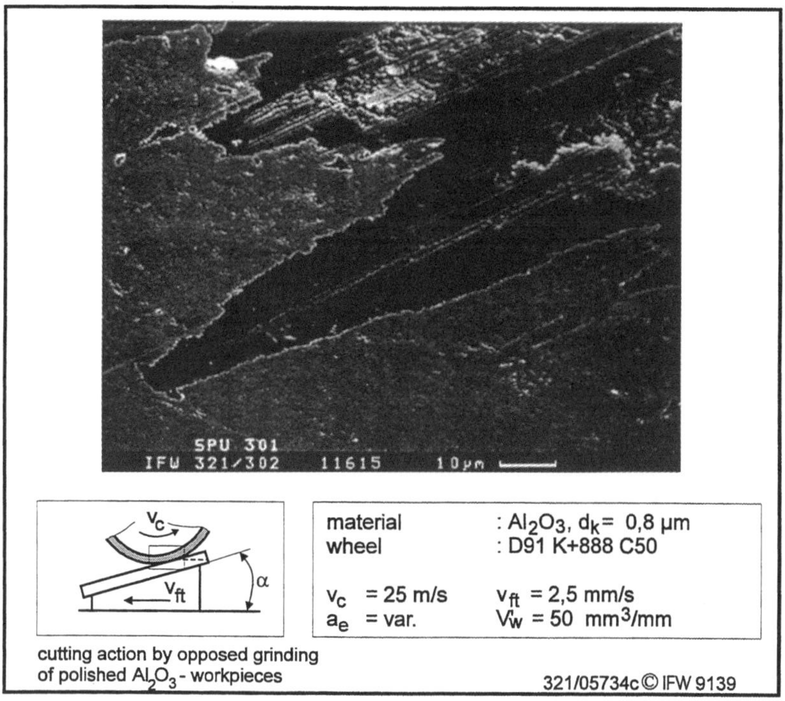

Figure 3.53: Grit-workpiece Interaction at Low Depth of Cut.

the depth of grit continuously, the polished specimen were tilted in a surface grinding process. As the grits contact the polished surface, sharp cutting grits immediately start removing material in the ductile mode (cutting). Duller cutting points first remove the sputtered gold coating in elastic contact (rubbing) and then displace material sideways in plastic deformation (ploughing). With increasing grit depth of cut, cracks begin to form, and material is removed by micro-brittle fracture.

In grinding ceramics the surface structure is not describable by the roughness itself. The material behavior of ceramics undergoes thermal and mechanical loading which leads to surfaces with different characteristics. Figure 3.54 shows micrographs of alumina ground by surface and creep-feed

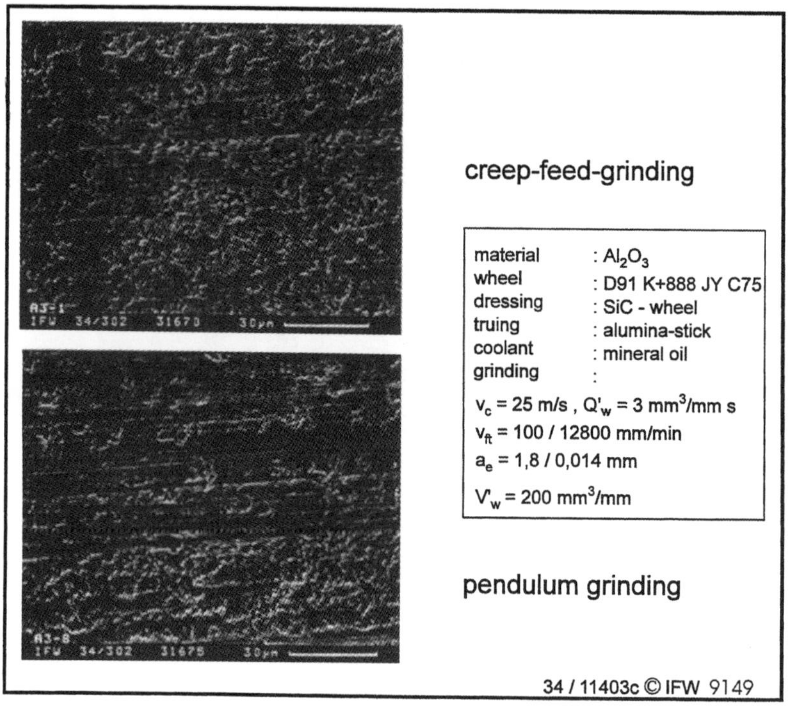

Figure 3.54: Surface Structure of Ground Alumina.

grinding processes with identically roughness R_a and R_z. The structure of the micrographs differs less; both plastic deformation as well as brittle material removal are visualized.

The differences in the material behavior are obvious at high magnification, Figure 3.55. Beside brittle-removed material, plastically-deformed material can be seen in both, surface and creep-feed ground surfaces. The plastic deformations are smooth, without excesses in creep-feed grinding, when compared to smeared and rugged, plastically-removed and rebounded material in surface grinding. This observation shows that different thermal

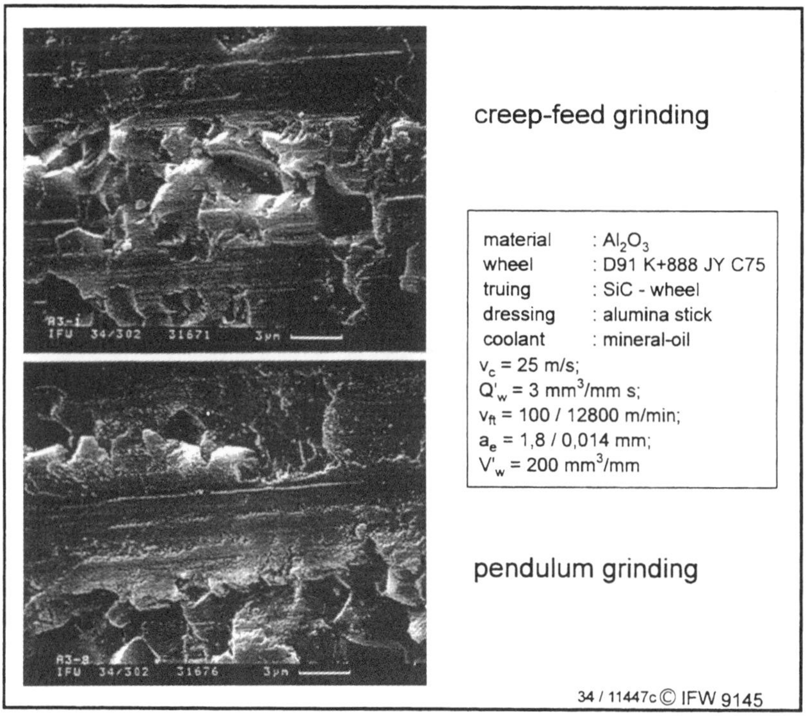

Figure 3.55: Plastic Deformations in Grinding Alumina.

and mechanical effects influence the microstructure of the machined ceramics. The different removal mechanisms assume influences on application of machined parts.

Besides the machining process itself, the surface structure is also influenced by the material. In view of their mechanical properties, grain size-varied alumina behave very differently in grinding. Figure 3.56 shows SEM micrographs of ground surfaces, which were ground identically at low grit depth of cut. For sintering the investigated alumina at temperatures ranging from 1300°C to 1700°C, no additives were used. The average grain size increases from 0.8 to 12 μm with the increase of the sintering temperatures.

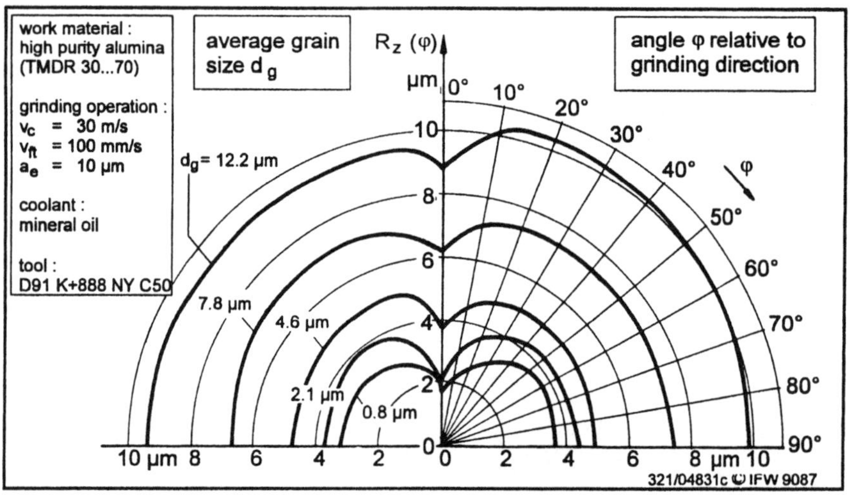

Figure 3.56: Surface Structure in Grinding of Different Grades of Alumina.

The ground surfaces exhibit different mechanisms of surface formation. With the increase in grain size, the surfaces become continuously rougher. The removal of single grains dominates over groove formation. The fine-grained variants TM30 and TM40 are covered with smooth grooves. TM50 and TM60 reveal many near-perfect grain facets and TM70 transgranular fracture is observed.

Figure 3.56 has already indicated the dependence of the grinding conditions on surface roughness in machining ceramics. In the grinding direction, the material is cut off. Perpendicular to the cutting direction, plastic deformations lead to higher roughnesses. Figure 3.57, in a polar diagram, shows various grades of alumina ceramic and their respective surface roughness R_z. The surface roughness increases with increasing grain size. The directionality

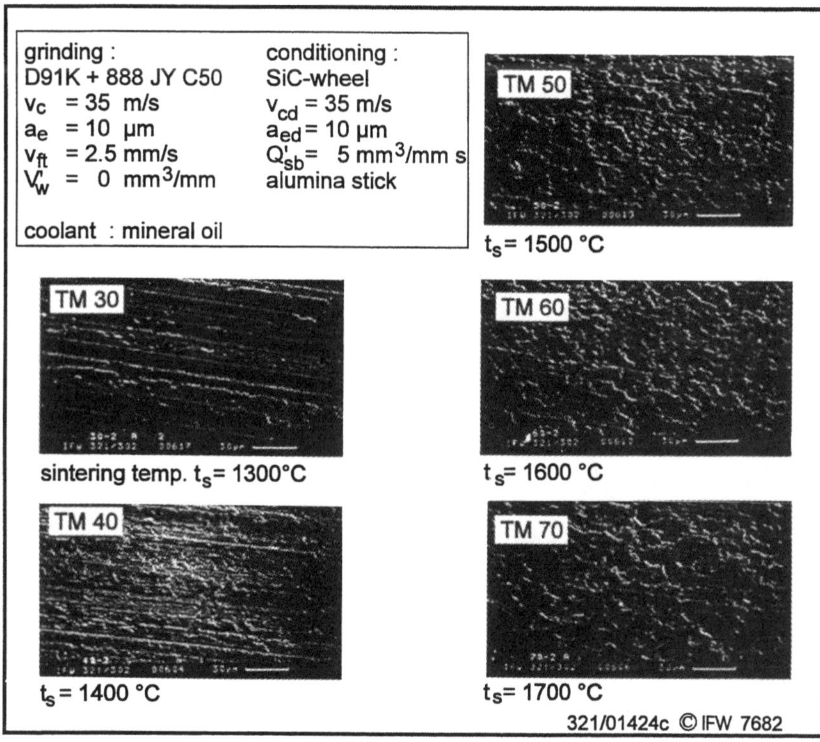

Figure 3.57: Roughness as a Function of the Grinding Condition.

of the roughness gradually disappears with larger material grain sizes because the material removal mechanism changes from ductile to brittle mode.

In addition to its influence on the grindability, the coolant affects the surface integrity of the ground ceramic. In the grinding of materials which are thermal shock sensitive, like aluminum oxide, the use of coolants with high heat capacity and convection coefficient can lead to a distortion of the surface layer. The influence of different coolants on the surface layer is shown in Figure 3.58. In case of oil, a plastically deformed surface with grinding grooves is shown. On the flanks of the grinding grooves, spalling occurs.

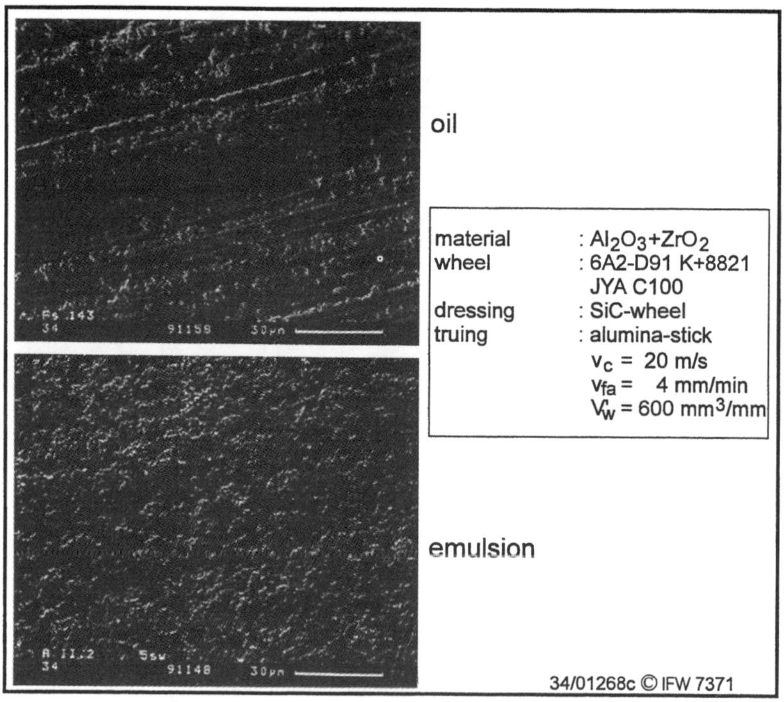

Figure 3.58: Surface Structure of Alumina after Grinding with Different Coolants.

These spallings may indicate the presence of brittle fracture during chip formation.

In the case of grinding with emulsion as coolant, no grinding grooves are visible. The surface is determined by unidirectional porosity. Besides the presence of brittle fracture in chip formation, thermal shock due to emulsion cooling seems to play an important role in material removal. The temporal and local variation, from high temperatures at the cutting edge down to ambient temperatures when the abrasive has passed and the coolant reaches the zone, causes thermal stresses inside the material. These thermal stresses can cause cracks and material spalls-off.

Subsurface Properties

Structural Transformations

In abrasive processes, such as polishing, lapping, honing, or grinding, each active abrasive grain generates an intense local stress field and a temperature rise. Thermal and mechanical effects influence the material removal mechanisms and therefore the surface structure.

The addition of finely divided zircon oxide (ZrO_2) increases the fracture toughness of ceramics because of crack deflection. Due to mechanical effects, ZrO_2 transforms from metastable tetragonal phase into monoclinic phase with a volumetrical increase. This induces stress fields in the matrix ceramic (AI_2 O_3) which lead to crack deviation and elimination. Figure 3.59 shows the principle of the dispersion composite and x-ray-diffractions of non-machined and ground Al_2O_3-ZrO_2-ceramic.

The stress fields of the cutting grains causes irreversible material deformation in form of dislocations, cracks, and void formations. Crushing, ploughing, and rubbing lead to surfaces as shown in chapter 'Surface topography'. The material removal mechanisms are usually classified into brittle fracture and ductile deformation. Other researchers define the ductile material removal as powder regime. The cut material is resintered on the surface. In these very thin outer layers of ground ceramics, dislocations as well as small grains could be found.

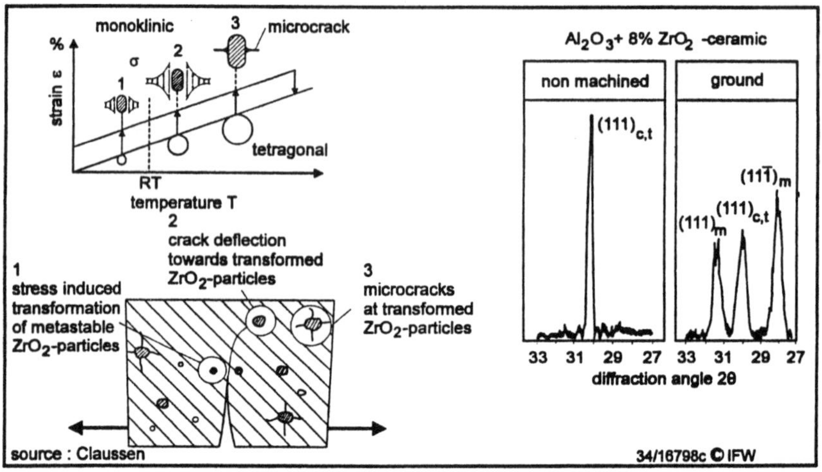

Figure 3.59: ZrO_2-transformation

TEM investigations of alumina show the nature of plastic deformations in ceramics, Figure 3.60. They reveal a concentration of dislocations in the cutting path of single grits.

Hardness

The influenced zone of machined ceramics is very thin. This causes difficulties in the description of the state of the surface properties, especially

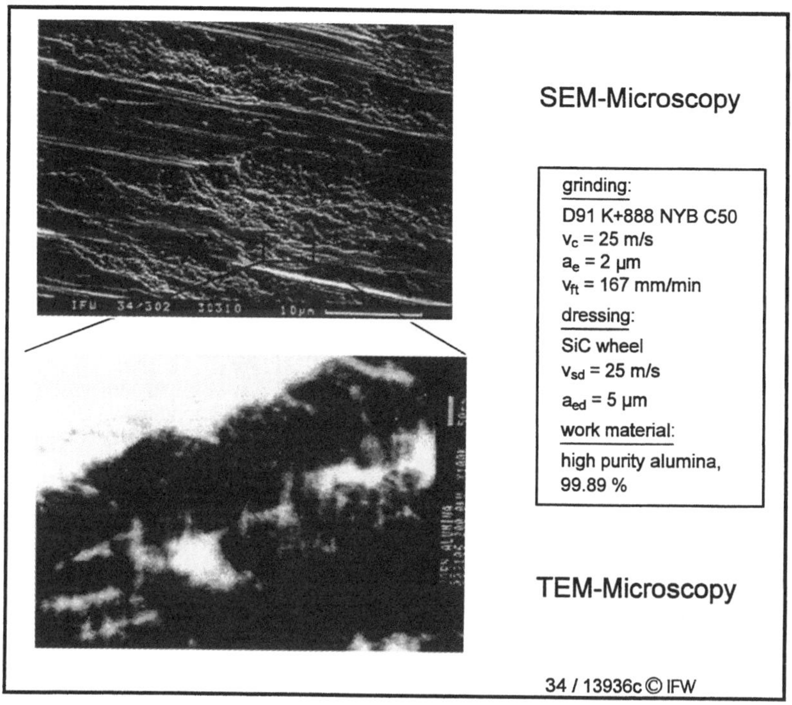

Figure 3.60: TEM Investigation of Alumina.

in hardness inspection. Most hardness measurement techniques are based on destructive measures. The thin surface layer cannot be measured without the influence of the base material. The only way is the microindentation.

Figure 3.61 shows results of microindention and observed near surface properties in ground Si_3N_4- and Al_2O_3/ZrO_2-ceramics. The results are presented as delta flow pressure curves. The flow pressure (yield stress) is the force divided by the projection of the area of the impression. The delta flow

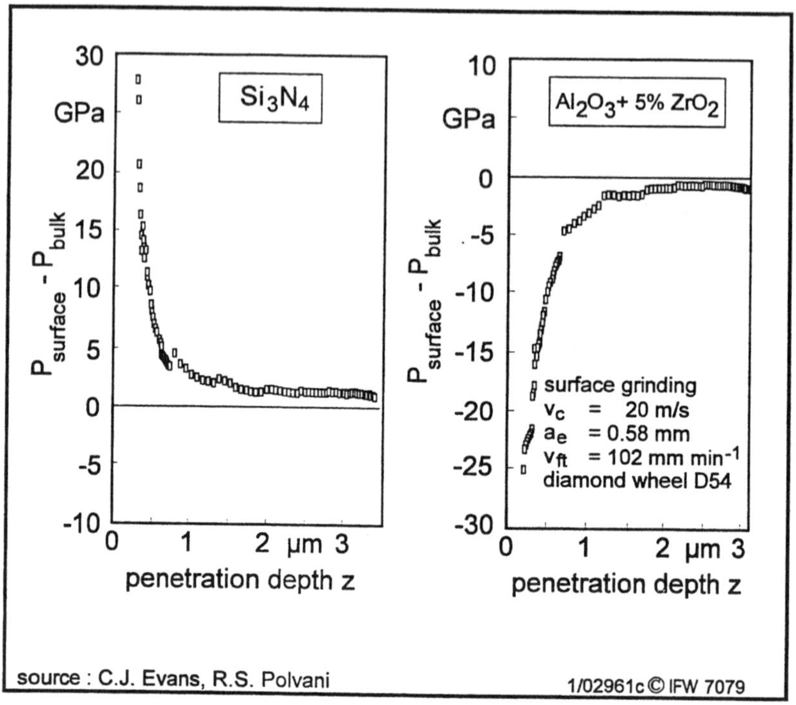

Figure 3.61: Delta Flow Pressure Curves for Ceramics.

pressure represents the difference between flow pressure in a defined depth of the indentation and unforced material.

Figure 3.61 shows increasing hardness in the case of ground Si_3N_4, and decreasing hardness for alumina ceramic. The hardness influenced zone extend for both materials equal to 2-3 µm. In grinding Si_3N_4, the increased surface hardness is caused by plastic deformations; however, the decreased hardness of the ground alumina is caused of material specific properties. Due to the low thermal shock resistance of alumina, fine cracks occur which leads to hardness softening.

Cracks and Flaws

The strength of materials are given by the microstructure and structure flaws of the material. Machined components are strength influenced by residual stresses and microcracks in the subsurface due to the process. Grinding processes lead to longitudinal, radial, and lateral surface cracks. For material strength, longitudinal and radial cracks are important because of the stress concentration of these notch effects.

The detection of cracks and flaws by dye penetration technique, non-destructive optical, or scanning electronic microscopically methods are not suitable for ceramics because of the size of the microcracks. Only macrocracks can be detected with these methods.

The most common technique for crack detection in ceramics is the cross section polishing technique. Therefore, the specimen were cut perpendicular to the machined surface. This cross-section will be polished with diamond grits of 3 to 1 μm to remove the influenced layer of the cross-section. Figure 3.62 shows lateral cracks underneath scratches of alumina in a cross-sectional view.

The amount of microcracks depends on the grinding conditions. Figure 3.63 shows different ground surfaces analyzed by scanning acoustic microscopy.[43] The pendulum ground surface show three times more microcracks than the creep feed ground alumina ceramic surface. This either causes increasing mechanical effects (grit depth of cut) due to higher workpiecespeed or high thermal alternation (gradients) due to the faster moving heat source.

Residual Stresses and Microstrain

For the exact description of the state of residual stresses in ground surfaces, it is necessary to determine the orientation of the main stress direction. From metallic materials, it is known that non-directional processes produce no preferred orientation of the residual stresses. In contrast, orientated processes like cutting or grinding are known to produce a distinctive stress orientation. This is also true for residual stresses in ground ceramics, Figure 3.64. The measurements revealed that the main stress direction is orie-

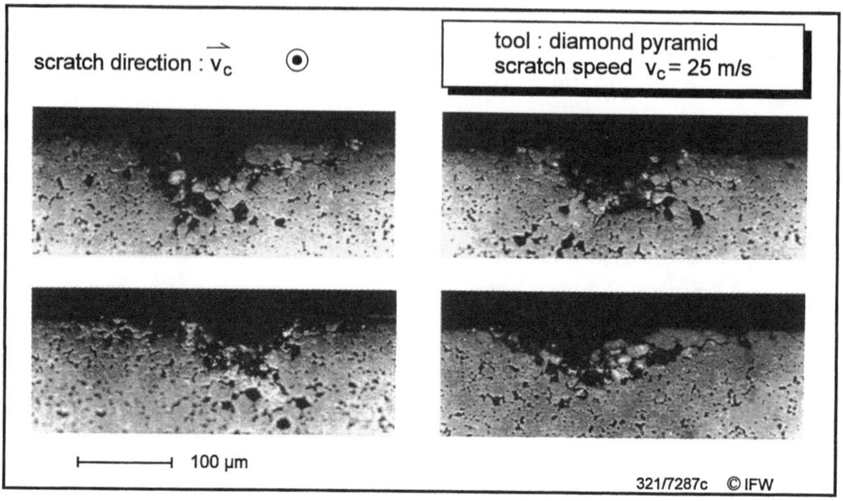

Figure 3.62: Lateral Cracks Underneath Scratches.

nted perpendicular to the direction of cut, whereas a minimum is found in the direction of cut.

The directional dependence of stresses in the surface layer correlates with the mechanical effects of the cutting edges. The largest amount of plastic deformation is found perpendicular to the cutting direction. The material remaining in the cutting path is deformed plastically to a smaller extent due to the crushing effects during material removal.

The influence of the mechanical effect of the grits can be verified by using different wheel specifications. With increasing grit size, the grit depth

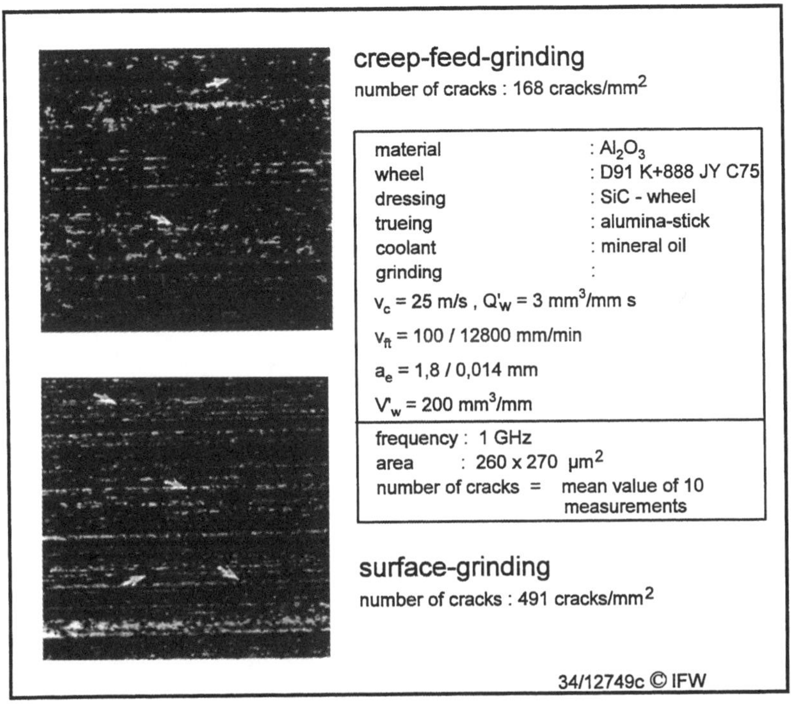

Figure 3.63: Crack Population for Different Grinding Conditions.

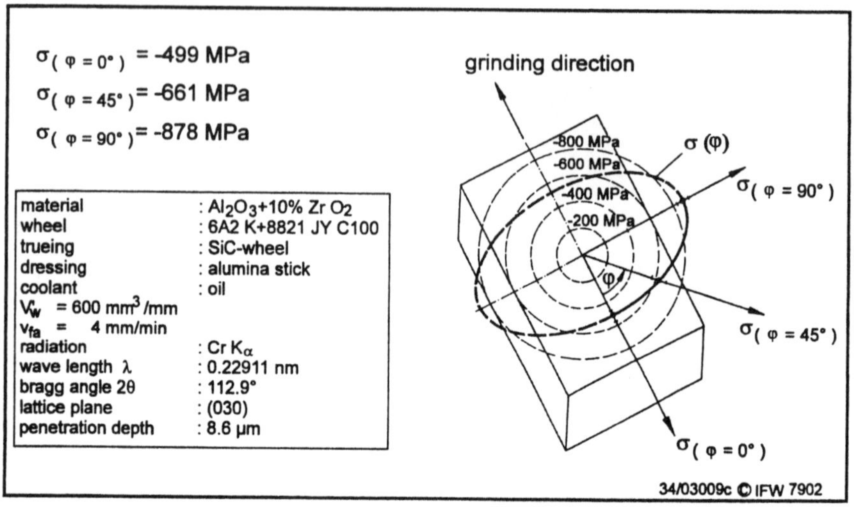

Figure 3.64: Influence of the Machining Direction on Residual Stresses.

of cut increases, whereas increasing number of cutting edges due to increasing concentration leads to decreasing grit depth of cut. Figure 3.65 shows the residual stresses as a function of the cutting speed, for wheels with different grit sizes and concentrations. At a defined cutting speed, a coarser wheel with fewer cutting edges (i.e. coarser grit size or lower concentration) showed higher compressive stresses. Furthermore, all wheel specifications showed the same tendency of lower compressive stresses at higher cutting speeds because of lower grit depth of cut with increasing cutting speed.

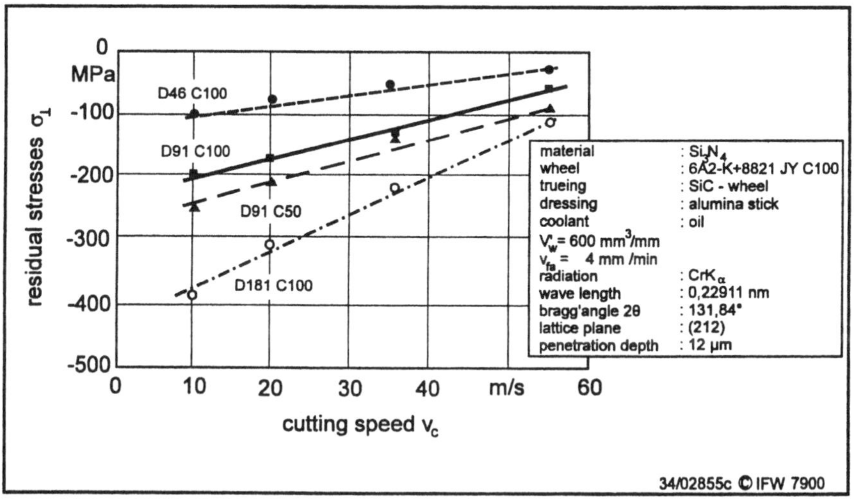

Figure 3.65: Residual Stresses for Different Machine Settings and System Quantities.

Residual Stress Gradients. The penetration depth of x-rays depends on their wave length. It is defined as the value where the primary intensity is reduced to l/e, or 37 percent of the initial value. In ceramics, the layer of residual stresses extend well below the surface, and there are high stress gradients within the penetration depth of the radiation. Figure 3.66 shows a typical profile of residual stresses in a surface ground alumina ceramic. Three different radiations were used to analyze the stresses at different lattice planes and different Bragg angels respectively. Directly underneath the surface, high co-

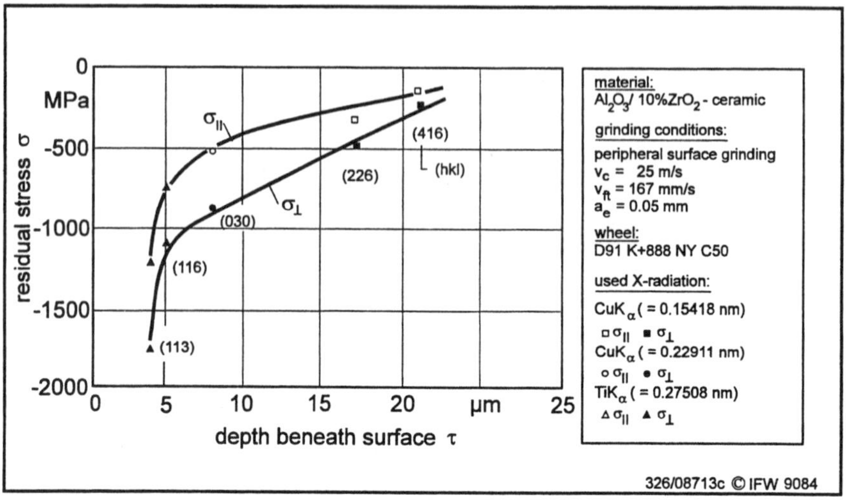

Figure 3.66: Stress Versus Depth Profile in Ground AL_2 O_3/ZrO_2-ceramic.

mpressive stresses were found with a decrease of lower compressive stresses with increasing penetration depth.

These gradients depend on the grinding conditions. Figure 3.67 shows the residual stresses of alumina as a function of the workpiece's speed in surface grinding measured with two different radiations. With increasing workpiece speed, the lower penetrating titanium radiation leads to greater compressive stresses than with chromium rays, which cause increasing stress gradients. For creep-feed grinding the stresses of both radiations are nearly the samc.

Investigations, using a method based on evaluating the curvature of the $\sin^2\psi$-2Θ-distribution (i.e. $\sin^2\psi$-ε-distribution) due to inhomogeneous stresses in the surface confirm the results described above. With increasing orientation angel ψ, the penetration depth of the radiation decreases. Therefore, information from different depth can be evaluated. Figure 3.68 shows the stress gradients in alumina ground with creep-feed and surface grinding conditions. The creep feed ground surface shows a very low stress

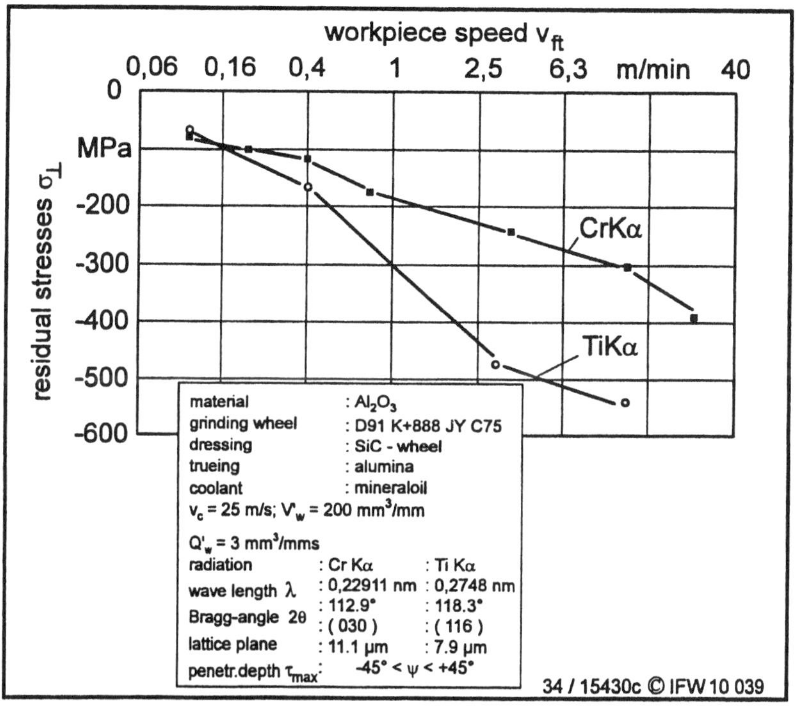

Figure 3.67: Residual Stresses with Different Radiations.

gradient compared to a high gradient for surface grinding conditions. The stresses perpendicular to the grinding direction are higher than the stresses parallel to the grinding direction. This technique calculates the stresses based on measuring data; the valid region reaches from 1.4 µm to 10.6 µm.

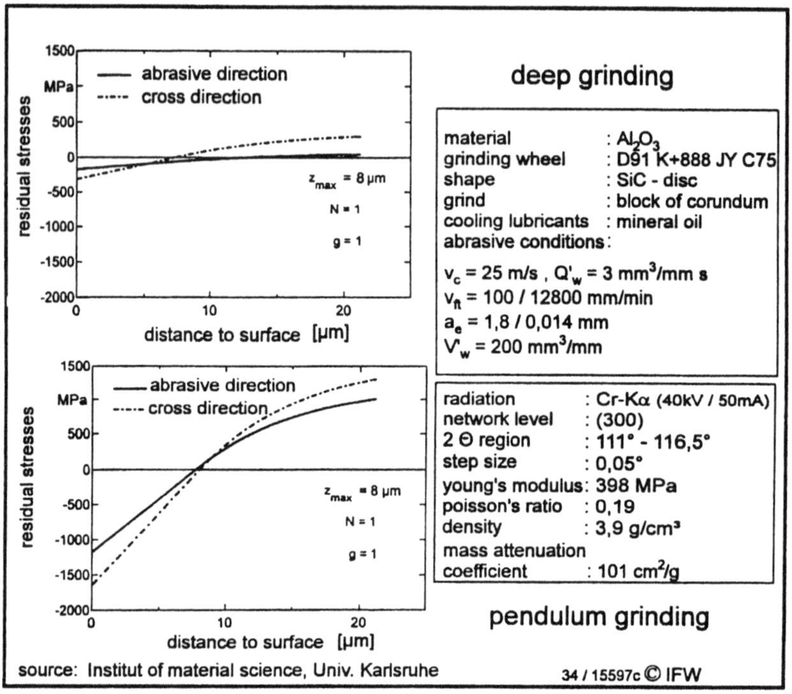

Figure 3.68: Residual Stress Gradients in Creep-feed and Surface-ground Alumina.

REFERENCES

1. John A Schey. 1977. Introduction to manufacturing processes. 271-278.
2. Robert H Todd, Dell K Allen and Leo Alting. 1994. Fundamental principles of manufacturing processes. Industrial Press Inc. New York. 32-36.
3. K Subramanian. 1996. Finishing, processing of ceramics. Materials Science & Technology, A Comprehensive Treatment. 217-257.
4. K Subramanian, P D Redington and S Ramanath. 1994. A systems approach for grinding of ceramics. Bulletin of the American for Ceramic Society. vol 73 No, 3.
5. K Subramanian, and T K Puthanangady. 1997. Diamond abrasive finishing of brittle materials - An overview. Finer Points. vol. 8 No.4, 18-28.
6. I D Marinescu. Tribological aspects of brittle material grinding. Proc. of Diamond 7 CBN Ultrahard Materials Symposium' 93. Windsor, Canada.
7. I D Marinescu. 1991. The grind ability of composite materials in the case of surface grinding with diamond wheels. Ph.D. Thesis. University of Galatzi, Romania.
8. S C Salmon. 1992. Modern grinding process technology. McGraw-Hill Inc. New York.
9. K-H Zum Gahr. 1987. Microstructure and wear of materials. Trib. Series 10. Elsevier, Amsterdam.
10. T G Bifano. 1988. Ductile-regime grinding of brittle materials. Ph.D. Thesis. North Carolina State University.
11. E A Almond, L A Lay and M G Gee. 1984. Comparison of sliding and abrasive wear mechanisms in ceramics and cemented carbides. Proc. Conf. Sci. Hard Mater. Rhodes.
12. M Ouadou and T G Mathia. 1991. Fully computerized sclerotopmeter for simulations and modeling of abrasive wear. Surface & Coat Tech., 45, 33-42.
13. S Yoshida and H Ito. 1990. The present and future of ductile-regime grinding of optical parts. Bul. Japan Soc. Pre. Eng. 24 (4), 239-243.
14. I D Marinescu, J A Webster and B Dimitrov. 1993. Brittle/ductile grinding regimes for brittle materials. Diamond & CBN Ultrahard Materials Symp.' 93. Windsor, Canada.
15. I D Marinescu and J A Webster. 1993. Tribo technological aspects of brittle materials grinding. Proc. of the 5th International Grinding Conference. Ohio, Society of manufacturing Engineers.
16. D Pavelescu, D Musat and A Tudor. 1988. Tribology, friction, wear and lubrication. Education Publishing House, Bucharest.

17. T Matsuo, S Toyoura and H Kita. 1984. Frictional behavior of diamond and cubic boron nitride abrasives on hard materials. Proc. of the 2nd International Conference on Science of Hard Materials. Rhodes. 897-905.
18. T Matsuo and E Oshima. 1984. Evaluation of superhard abrasive grains by overcut fly-milling test. Annals of the CIRP, vol 33/1, 185-188.
19. A Broese van Groenou and J D B Veldkamp. 1978/79. Grinding brittle materials. Philips Technical Review. vol 38, 131-144.
20. A W Ruff, H Shin and C J Evans. 1995. Damage processes in ceramics resulting from diamond tool indentation and scratching in various environments. Wear. 181-183, 551-562.
21. S Malkin and J E Ritter. 1989. Grinding mechanisms and strength degradation for ceramics. Journal of Engineering for Industry, Trans. of the ASME. vol 111, 167-174.
22. H H K Xu, S Jahanmir and Y Wang. 1995. Effect of grain size on scratch interactions and material removal in alumina. J. Amer. Ceram. Soc. vol 78, 881-891.
23. J E Mayer Jr and G P Fang. 1993. Diamond grinding of silicon nitride. NIST SP 847, 205-222.
24. S Malkin. 1989. Grinding technology: theory and application of machining with abrasives. John Wiley & Sons. New York.
25. S Malkin and T W Hwang. 1996. Grinding mechanisms for ceramics. Annals of the CIRP. vol 45/2, 569-580.
26. C Chen, Y Jung and I Inasaki. 1989. Surface cylindrical and internal grinding of advanced ceramics. Grinding Fundamentals and Applications. Eds., S Malkin and J A Kovach. PED-Vol 39, ASME, 201-211.
27. S Malkin. 1995. From steels to ceramics in grinding. Finer Points. Vol 7 No 3, 18-23.
28. G Lutz. 1995. Grinding technical ceramics. Finer Points. Vol 7 No 3, 6-10.
29. T G Bifano, T A Dow and R O Scattergood. 1991. Ductile regime grinding: a new technology for machining brittle materials. Journal of Engineering for Industry, Trans. of the ASME. vol 113, 184-189.
30. B Zhang and T D Howes. 1994. Material removal mechanisms in grinding ceramics. Annals of the CIRP. 43/1, 305-308.
31. I D Marinescu. 1996. Double fracture model in lapping of ceramics. Proc. of Precision Grinding of Brittle Materials 1996 Spring ASPE Topical Meeting. Maryland. 88-94.
32. S Shimada, N Ikawa, T Inamura, N Takezawa, H Ohmori and T Sata. 1995. Brittle-ductile transition phenomena in microindentation and micromachining. Annals of the CIRP. vol 44/1, 523-526.
33. Z Zhong and V C Venkatesh. 1995. Semi-ductile grinding and polishing of ophthalmic aspherics and spherics. Annals of the CIRP. vol 44/1, 339-342.

34. Hasselt, R. van: The functional behavior of surfaces, Annals of the CIRP 17 (1968), p. 217
35. Kahles, J.F. et al.: Survey on surface roughness and surface integrity requirements for machined components, Annals of the CIRP 23 (1974), p. 191
36. Tönshoff, H.K.: Spanen-Grundlagen Springer Verlag Berlin 1995
37. Weingraber, H. von, Abou-Aly, M.: Handbuch Technischer Oberflächen Friedr. Vieweg Verlag, Braunschweig 1989
38. Field, M,. Kahles, J.F.: Review of surface integrity of machined components. Annals of the CIRP 20 (1971) 2, pp. 153-163
39. Tönshoff, H.K., Brinksmeier, E.: Determination of the mechanical and thermal influences on machined surfaces. Annals of the CIRP 29 (1980) 2, pp. 519-529
40. Brinksmeier, E.: Prozeβ- und Werkstückqualität in der Feinbearbeitung Habilitationsschrift Universität Hannover 1991
41. Tönshoff, H.K.: RandzanenbeeinRussung durch Spanen und Abtragen Annals of the CIRP 23 (1974) 1, S. 187-188
42. Tönshoff, H.K., Lierse, T.: Thermomechanik des Schleifens und Prijfung der erzeugten Funktionsflächen, Proc. of the conference "Werkstoffwoche 96", Stuttgart, 28.-31.05.1996
43. Arnold, W. et al.: Characterization of surfaces by acoustic imagining techshort fibers

4

Grinding

Ichiro Inasaki, Hans R. Meyer, Fritz Klocke, Junji Shibata, Gunther Spur, Hans K. Tonshoff, and Hans G. Wobker

1.0 FUNDAMENTALS OF GRINDING

Ichiro Inasaki

Material Removal Mechanism in Grinding

Grinding is a representative abrasive process in which a grinding wheel made of abrasives and bonding material is used. The grinding wheel is conditioned, i.e., trued and dressed, on the grinding machine before grinding as shown in Figure 4.1. This is very characteristic of grinding because the conditioning process has a decisive influence on the final result. The mechanism of material removal with abrasive cutting edges on the wheel surface is basically the same as that with cutting tools. In metal machining, for example, material is removed mainly through shear deformation in both material removal processes. The size of chips removed in grinding is, however, much smaller than the case of cutting providing better surface finish and machining accuracy. The order of chip thickness in grinding is far less than 0.1 mm, whereas it is larger than 0.1 mm in cutting.

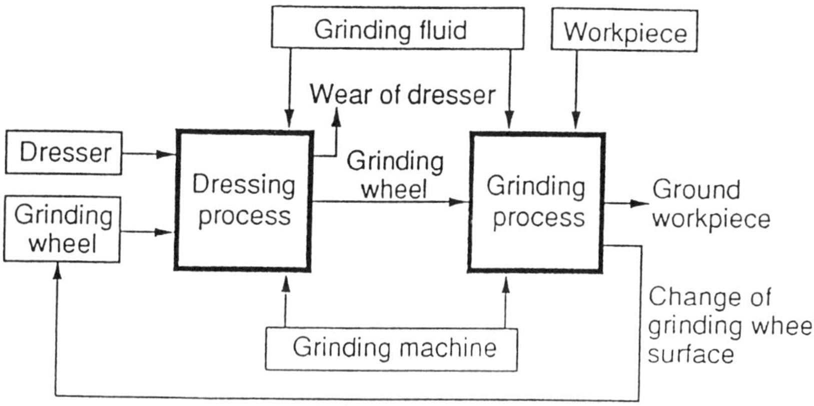

Figure 4.1: Factors Affecting the Grinding Process

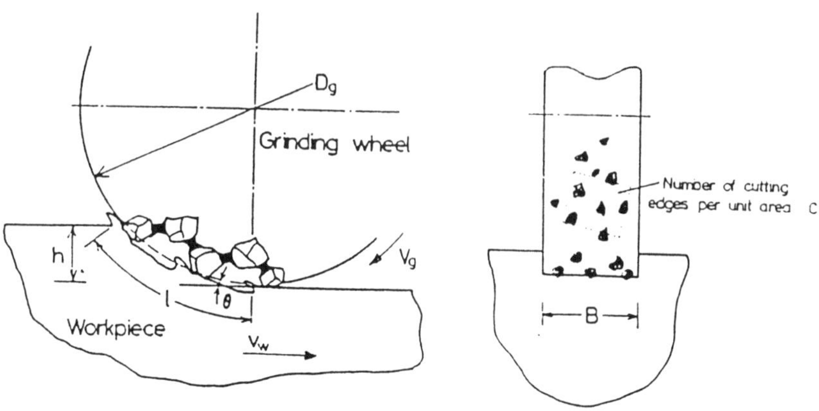

Figure 4.2: Parameters in Surface Grinding

The average cross-sectional area of chips A in grinding can be estimated as follows taking an example of surface grinding (Figure 4.2).[1]

The material removal rate q is where b is the grinding width, v_w is the workpiece speed, and a is the depth of cut. Number of chips produced per unit time N is given by:

$$q = bv_w a \tag{4.1}$$

$$N = Cbv_s \tag{4.2}$$

assuming that each culling edge on the wheel surface produces a chip where, v_s is the grinding wheel surface speed and C is the number of cutting edges per unit area on the wheel surface. The average chip volume V is, therefore:

$$V = \frac{q}{N} \tag{4.3a}$$

$$= \frac{v_w a}{Cv_s} \tag{4.3b}$$

As far as it can be assumed that the chip length is equal to the contact length between the grinding wheel and the workpiece, the chip length is given by:

$$l = \sqrt{d_s a} \tag{4.4}$$

where d_s is the wheel diameter.

From Equations (4.3a), (4.3b), and (4.4), the average cross-sectional area of chips is obtained as:

$$A = \frac{V}{l} \tag{4.5a}$$

$$= \frac{v_w}{Cv_s}\sqrt{\frac{a}{d_s}} \tag{4.5b}$$

Equation (4.5b) can be generalized for cylindrical as well as internal grinding using the equivalent wheel diameter given by:

$$\frac{1}{de} = \frac{1}{ds} \pm \frac{1}{dw} \tag{4.6}$$

Where:

dw :Workpiece diameter
+ :Cylindrical grinding
- : Internal grinding
dw → ∞ : Surface grinding

Consequently,

$$A = \frac{v_w}{Cv_s}\sqrt{\frac{a}{de}} \tag{4.7}$$

Although the average cross-sectional area of chips given by Equation (4.7) has a simple form, it is very useful for understanding qualitatively the influence of set-up parameters on the grinding process. When *A* is increased through the increase of the workpiece speed or the depth of cut, the cutting force acting upon each cutting edge will increase resulting in the breakage or dislodging of the abrasives from the binder. Such phenomenon is called "self-sharpening." The sharpness of abrasive cutting edges can be maintained by self-sharpening; consequently, the increase of normal grinding force versus grinding time is suppressed. The surface roughness of the workpiece is, however, deteriorated. To the contrary, A is decreased through, for example, the reduction of the wheel surface speed if the attritious wear of cutting edges becomes significant. The heat generation, therefore, becomes marked whereas the surface roughness becomes smaller. It can be generally said from Equation (4.7) that A increases as the increase of material removal rate q. To find out the most appropriate value of A which can meet the output requirement is, therefore, crucial in grinding.

The average cross-sectional area of chips can also be changed by varying the number of cutting edges per unit area C. This can be achieved by applying different grinding wheels or by changing the dressing condition. If rough dressing with the large lead is performed, C will be decreased causing the increase of the cutting force upon each cutting edge. The change of wheel surface topography versus time is schematically illustrated in Figure 4.3 as being related to the chip cross section A.

In addition to the number of cutting edges, what has decisive influence on the grinding process is the chip pockets on the wheel surface. If there are not enough chip pockets, swarfs will be loaded on the wheel surface and consequently, troubles such as grinding burn will occur. The volume of chip pockets, which is the function of the porosity of the grinding wheel is, roughly speaking, inversely proportional to the number of cutting edges on the wheel surface. In order to increase the material removal rate, therefore, it is crucial to give enough volume of chip pockets on the wheel surface.

Grinding Stiffness, Contact Stiffness, and Time Constant of the Grinding Process

For achieving high accuracy grinding, which is the most important task of grinding, it is necessary to take the grinding force, elastic deformation of the wheel, and the stiffness of the grinding machine into consideration.

The grinding force can be theoretically derived utilizing the average cross-sectional area of chips given by Equation (4.7). The specific grinding energy, which is defined as the energy per unit volume of material removal, is:

$$E = E_0 a_m^{-\varepsilon} \tag{4.8}$$

where E_0 and ε (= 0.2 ~ 0.5) are constants. It is assumed in Equation (4.8) that there is so-called "size effect" which implies that the smaller the chip cross-section, the larger the specific grinding energy becomes. The grinding power *P* can be obtained through the specific grinding energy multiplied by the material removal rate. Thus:

$$P = Eq \tag{4.9a}$$

$$= E_0 b c^{\varepsilon} v_s^{\varepsilon} v_w^{1-\varepsilon} a^{1-\frac{\varepsilon}{2}} d_e^{\frac{\varepsilon}{2}} \tag{4.9b}$$

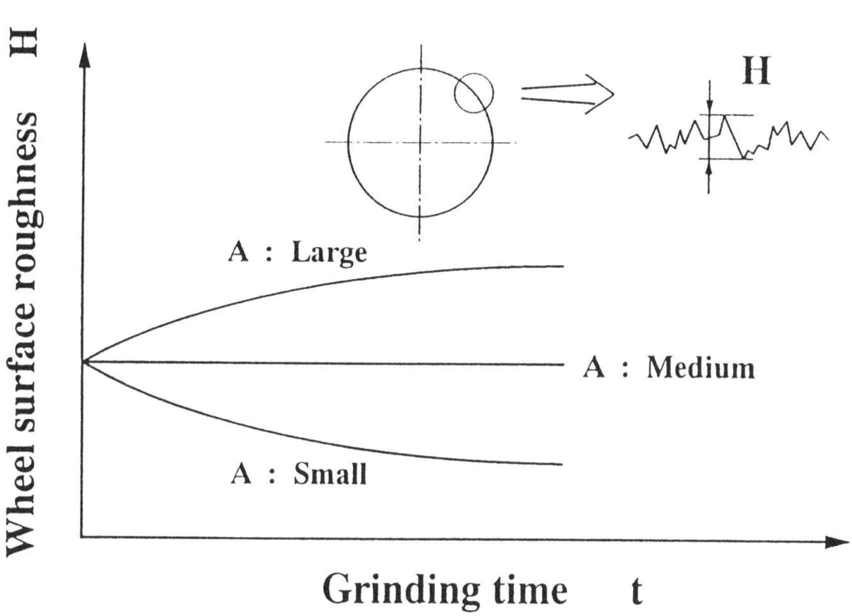

Figure 4.3: Change of Wheel Surface Topography

The tangential grinding force is, therefore, obtained by P/v_s. If it is possible to assume that the exponent ε is equal to 0 (i.e. no side effect) tangential grinding force F_T is given by:

$$F_T = E_0 b \frac{v_w}{v_s} a \tag{4.10}$$

The normal grinding force F_N is given by:

$$F = \frac{1}{\lambda} F_T \tag{4.11}$$

where λ is the constant.

From Equations (4.10) and (4.11), an important parameter, which has a decisive influence on the grinding accuracy, can be derived. If we differentiate the normal grinding force with respect to the depth of cut, we can obtain the increase of the grinding force due to the increase of the depth of cut. Let us call this parameter the "grinding stiffness kg" unit because it has the dimension of the stiffness. The concept of the grinding stiffness makes it possible for us to assume a spring which has a spring constant of kg between the grinding wheel and the workpiece. The grinding stiffness actually has a non-linear characteristic of soft-spring type as understood from Equation (4.9b), although we assume its linearity as in Equation (4.12) in the following discussions.

$$K_g = \frac{dF_N}{da} \tag{4.12a}$$

$$= \frac{1}{\lambda} E_0 b \frac{v_w}{v_s} \tag{4.12b}$$

Assuming the surface grinding process, which is an intermittent infeed grinding, the actual depth of cut a is given by:

$$a = a_n - \delta \tag{4.13}$$

where a_n is the nominal depth of cut and δ is the elastic deformation of the mechanical system. The elastic deformation is obtained by dividing the normal grinding force by the static stiffness of the mechanical system k_m. Thus:

$$\delta = \frac{F_N}{k_m} \tag{4.14}$$

Combining Equations (4.12b), (4.13), and (4.14), the relationship between the actual depth of cut and the nominal depth of cut is obtained as:

$$\frac{a}{a_n} = \frac{1}{1 + \frac{k_g}{k_m}} \tag{4.15}$$

From Equation (4.15), it should be recognized that the increase of the mechanical stiffness is crucial for improving the grinding accuracy.[2] The grinding stiffness also affects the occurrence of the self-excited chatter vibration in grinding.

In addition to the grinding stiffness, it is necessary to take the contact stiffness k_c between the grinding wheel and the workpiece into consideration when we discuss the grinding process. The contact stiffness determines the elastic deformation of the grinding wheel surface due to the normal force. Needless to say, it affects the actual depth of cut together with the stiffness of the mechanical system. Its influence can be taken into account by considering that the contact stiffness is connected in series to the stiffness of the mechanical system as follows:

$$\frac{1}{k} = \frac{1}{k_c} + \frac{1}{k_m}. \tag{4.16}$$

The contact stiffness is affected mainly by the bonding material and has a nonlinear characteristic of hard-spring type. It also has a decisive influence on the stability of the grinding process, i.e., the higher the contact stiffness, the more unstable the grinding process becomes.[3]

For the continuous infeed grinding such as cylindrical and internal grinding, it is useful to introduce the parameter "time constant" for discussing the process behavior.[4] Let us consider the simplest grinding cycle consisting of a rough grinding and a spark-out grinding as shown in Figure 4.4. The actual change of the workpiece radius is different from the nominal infeed motion due to the elastic deformation of the mechanical system. This phenomenon affects not only the grinding time but also the quality of the products in terms of the surface roughness and the geometrical accuracy, because the actual infeed rate changes during the grinding time.

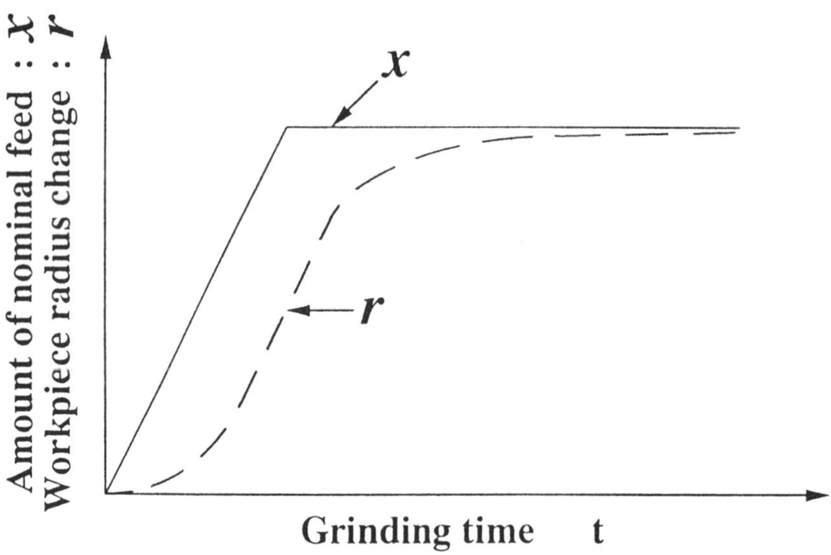

Figure 4.4: An Example of Grinding Cycle

From the equations for actual radius change of the workpiece *r*:

$$r = x - \delta \tag{4.17a}$$

$$= v_f t - \delta, \tag{4.17b}$$

where x is the amount of the nominal infeed, v_f is the nominal infeed speed, and δ is the elastic deformation of the mechanical system which is given by:

$$\delta = \frac{F_N}{k_m}, \tag{4.18}$$

Thus, the following equation can be obtained:

$$r + \frac{F_N}{k_m} = v_f\ t. \tag{4.19}$$

Assuming the plunge grinding process, in which the material removal rate q is given by:

$$q = \pi d_w \frac{dr}{dt}, \tag{4.20}$$

where dr/dt is the reduction rate of the workpiece radius, and the normal grinding force is proportional to the material removal rate, Equation (4.20) becomes:

$$r + \sigma\pi \frac{d_w}{k_m} \frac{dr}{dt} = v_f\, t, \tag{4.21}$$

where α is the constant. The solution of this first order linear differential equation is

$$r(t) = v_t - v_f\, T\, \{1 - e^{\frac{t}{T}}\}, \tag{4.22}$$

where

$$T = \alpha\pi \frac{d_w}{k_m}, \tag{4.23}$$

is the time constant of the grinding process.

The time constant is an important parameter for assessing the grinding process performance. Generally speaking, *T* should be as small as possible for achieving the reliable grinding process. The time constant increases as the decrease of the mechanical stiffness and the increase of the constant α. The materials that are difficult to machine have a large value of α, and consequently, an increased time constant. The time constant is not necessarily constant during the grinding process; it generally increases due to the increase of the parameter α.

Characteristics of Ceramic Grinding.

Being different from metal grinding, material removal through brittle fracture plays an important role in ceramic grinding. This is due to the fact that ceramics possess a high degree of brittleness. Figure 4.5 shows how the average cross-sectional area of chips affects the specific grinding energy in

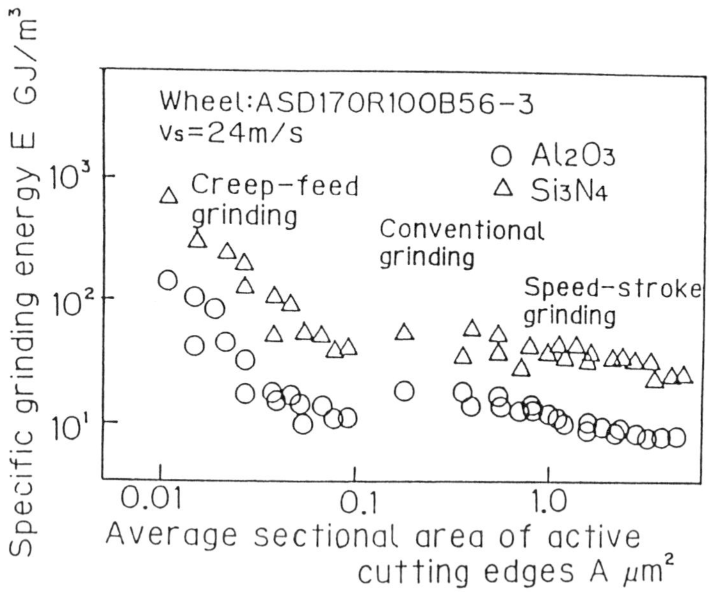

Figure 4.5: Specific Grinding Energy of Advanced Ceramics

surface grinding.[5] The specific grinding energy *e* is calculated using the following equation:

$$e = \frac{F_T\ v_s}{bv_w a}. \tag{4.24}$$

What is evident from the figure is that the specific grinding energy starts to increase significantly in the range of small average cross-sectional area of chips *A*, whereas the energy is almost constant in the large value of *A*.

The influence of *A* on the surface roughness is shown in Figure 4.6. The surface roughness increases as *A* increases up to about 4 µm, where it has a tendency to become constant. In addition, the difference in the surface roughness between longitudinal (R //) and trans-grinding direction ($R_\perp$) becomes insignificant in the range of large chip cross-section.

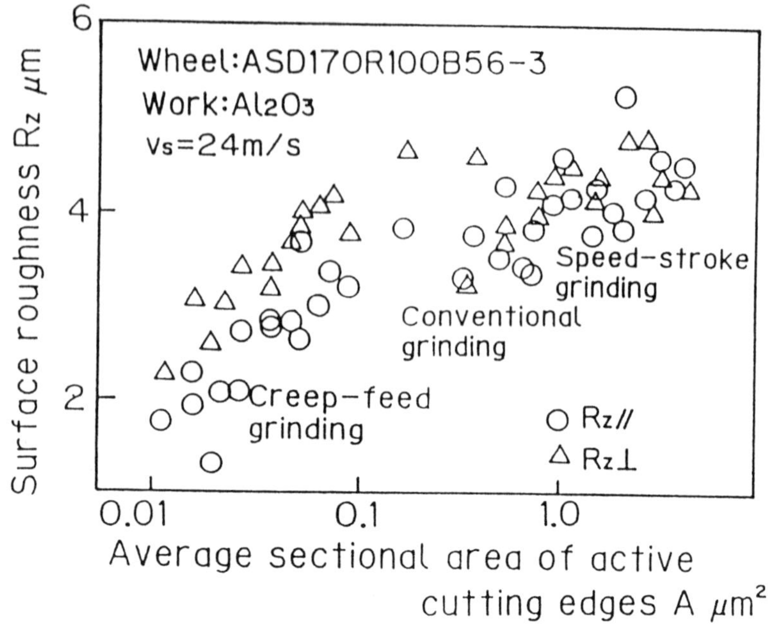

Figure 4.6: Surface Roughness of Ground Ceramics

Through investigation of the specific grinding energy and the surface roughness shown in Figure 4.5 and Figure 4.6, it can be pointed out that the material is mainly removed by the brittle fracture of the ceramics in the range of large cross-sectional area of chips, whereas the material is removed by the shear deformation in the range of small chip cross section. From the viewpoint of ground surface quality, not only surface roughness but also subsurface layer with cracks, it is recommended to grind ceramics with set-up parameters which give small values to *A*. However, such a condition results in the large grinding power and the force under the same material removal rate because of the increased contact length between the wheel and the workpiece. Besides, increase of the grinding power and the force versus time are significant when the chip cross-section *A* is small (Figure 4.7). Excessive attritious wear of cutting edges due to low cutting force upon each cutting edge is considered to be the reason for this.

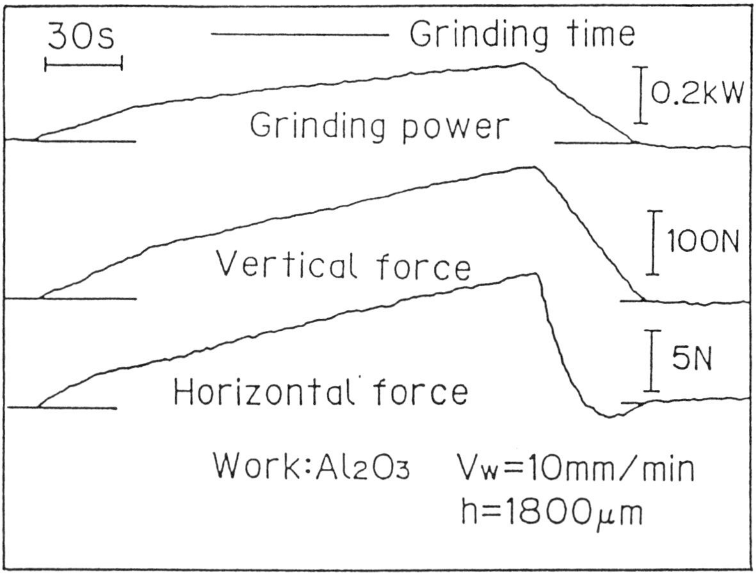

Figure 4.7: Increase of Grinding Force and Power

Guidelines for High Efficiency Grinding of Ceramics

Experimental results shown in Figures 4.5, 4.6, and 4.7 provide us with some useful information for improving the efficiency of ceramic grinding. To achieve the high material removal rate, it is recommended to take advantage of brittle mode material removal which is promoted by the large value of the chip cross-section *A*. As shown in Figure 4.8, the specific normal grinding force is reduced by increasing the area of chip cross-section. In this grinding test, the material removal rate is kept at a constant value although the average cross-sectional area of chips is changed by changing the combination of the workpiece speed and the depth of cut. The average cross-sectional area of chips becomes large when the depth of cut is small and the workpiece velocity is high under a constant material removal rate.

The increase of grinding force versus time is compared for different values of *A* in Figure 4.9. In this case again, the material removal rate is kept at a constant value. The advantage of grinding with a large value of the chip

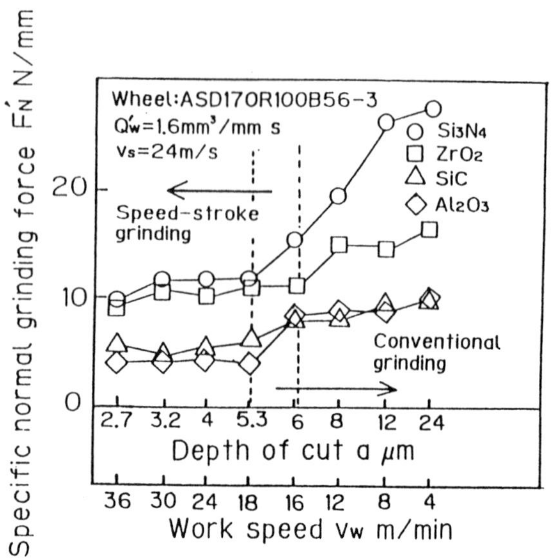

Figure 4.8: Specific Normal Grinding Force

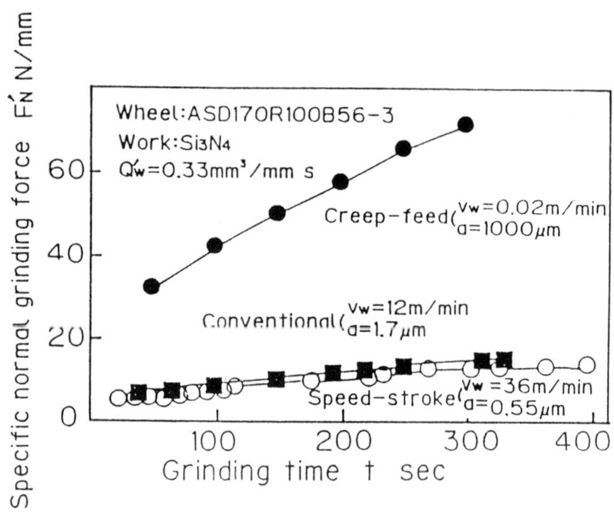

Figure 4.9: Increase of Specific Normal Grinding Force

cross-section is obvious. The increase in the grinding force can be markedly suppressed by applying the combination of the small value of depth of cut and high workpiece velocity. The activated self-sharpening of abrasives due to the large cutting force acting upon each cutting edge must be the reason why the increase of the grinding force is suppressed.

Consequently, the following procedure is recommended for improving the grinding efficiency of ceramics:

(1) In the rough grinding process for removing most of the allowance, the average cross-sectional area of chips should be set at large value to take advantage of brittle mode machining.

(2) The subsurface layer introduced by the rough grinding process should be removed by the finishing process featured by the small cross-sectional area of chips.

2.0 GRINDING TOOLS

Hans R. Meyer, Fritz Klocke

Introduction

Grinding is practically the only way of machining ceramic materials in the fired state. Their great hardness and high wear resistance necessitate production systems that work with loose or bonded abrasives. The physical properties of diamond make it particularly suitable as an abrasive for these applications. Assuming that the practical issues of technological application are resolved, the economics of operation are the decisive factor. The key factors in grinding with diamond wheels are the selection of the machine and cooling conditions, the optimization of the diamond grinding wheel, and the associated dressing and trueing strategy.

Diamond wheels are different from conventional wheels in several respects. Basically, the tool system may be broken down into three sub-systems: the abrasive system, the bond system, and the core system (Figure 4.10). The diamond wheels themselves feature high macro-geometrical accuracy as delivered.

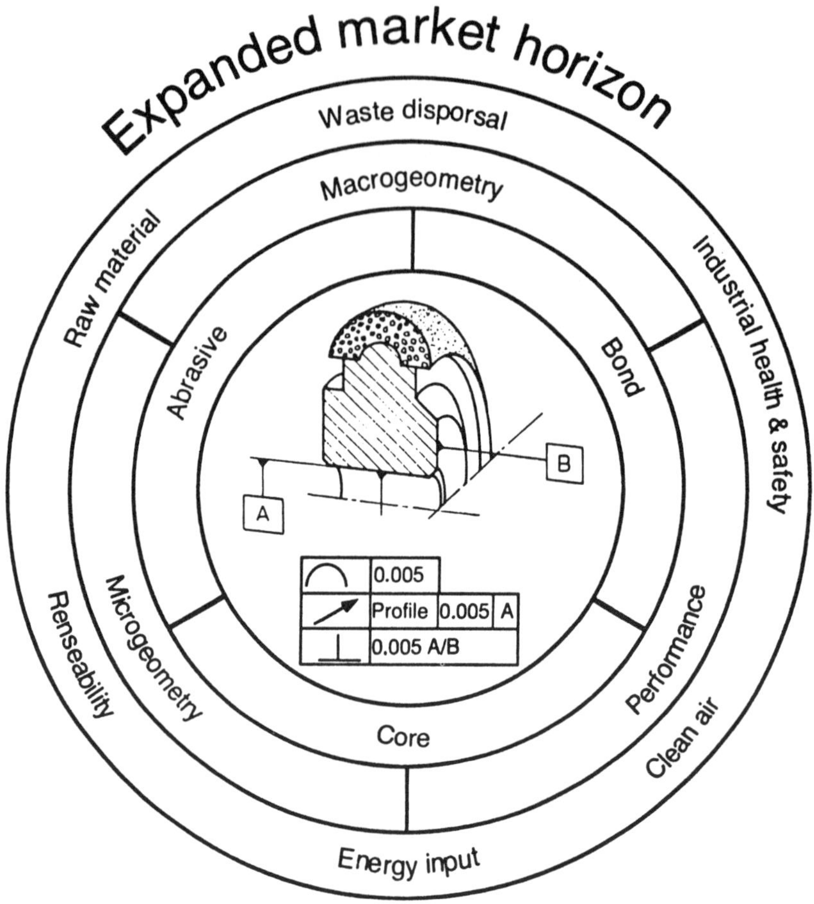

Figure 4.10: Expanded Market Horizon

Grinding behavior in application is largely determined by the micro-geometrical characteristics of the chip space of the grinding wheel. The initial state of the grinding layer determines the initial wheel cutting behavior, the initial wear, and thus also the non-steady-state break-in phase. The grinding wheel topography reflects wheel composition and any dressing processes that have taken place.

In the past, the application of diamond grinding wheels was mainly determined by technical and economic features. Today, the selection of diamond grinding wheels is largely determined by other, additional criteria (Figure 4.10). These include aspects such as recycling capability of the raw materials and cores, industrial health and safety, the raw materials used and waste disposal, and aspects of energy input per volume unit of the material to be machined. Ceramics machining, in particular, is characterized by high-performance grinding processes such as creep-feed grinding and high-speed grinding, which open up technological and economic perspectives; this also means new requirements for grinding wheel design and strength and mounting systems.

Grinding Wheel Design

Abrasive Materials

An important criterion for the specification of an abrasive is the hardness of workpiece material. A hardness comparison between ceramic and metallic materials shows that diamond, in particular, is suitable for hard machining ceramics. The diamonds used for this purpose are mainly synthetic diamonds, which are manufactured by lattice conversion from the soft hexagonal layer-lattice structure of graphite. This gives a considerable increase in density and an extreme increase in hardness (Figure 4.11).

Diamonds differ in their external morphology. Basically, crystals are built up from initial cubic and octahedral shapes and their combinations. Grinding applications use either grown crystals or crushed grits. These basic shapes may be influenced by the synthesis parameters. The synthesis process permits control of diamond characteristics, generating either blocky grit sha-

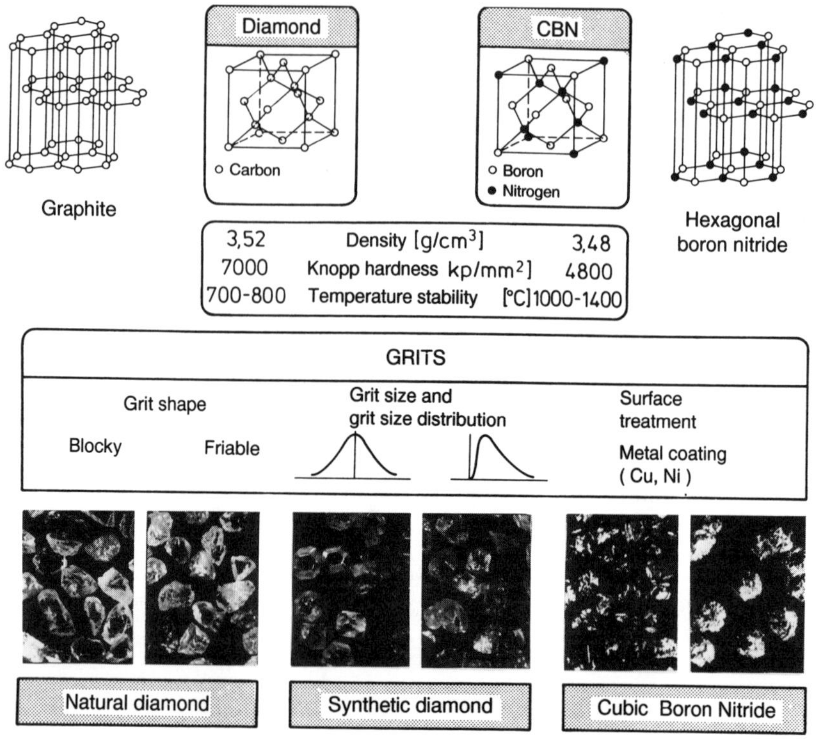

Figure 4.11: The density and hardness of natural diamond, synthetic diamond, and CBN.

pes with very high impact strength, or friable grits with low impact strength. Diamonds are also classified by grit size and grit size distribution. Micron grits, in particular, may be specified by asymmetrical grit size distributions, i.e., the size range is limited in one direction.

Thermal transition behavior between diamond, bond, and cohesion in specific bond types may be improved by giving the diamonds metal coatings, which are generally copper alloys or nickel phosphorus alloys. Thermal conductivity is an important physical quantity (Figure 4.12), that is very much dependent on the purity of diamonds. Thus, the thermal conductivity of high-purity diamonds is around 2000 W/mK, which is far above the thermal conductivity of all other materials. Even synthetic diamonds used for grinding applications have higher thermal conductivity than silver or copper.

The metal coatings used basically have two functions to fulfill. Firstly, their external structure can be varied. They can have very rough, jagged, or even spiky surface structure. This gives good mechanical anchoring of the grits in the resin bond. Secondly, the nickel-phosphorus layer has considerably lower thermal conductivity than diamond, and thus acts as a thermal retarder as long as thermal breakdown of the resin bond is avoided. The wide variety of capabilities shows that the grit system can be used to control the desired working result within a wide range.

Diamond is the hardest known cutting material, but this hardness is very much dependent on temperature. It decreases considerably above 400°C in an oxidizing atmosphere (Figure 4.12). At higher temperatures, the diamond lattice is converted to a hexagonal layer lattice. It is important to be aware of these factors in order to prevent thermal overloading of the diamond by providing adequate cooling/lubrication and by proper specification of the other process parameters. There is also a danger that carbon atoms from a partly graphitized area may diffuse into the workpiece and cause chemical wear if there is chemical affinity with the workpiece and if the activation energy is exceeded.

Bonds

Multilayer Wheels. With multilayer diamond grinding wheels, the grits are held in a metal, resin, or vitrified bond. These grinding wheels can be dressed.

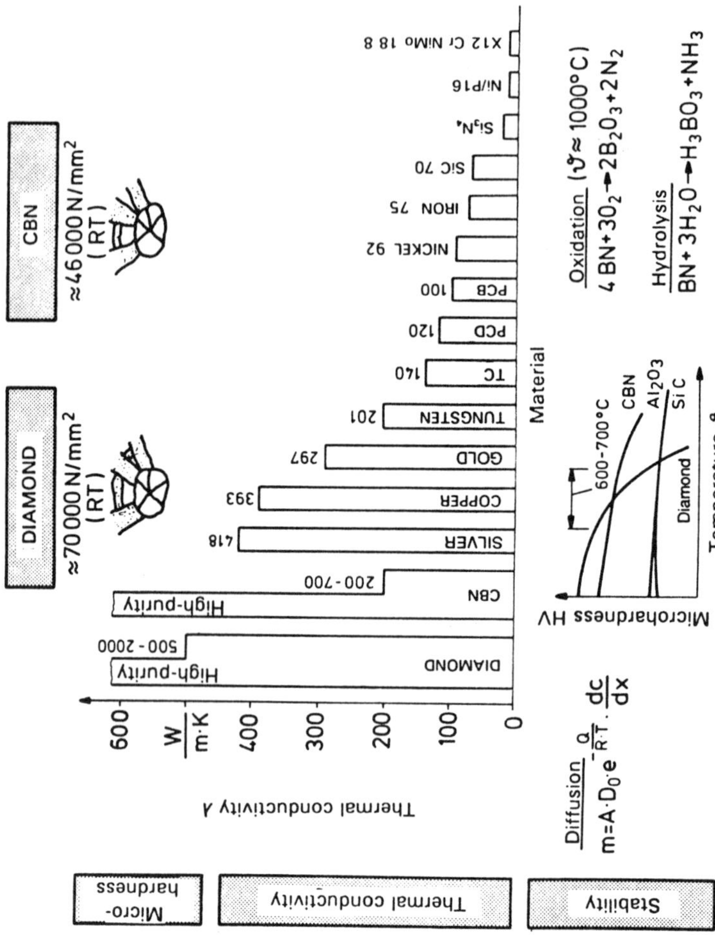

Figure 4.12: Properties of diamond and CBN.

Metal Bonds. Metal bonds mainly use modified copper/tin and cobalt/bronze materials. Iron/copper/tin systems are also used in special applications. There are reports from Japan on the use of grey cast iron bonds in diamond wheels for ceramics grinding.[6][7] The composition of the bond and the manufacturing parameters basically produce changes in abrasive wear resistance and ductility/brittleness of the bond. The main feature of metal bonds is that they key in the diamond very intensively, and give very high abrasive wear resistance to small detritus that occurs when machining short-chipping materials. This means that additional sharpening processes are normally necessary when grinding ceramics with metal-bond diamond wheels. Wheel manufacture is done by sintering at high temperature; the diamonds used for metal bonds must have high heat resistance in order to avoid thermal damage from this process.

Resin Bonds. The material used to bond the diamond in the resin matrix is mainly phenolic, and fillers may be added to control the bonding characteristics. These fillers may be used to change wear resistance and thermal conductivity or may serve as solid lubricants. Polyamide resins may be used, instead of phenolic resins. These polyimides have higher strength, thermal stability, and elasticity. Resin bonds are mainly used with metal-coated diamond grits, in order to improve mechanical anchoring in the bond matrix. Wear resistance against abrasion is considerably lower than with metal bonds.

Vitrified Bonds. Resin bonds and metal bonds are generally manufactured free of pores (as determined by bond formulation and manufacturing conditions), unlike normally fired vitrified bonds, which have a defined pore content.

Thus, their structure is basically the same as that of conventional abrasives. However, due to the mechanical/thermal properties of the diamond, these bonds are designed quite differently; there are also differences in firing temperatures and firing characteristic. Vitrified bonds can also be manufactured free of pores by the hot pressing technique.

Monolayer Wheels

In electroplated monolayer grinding wheels, the layer thickness corresponds approximately to the mean grit size used. Monolayer wheels cannot be dressed, and wheel life ends when the layer no longer provides the required performance. Monolayer vitrified-bond wheels are used to true profiles, particularly for machining ceramics in green state. The service life of the wheels is long, and tool change is not required at short intervals. But it is important to remember that the behavior of monolayer wheels changes during the grinding process. It is not possible to achieve a virtually steady-state in the grinding process, because the wheel has only one layer of grit. This is not so important in the machining of ceramics in green state, because the wheel layer components are subjected to only minor abrasive attack by the chips removed from the workpiece. This means long wheel life due to low wear gradient over grinding time.

The available chip space is determined by grit protrusion over the bond and also by the usable grit volume available for the chip-removal process. There is a trade-off here between sufficient retention strength of the grit in the bond (which increases with increasing layer thickness) and maximum chip space (achieved by small layer thickness). Bond material is usually an electrolytically deposited nickel layer, with a thickness of about 50% of mean grit diameter.

There are two other bonding systems possible for monolayers. The first alternative is chemically-deposited, nickel-phosphorus alloys. Following further thermal treatment, these have considerably higher hardness than electrolytically-deposited nickel layers. They also give a more uniform layer thickness, especially with large profile depths. The drawbacks of this process are that it requires more extensive production facilities and considerably higher deposit temperatures, and there is greater crack sensitivity because the deposited bond layer is more brittle. The mean layer thickness is likewise 50% of the mean grit diameter.

The second alternative for metal-core diamond wheels is high-wetting active brazing materials.[8][9] If titanium-containing brazing materials are used, production requires either a vacuum of $<10^{-4}$ bar, or an argon atmosphere; other complex brazing materials on a Ni-Cr-Bo-Si base also require an inert gas atmosphere. The use of active brazing materials and high-wetting capability makes it possible to achieve mean layer thicknesses of about 20 to 30% of mean grit diameter. This makes it possible to achieve correspondingly larger chip spaces. This production technique requires complicated equipment. It also requires steel cores due to the high manufacturing temperatures, and it is not possible to manufacture highly accurate profiles because of thermal expansion. Reproducible production of uniform cutting layer properties is particularly difficult with profiles, because it is practically impossible to control grit size distribution accurately enough. In addition, the greater chip space has only slight influence on grinding and wear behavior when machining ceramics. All these factors mean that the benefits which are theoretically possible are very hard to achieve in practice in ceramics machining. That is why these bond systems have not yet achieved wide application.

Cores

Unlike conventional grinding wheels, diamond layers can be combined with a wide range of different core materials. This gives a wide parameter field to adapt the diamond wheel to the specific machining operation. This is particularly important in the machining of brittle ceramics, permitting optimal adaptation of the wheel to the specific properties of the material, especially with respect to vibration behavior thermal conductivity (Figure 4.13).

For monolayer diamond wheels, an electrically-conducting core is necessary. Metallic materials such as steel and aluminum are mostly used here. Carbon fiber reinforced cores with resin matrix may be used in special applications requiring extremely high peripheral speeds.[10]

For multi-layer grinding wheels, there is a wide range of different grinding layer/core combinations, a selection of which are shown in Figure 4.14. When machining ceramic materials, it is important to minimize external vibration sources because of the brittle nature of these materials. The grinding wheel system should therefore be optimized to reduce run-out error, clamping error, and grinding wheel unbalance. The vibration effect of these possible error sources can be minimized by keeping the weight of the grinding wheel low. Resin composite cores are often used in these applications be-

Property	Resin with Filler		Resin with Fiber Reinforcement	Metal	Vitrified
	Metal	Non-Metal			
Thermal Conductivity	◐	○	◐	●	○
Damping	◐	●	◐	○	◐
Specific Young's Modulus	○ low	○ low	● high	◐	○ low
Density	● low	● low	● low	○ high	● low
Strength Bursting	○ low	○ low	● high	● high	○ low
Machinability	◐	◐	○	●	○
Material Cost Manuf. Cost	◐	◐	○ high	● low	● low

Assesment: ● good ◐ medium ○ poor

Figure 4.13: Cores

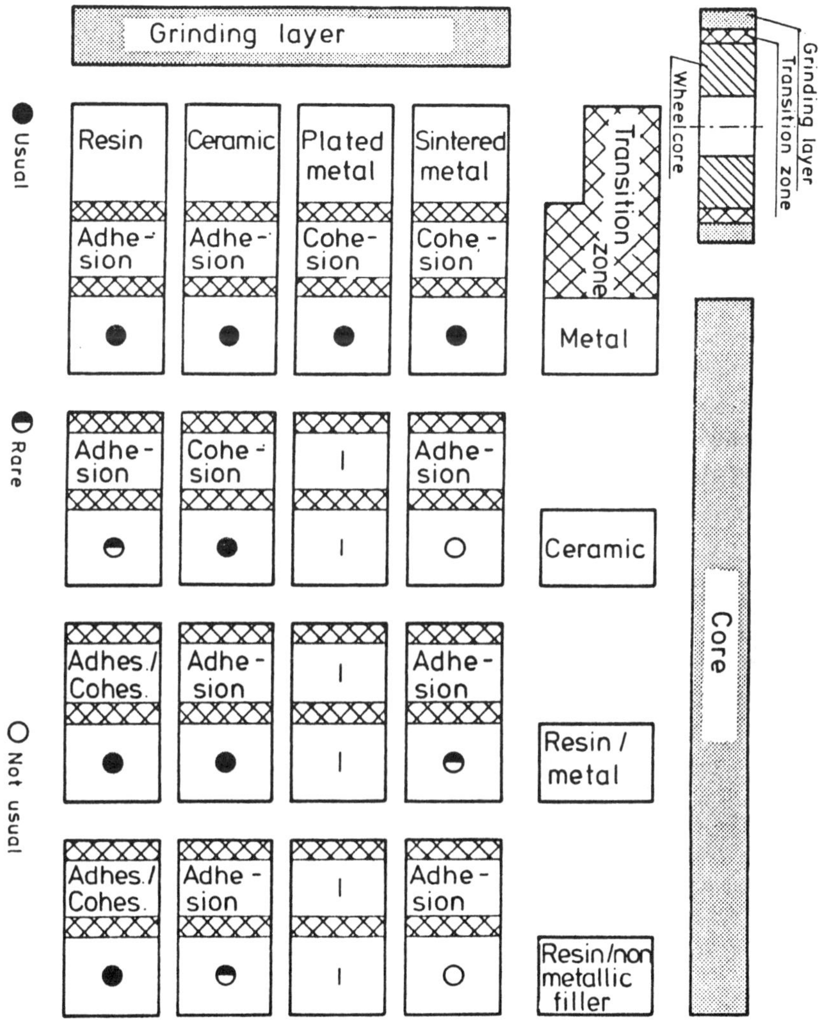

Figure 4.14: Grinding Layer

cause they feature low weight, high intrinsic damping, medium strength, and thermal conductivity.

Wheel Description

The full description of a grinding wheel must give information on geometry, layer composition, and core (Figure 4.15). The most important wheel shapes are codified in the FEPA standard and have been incorporated in ISO 6168-1979. In Europe, the description of micron powers mostly uses the FEPA standard, which has likewise been taken over in ISO 6106-1980 and in DIN 848. In the USA, the sieve sizes are mainly indicated to U.S. standard ASTM E-11-70. Micron powders smaller than D46 (FEPA) or 325/400 mesh (ASTM) are currently not standardized. For micron powders, the classification is no longer done by sieving but by sedimentation.

The grit concentration describes the volumetric diamond content in the layer volume. The baseline value is 4.4 carats per cm^3 layer volume, which is generally indicated as C100. The calculation with diamond density shows that C100 gives a percentage by volume of 25%. Thus, C50 means that diamond content is 2.2 carats per cm^3.

Tailoring the Wheel to the Material and Process

Grinding Ability

The grinding behavior or grinding ability of the wheel are general terms, referring both to the physically-defined parameters and the empirical values from practical operation. They describe the ability of a wheel to grind a given material under defined conditions. It is not possible to quantify this parameter; at most, this can be done with reference to a specific application. For this reason, the grinding ability of a wheel is not given exclusively with the wheel parameters but always takes account of specific cutting properties of the material, the dominant kinematic parameters of the grinding process,

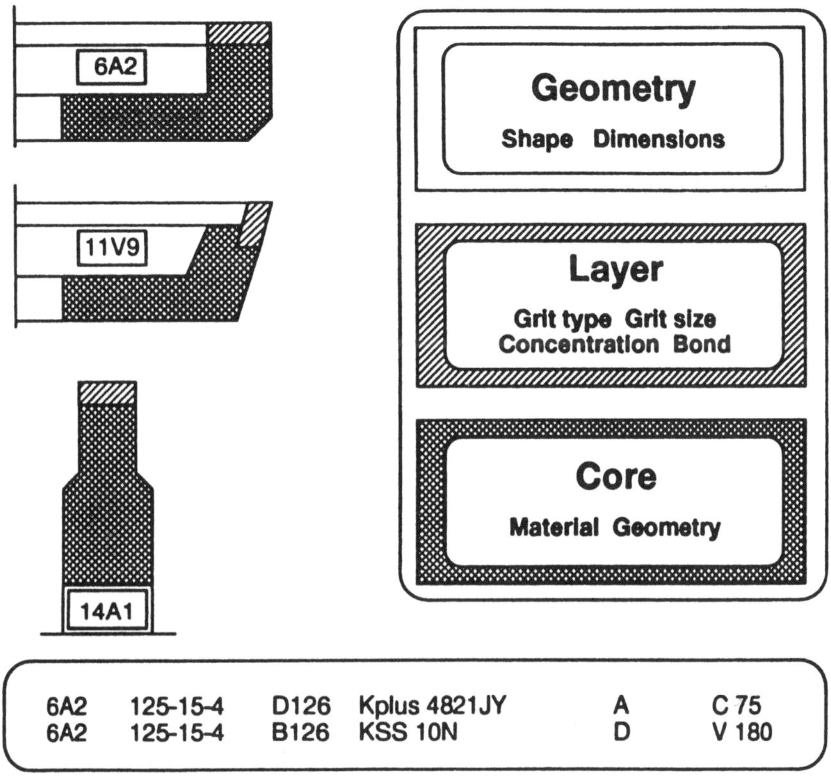

Figure 4.15: Grinding wheels.

and grinding wheel data.

Despite this complexity, it can make sense to distinguish the various families of bonds from one another, as an initial guideline to help select the bond for a given application (Figure 4.16). Bonds with a wide fluctuation in range of properties make particularly severe demands on application preparation and constancy of the grinding process; on the other hand, they make it possible to increase performance by controlled use of outstanding performance features. Performance capability can be optimized by a wheel tailored to the specific application, although a wheel with slightly different properties may be easier to use and may not react so sensitively to process variations.

Precision Machining, Finish Machining of Contours

The main feature in wheel specification for finish machining of ceramics is the avoidance of local stress peaks, which may lead to sudden unpredictable brittle fracture of the component. The most important rule in wheel specification is therefore to avoid or suppress crack growth which might cause a rise in maximum fracture stress.[11][12][16] Even in brittle materials, it is possible to achieve ductile behavior under certain conditions, as shown by extensive and detailed crack tests with single-grit wheels and systematic studies of process parameters.[12][13][14][15][16][17][18][19] These conditions are as follow:

- Low force per individual grit
- Small individual chip thickness, not exceeding a certain material specific value
- Large cutting point radii generate compressive stress fields and promote ductile grinding. However, this only takes effect if there are no simultaneously higher normal forces and considerable heat development to offset the positive effect described above.

This means that creep-feed grinding and high cutting speeds may be favorable, provided that this is not offset by thermal damage, which may occur directly on the grinding wheel and also directly on the workpiece.

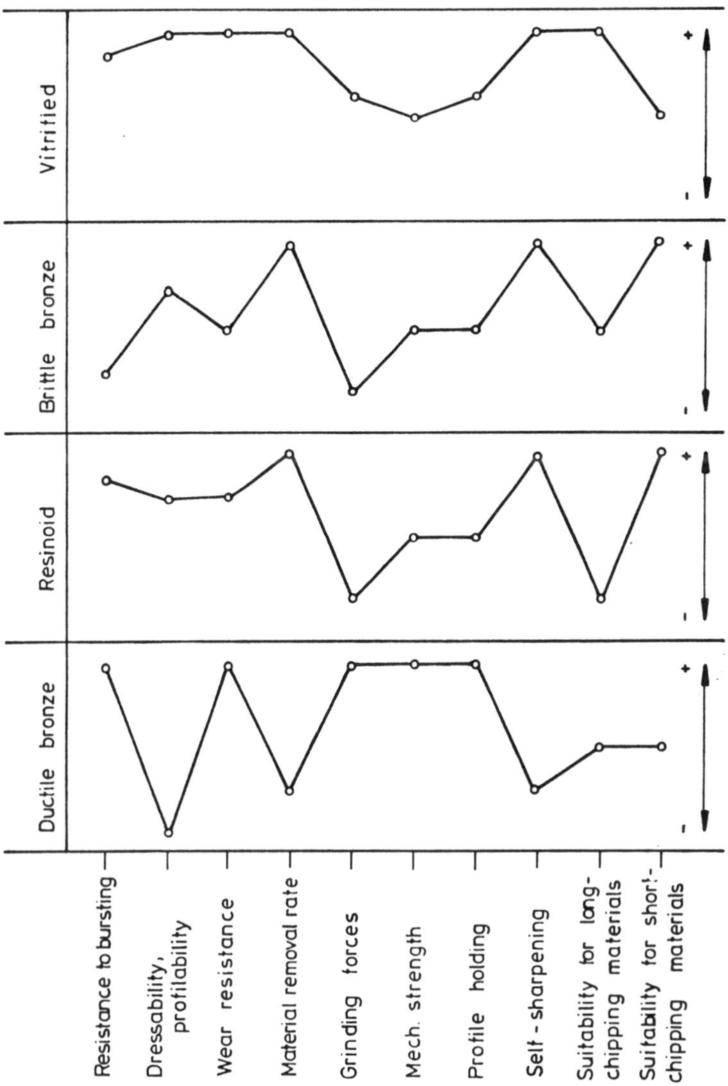

Figure 4.16: Guidelines for Bonding Selection

Rough Grinding

The optimization criteria are different in rough grinding - in this case, there are subsequent operations for removal of further surface layers. The allowances for final machining must be greater than the surface zone subject to damage from rough machining. The following optimization criteria must therefore be included in wheel design:

- Low energy requirement for removal of one material volume unit. This is favored by brittle fracture chip formation. The chip thickness must exceed a material-specific thickness.
- Low increase in force vs. material volume removed.
- Low specific wheel wear.

Wheel Specification

In keeping with the target criteria mentioned above, diamond wheels for machining brittle ceramic materials must be specified in two basic stages.

Rough Design

Based on the material to be machined, the emphasis is on the following:

- Part geometry
- Intended dimensional, shape and positional accuracies, and
- Surface qualities.

The machine parameters such as dressing and trueing processes are then taken into consideration to specify the following:

- Wheel diameters,
- Wheel bond, and
- Diamond.

Fine Design

The fine adjustment phase is to optimize wheel composition, based on the thermomechanical properties of the material to be ground, with the following goals:

- Finish grinding: to minimize the surface zone influenced by grinding
- Rough grinding: to achieve an energy-efficient brittle fracture process.

The main parameters here are process technology (such as creep-feed grinding, reciprocating grinding, up/down grinding, high-speed grinding) and coolant/lubricant parameters. The goal is to optimize the following:

- Diamond grit type,
- Diamond grit size, and
- Diamond concentration.

Wear Mechanisms and Diamond Grit Type

There are different forms of wear of the diamond grit, depending on wear state (Figure 4.17). Wear on the grinding grit starts where there is mainly thermal load in the outer grit regions. If the temperatures are too high, the diamonds may be graphitized in some areas, which causes considerable reduction of wear resistance. This causes wear flats on the abrasive grit (Figure 4.17b). Flats due to abrasion and thermal effects may also be caused by the use of wear-resistant bonds with high grit retention force (e.g. metal bonds), high-impact-strength diamond types which do not splinter easily, high wheel peripheral speeds, and creep-feed grinding. With friable grit types in particular, even small mechanical overloads may be sufficient to form new, sharp secondary cutting points by partial breakout of diamond fragments (Figure 4.17d). With blocky diamonds, if there is mechanical overload under conditions of sufficient retention in the bond, this may cause partial grit breakout, but the number of new secondary cutting points generated is smaller; the cutting points are elongated, and not so pointed and sharp-edged (Figure 4.17e). Complete grit breakout occurs if the retention force of com-

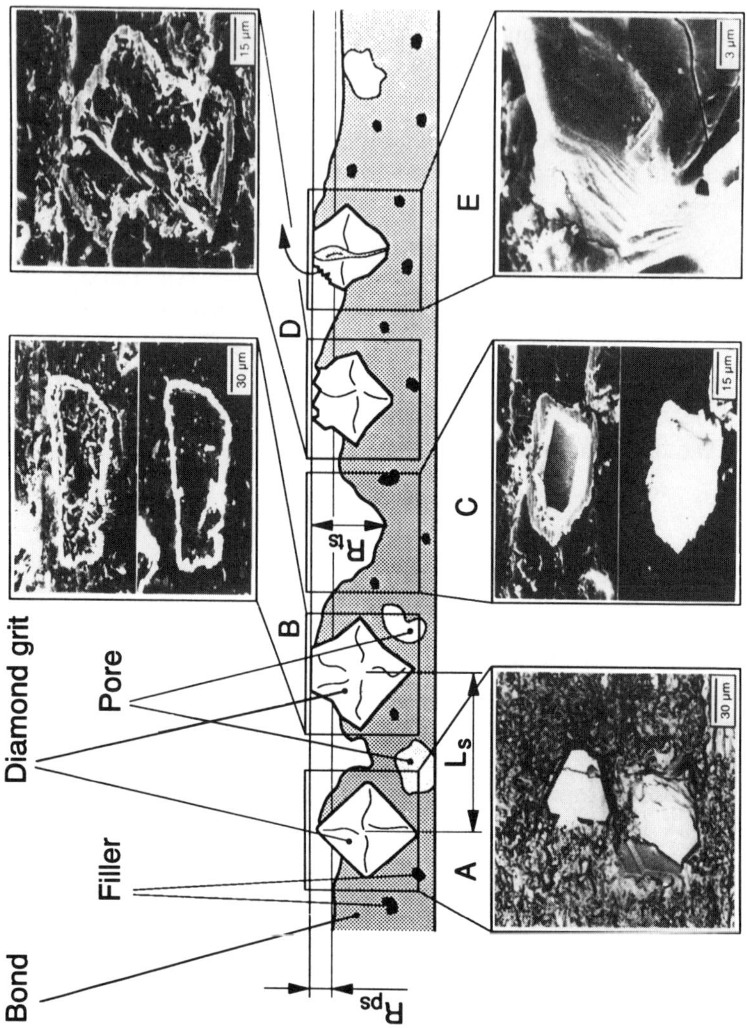

Figure 4.17: Wear Mechanisms and Diamond Grit Type

plete grits or groups of grits is exceeded (Figure 4.17d).

The wear forms on the diamond grit cause a direct influence on the individual grit contact forces and total grinding forces, which in turn act as reaction forces on the workpiece. The wear forms on the diamond are determined not only by the process parameters occurring in the cutting zone, but also by retention capability in the bond, so assessment of diamond grit types must also take account of the bond.

The small size of chips and the high abrasive wear resistance of the bonds means that bond wear plays only a minor role in the grinding of most engineering ceramics. On the other hand, the hardness difference between the material to be ground and the diamonds is relatively small, so the change in cutting edge topography is mainly determined by grit wear. This in turn causes the grinding wheel surface to be smoothed, and the cutting edges further down are soon engaged in the cutting process. This increases the grinding forces. This in turn is reinforced by the reduction in available chip space.

Thus, both the grit type and the bond are vitally important in wheel specification. A calculation model has been developed showing that with friable diamonds, the wear on the diamond cutting edges is on average evenly distributed over the material volume. This is demonstrated by the fact that there is only a slight change in the cutting edge rounding radius and only a slight increase in cutting point density.[15] With blocky diamonds, the formation of pronounced wear flats tends to cause a considerable increase in rounding radius versus material volume removed (Figures 4.18 and 4.19). Under otherwise identical conditions, the radii generated on the grits correspond to the average grit diameter.[15]

Blocky diamond grades with low grinding wheel wear should be considered where: rigid machine systems are used, there are small contact lengths in engagement, and sufficient coolant/lubricant can be applied. Together with suitable bonds, the conditions then may be such that damage-free workpiece surfaces can be achieved only with blocky, high-impact-strength grit types.[12] Blocky diamonds are likewise preferable to friable types where high material removal rates have to be achieved by high-speed grinding in the energy-efficient brittle fracture range.[15] Metal bonds have only limited suitability for high-speed processes, due to the high normal forces occurring; favorable cutting force characteristics and wheel life between dressings can be achieved with vitrified-bond and resin bond wheels (Figure 4.20).

In creep-feed grinding of HPSN with phenolic resin diamond wheels,

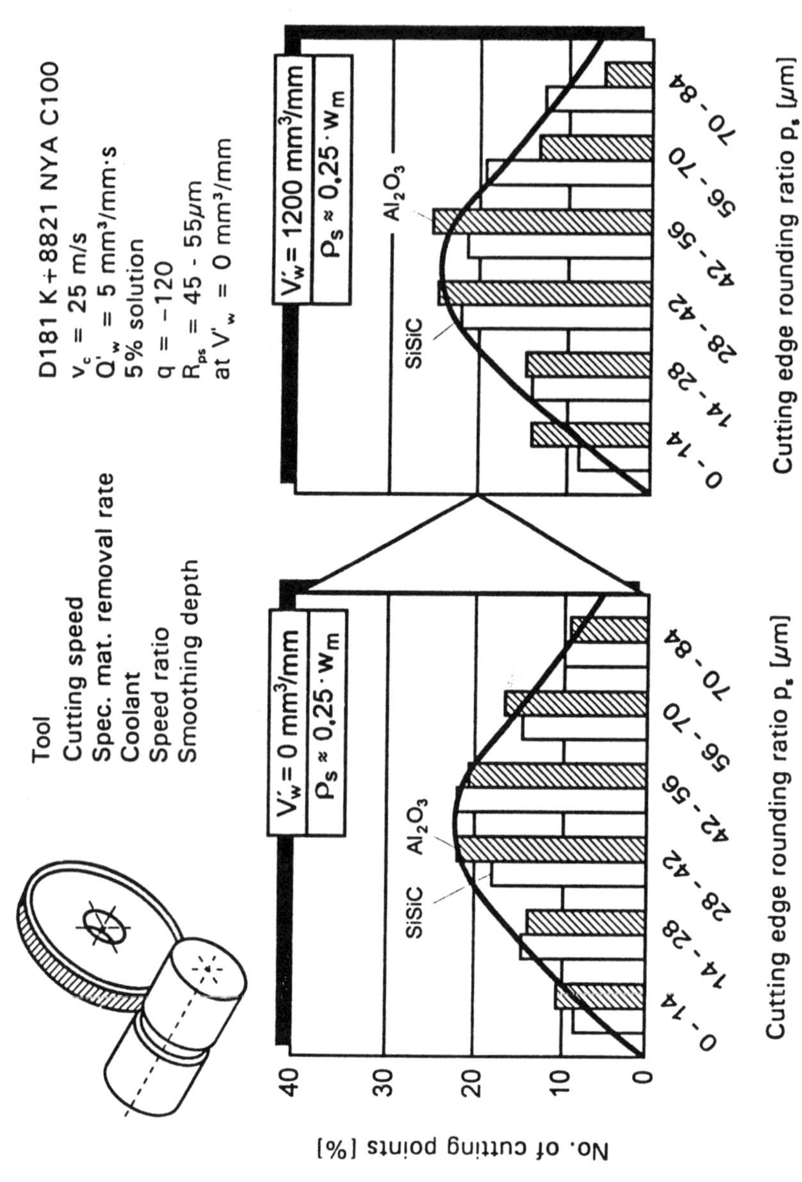

Figure 4.18: Cutting Edge Rounding Ratios of SiSiC and Al_2O_3

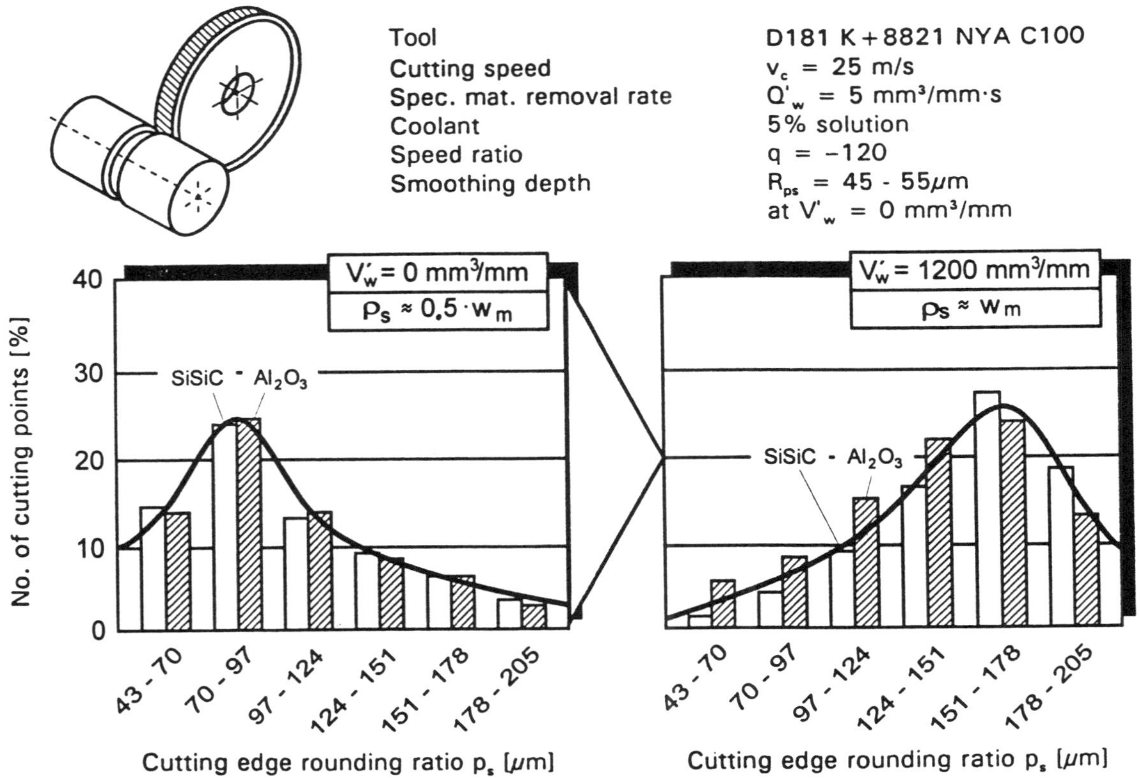

Figure 4.19: Cutting Edge Rounding Ratios of SiSiC and Al_2O_3

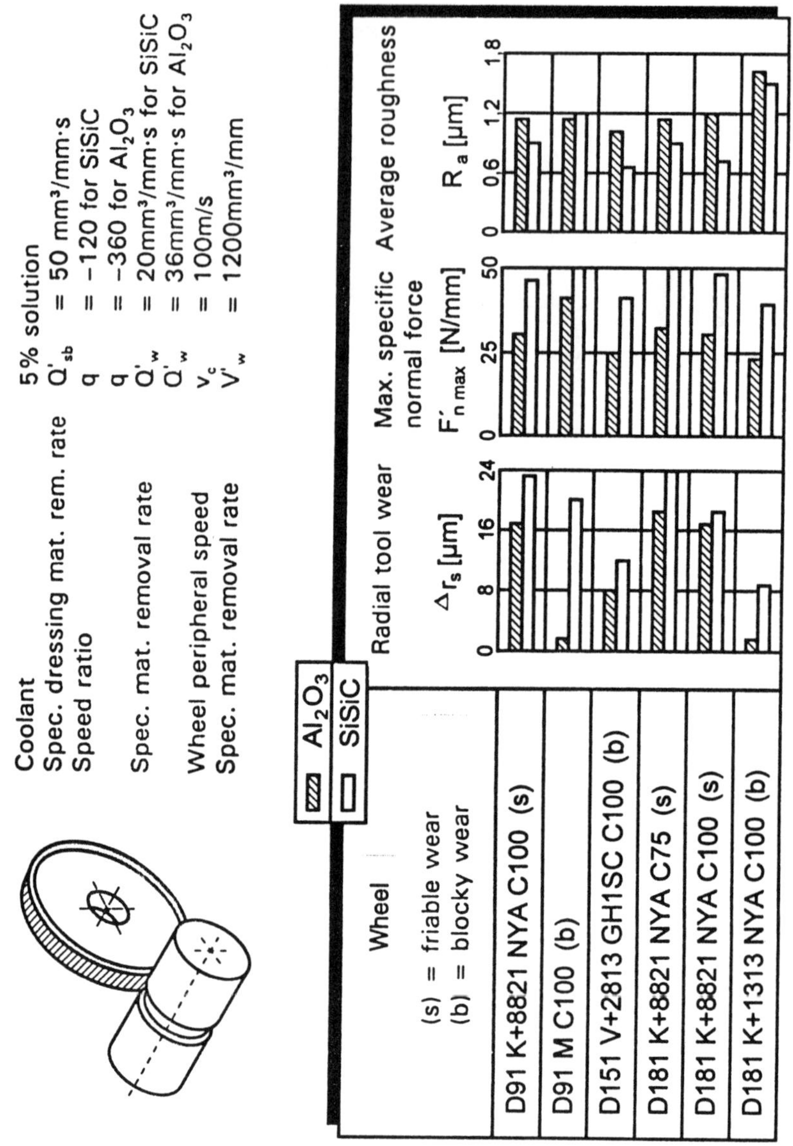

Figure 4.20: Cutting Force Characteristics and Wheel Life

Figure 4.20:

variation of the diamond grit type under otherwise identical parameters showed that blocky diamond grit types give considerably higher normal forces in the grinding process. This is because flats are generated during the grinding process, with simultaneously reduced ability to form new sharp cutting edges by splintering (Figure 4.21). In order to achieve lower mechanical and thermal stress on the ground surface zone, it is preferable to use sharp-edged, friable diamonds to machine HPSN, even though this may cause a slight increase in wheel wear and coarser surface quality (Figure 4.21). Extremely friable diamond grit types generate the lowest normal forces, but also the highest wheel wear.

Grit Size

Grit size has an influence on the attainable surface quality, the wear behavior, and the grinding forces occurring in the cutting process. It has been found in reciprocating surface grinding with small infeeds up to $a_e = 0.1$ mm[17][19] that finer grit sizes generate better surface qualities and higher normal forces. These conditions may be reversed in creep-feed grinding of HPSN with phenolic bond grinding wheels (Figure 4.22). Here again, larger grit sizes produce higher grinding forces. This is because large grits are retained more firmly in the bond and form larger wear flats. The wear influence on the grinding forces that occur in the cutting process is thus considerably stronger than the decrease in the number of cutting edges due to the increase in the size of the diamond grit. But this also shows that grinding wheel topography, and thus grinding behavior, is decisively dependent on cutting edge protrusion over the bond. Sharpening and optimal grit protrusion are decisive both for wear and for the grinding forces generated. If the chip space preparation by sharpening processes or other suitable processes is not carried out in the manner appropriate to the material and machining parameters, it is not possible to achieve satisfactory working results even with optimized grinding wheel composition.

The selection of suitable grit size is also influenced by the material to be ground. The lower the hardness of the ceramic material to be machined, and the more the process operates with brittle fracture, the less pronounced will be the wear flat on the cutting edge. This means that smaller total grin-

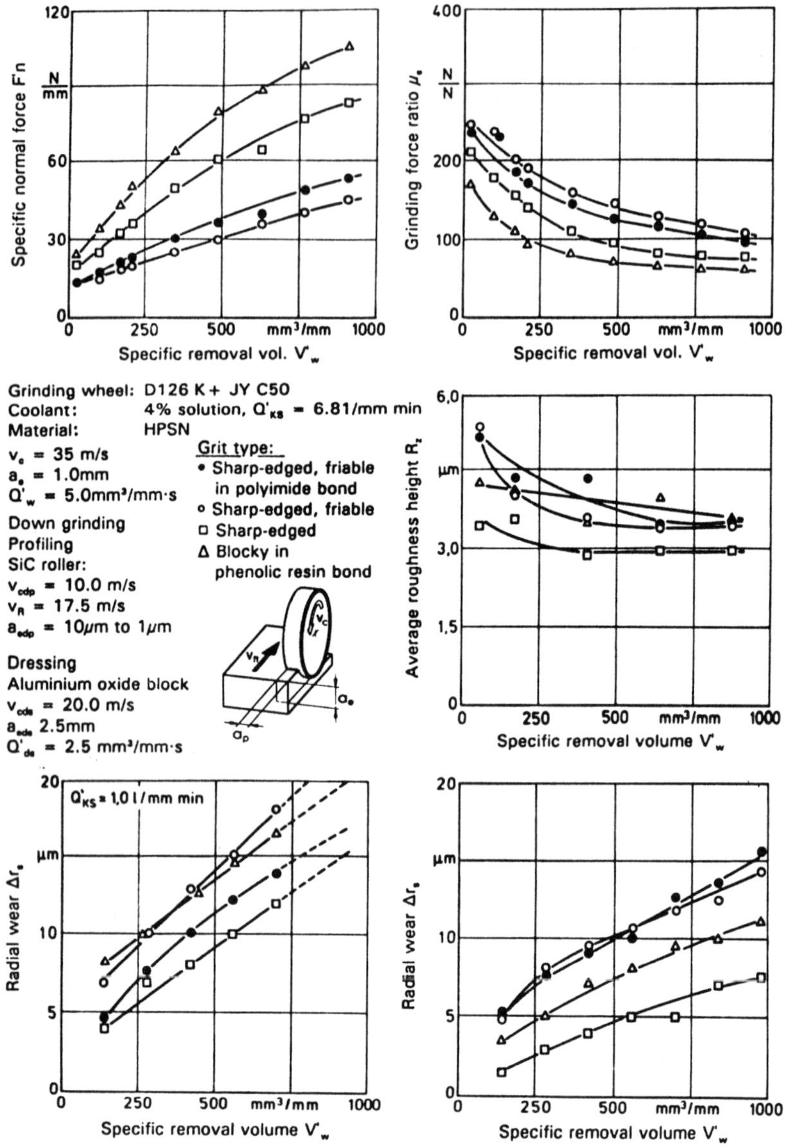

Figure 4.21: Diamond Grit Types vs. Effect on Wheel Wear

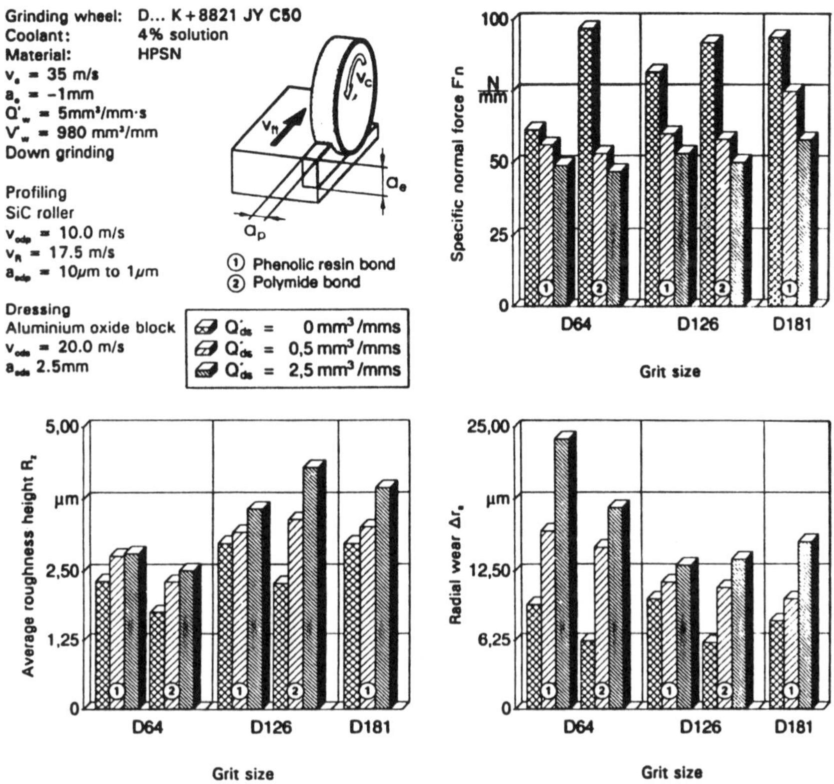

Figure 4.22: Finer Grit Sizes generate Better Surface Qualities

ding forces will occur in cases with larger grit sizes. This has been demonstrated in grinding zircon oxide, compared with HPSN and SiC.[14]

Grit Concentration

An increase in grit concentration causes an increase in the number of active cutting points. There is a decrease in the average distance from one cutting edge to another, and the average maximum chip thickness decreases under otherwise identical kinematic contact conditions. The result is a decrease in the individual grit contact forces.

However, this does not mean that a higher grit concentration will always give a better working result. The volumetric grit concentration must be adapted to the machining application.

In reciprocating grinding with small infeed increments, the grit concentration has only minor impact on forces and surface qualities.[17][19] Closer correlations have been observed in face grinding with large contact surfaces and in surface creep-feed grinding.[14][18] In creep-feed grinding of various ceramics, it has been found that the volumetric grit concentration also has to be adapted to the workpiece material. In grinding HPSN, there is a considerable force increase with an increase in grit concentration from C50 to C100 (Figure 4.23). However, this is not true to the same extent for radial wear. This suggests that flat formation on the cutting point is a major factor in force increase. Wheels with high concentration are therefore not advisable for creep-feed grinding of HPSN. With Si-SiC, which tends more towards brittle fracture, the thermal load on the grinding layer components is lower; there is less wear flat formation. There is only a slight increase in total grinding forces as concentration rises. In this example, this also means that higher grit concentration causes lower individual grit engagement forces. Increased concentration thus gives considerably lower wheel wear (Figure 4.23).

In summary, materials that exhibit extreme brittle fracture and good thermal conductivity reduce thermal loading of the wheel bond and of the diamond grit. Higher grit concentration may be expected to give benefits if these conditions are met due to the material or to other process parameters.

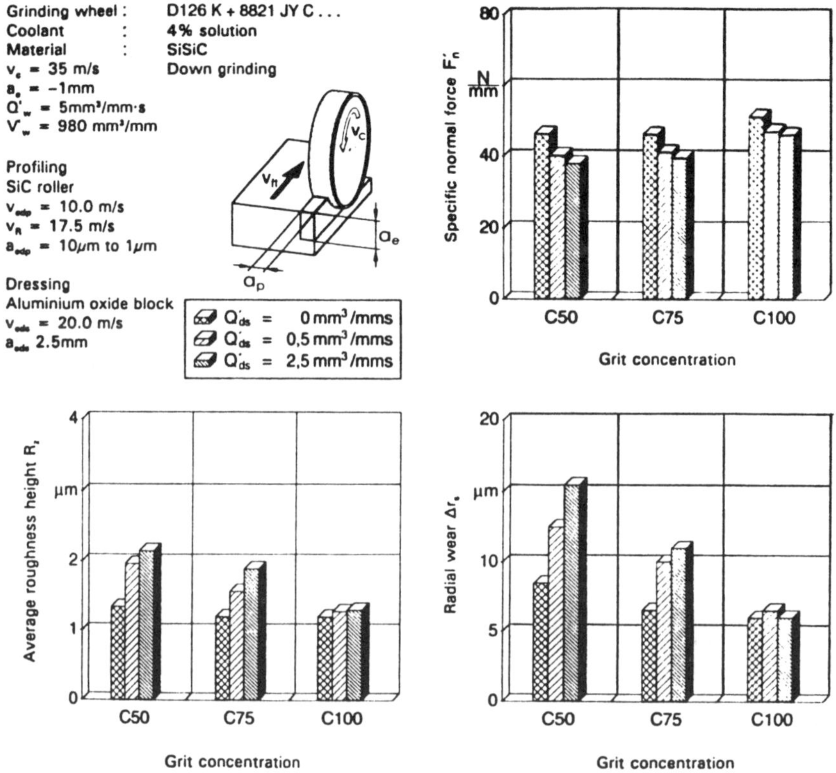

Figure 4.23: Grit Concentration vs. Wheel Wear

Dressing Strategy and Bond Specification

The decisive factors for grinding behavior include not only grit type, grit size, and concentration, but also specification of the bond system and conditioning of the grinding layer. The interplay between wear processes on the grit and on the bond determine grinding behavior and the working result. This is shown by the example of creep-feed grinding HPSN with various diamond wheels (Figure 4.24). The least sensitive reaction is that of vitrified-bond diamond wheels to different dressing conditions. The vitrified bond has brittle fracture behavior, and the bond webs are weakened by the existence of pores in the cross-section. This results in low grinding forces during grinding because wear causes the diamond grits to break out of the bond at an early stage without forming large wear flats. This results in low normal force and high wheel wear. Metal bond wheels in particular react strongly to differences in dressing. Normal forces are significantly reduced by wheel dressing, and also in wheel wear, but not to the same extent. This shows that even with metal bond wheels, the use of suitable dressing conditions can produce process parameters comparable to those of vitrified bond and resin bond wheels. This requires knowledge of dressing behavior, the use of suitable dressing units, and knowledge of the appropriate dressing strategy.

Figure 4.25 gives a summary of the various study results on the influence of wheel composition on process parameters and working result.

Bending Fracture Strength

The mechanical and thermal load on the surface zone is influenced by wheel composition, i.e. grit type, grit size, concentration, and bond. This is overlaid by changes in wear condition of the active cutting edges over time and by the number of active cutting edges.

An example of this is shown in Figure 4.26, with the results obtained in creep-feed grinding of HPSN. The use of sharp-edged, friable diamonds in phenolic resin bond with a concentration of C50 evidently produced considerable damage in the surface layer. The characteristic bending fracture stre-

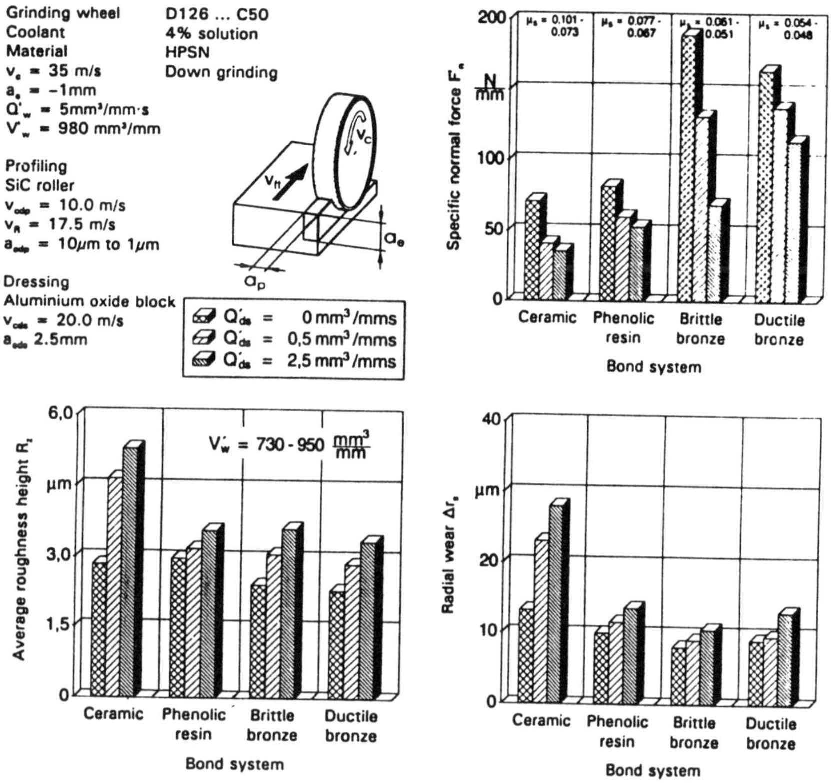

Figure 4.24: Creep-feed Gringing HPSN with Various Diamond Wheels

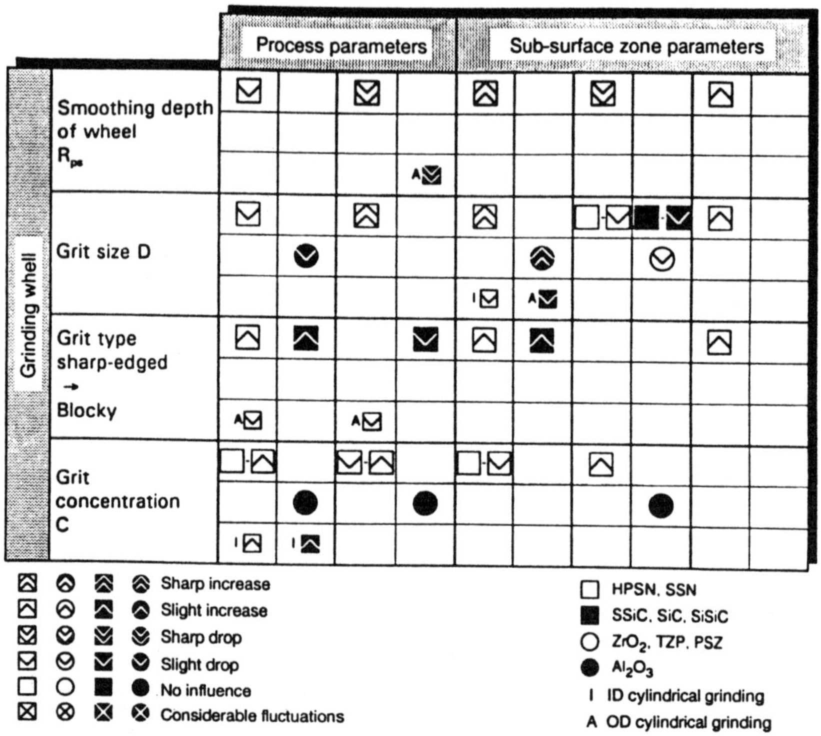

Figure 4.25: Influence of Wheel Composition on Process Parameters

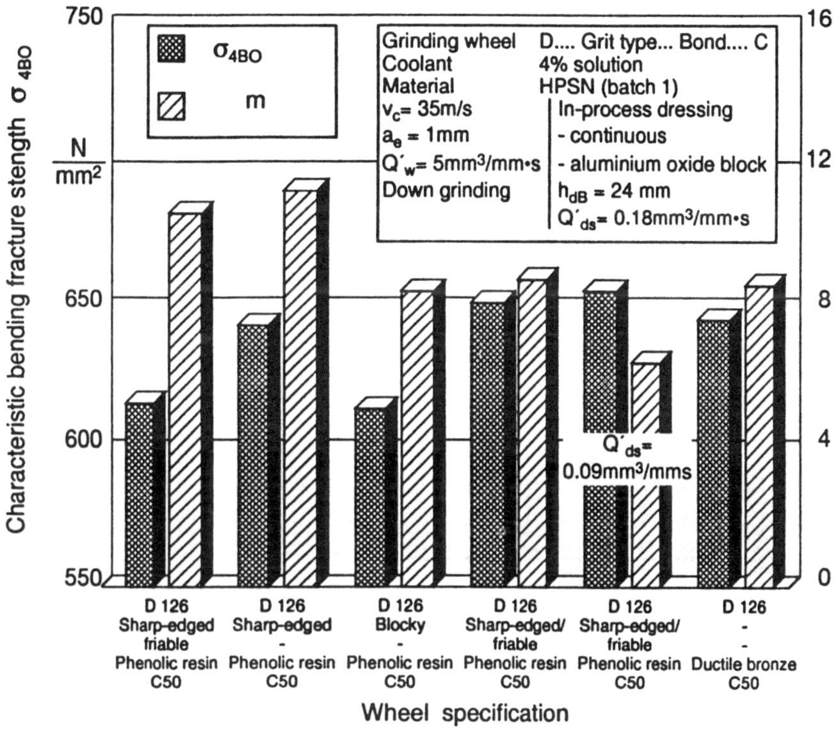

Figure 4.26: Influence of Wheel Composition on the Surface Zone

ngths have the lowest values at = 618N/mm². A less friable diamond grit type produces a considerable increase in characteristic bending fracture strength by the formation of new sharp cutting edges and the establishment of inherent compressive stresses. The Weibull modulus also increases. A more blocky diamond causes a decrease in bending fracture strength, and the low Weibull modulus levels also show greater scatter in the strength data. High individual grit contact forces and hence higher thermal loads and increased wheel wear now lead in turn to greater damage to the sub-surface zone. An increase to concentration C100 with the use of sharp-edged, splintery grit types likewise fails to produce an increase in strength values. This is due to increased wheel wear (Figure 4.23).

The conditions shown in Figure 4.26 also suggest that the influence of the bond can be largely eliminated by using suitable dressing procedures, so that strength behavior is mainly determined by the characteristics of the diamond grit, even where different bonds are used. Wheels D64/phenolic resin and D126/ductile bronze were dressed continuously during the grinding process, using different dressing material removal rates adapted to the bond and the grit size. This means that the same percentage protrusions were generated on both grinding wheel types. This made it possible to avoid thermal overload even with blocky diamonds in a ductile bronze bond. The working result is largely determined by the wear characteristic of the diamonds and by the chip formation process. The attainable bending fracture strengths are comparable to those achieved with phenolic resin bond wheels of the same grit size and with blocky diamonds (Figure 4.26). In proper continuous dressing, the working result is mainly determined by the diamond grit type.

Note that different bond systems move the position of minimum damage to different grit sizes.[14] An increase in concentration from C50 to C100 can likewise produce an increase in bending fracture strength. Chip thickness per individual grit is smaller, and provided that thermal surface damage can be prevented by adequate cooling/lubrication, higher grit concentrations also permit higher bending fracture strengths and also smaller scatter range of the strength data. In principle, this can also be achieved by a reduction in grit size from D126 to D64 (Fgure 4.26). The conditions shown in this figure are characteristic of grinding with HPSN, and have also been demonstrated for grinding with zircon oxide.[14]

3.0 THE CONDITIONING OF GRINDING WHEEL —DRESSING AND TRUEING—

Junji Shibata

Introduction

Grinding performance with super-abrasive wheels depends basically on material properties of the abrasives and the bond of which they are composed. However, at the same time, geometrical accuracy of the wheel, the protrusion height of the individual cutting edge, its distribution patterns, etc. will also play a valuable role. Therefore, in advance to grinding operations, the proper initial preparation of the wheel and its surface is an absolute necessity. Thus, the surface conditioning of the wheel is divided into dressing and trueing.

Dressing is such an operation that worn (dulled) cutting edges and the loaded wheel face with swarfs are regenerated by fracturing worn ones and removing impregnated swarfs from the face with a dressing tool.

Trueing, on the other hand, is defined mainly as a correcting operation of the wheel 'whirling.' Correction of a wheel shape, or generation of a wheel form for the 'form grinding' operations are also included in the trueing operation.

For electro-plated super abrasive wheels, the dressing or trueing operation is not usually required, because they have only a single abrasive layer which is attached by an electro-plating process to the peripheral surface of the metal base plate. Though it could be possible to control the grain height of the electro-plated wheels to some extent (in the order of a few μm) by means of the dressing operation (sometimes called shaving), such an application would be very rare. In a word, these conditionings are only applied to resin, ceramic (vitrified), and sintered-metal bonded super-abrasive wheel.

Though the dressing and trueing operations are believed to be the key technology which enables us to fully use the excellent ability of super abrasives, this has not yet been made clear. Rather, in contrast to easy-to-use conventional silicon carbide (SiC) wheels and aluminum oxide ($A1_20_3$) wheels, this still remains unsolved and is a difficult problem in technical operations. The primary reason for this difficulty lies with the cutting edges on the wheel face. How should we estimate and control these edges? It seems to present a problem which today's technology cannot yet solve.

This industrial situation means that we cannot use super-abrasives as effectively and efficiently today as we would wish to. This is primarily due to a relative criteria of grinding characteristics: the uncertainty, unexpectedness and insufficiency of constant stock removal rate, better surface finish, and overriding justification for the expense of the wheel.

Dressing of the Wheel

Function and Significance of Dressing

The primary purpose of the dressing should be generation or regeneration on the desired surface condition of the grinding wheel, necessary for assuring the quality of the finished part. A balance between the high quality of the finished surface, and the high efficiency and reliability of the operations must also be addressed, bearing in mind cost constraints.

To create such a favorable condition, one may suggest that a redesign or a remodel of the said tool by the dressing is as follows:

- By removing the bonding material, cutting edges are protruded above the bond, and places for chip clearance (chip pocket) are formed effectively on the cutting surface of the wheel, especially after the trueing by which (as mentioned later) the wheel face becomes very smooth due to its insufficient exposure to poreless (no chip pocket) bond structure of resin-bonded and metal-bonded wheels.
- Worn (dulled) cutting edges with 'wear flats' on their tips are fractured, extracted (shedded), and resharpened to cxposc ncw sharp grains. In addition, their distribution pattern could be adjusted for a more desired one.
- Cutting surface impregnated with smarfs (loaded) is removed and a clean face generated.

As a result, the dressed wheel surface allows the individual grain cutting edge to penetrate the workpiece and create swarfs more easily.

The wheel needs to be dressed or trued not only for the use of a new grinding wheel but also when such unfavorable phenomena like chatter vibration, grinding burn (metallurgical burn), etc., are present; this could result in damage to the ground surface or, when tolerance of the works lies toward unacceptable limits, with the wheel wear. However, we believe that these operations should be avoided as much as possible, because they oblige us to interrupt a work flow in the production line.

For example, the grinding time per a work "t_u" can be given approximately by the next well-known equation:

$$t_u = (M_w/M) + t_w + (M_w/M_T)t_D \tag{4.25}$$

where, M: stock removal rate (mm³ / sec.)

M_T: total stock removed in a dressing interval (mm^3)

M_w: stock removed per a work (mm^3)

t_D: time required for changing and dressing the wheel (sec)

t_w: time required for work preparation (loading and unloading of a work) (sec)

In order to improve grinding efficiency and reduce grinding cost, the following requirements can be easily estimated on the basis of the above relation:

(a) To increase the dressing interval as long as possible, which brings about the increase of M_T.
(b) To decrease the time spent for indirect operations (t_D,t_w), e.g., by adopting an on-the-machine dressing procedure.

The dressing interval for the super-abrasive wheel is indeed far longer in comparison with conventional ones. In other words, the value M_T of the super-abrasive wheel is extremely large. However, nowadays, some of this advantage might be canceled out by the long dressing time t_D required for them. In addition to this, loss of expensive abrasive particles from the wheel surface due to the dressing action could not be neglected in order to ensure economic results because initial super-abrasive wheel acquisition cost may be several hundreds times as conventional ones. Consequently, it is very important for the effective utilization of the super-abrasive wheels to save in the waste of the wheel due to the dressing as well as to reduce the dressing time.

Dressing Method and its Principle

The common type of dressing tools available today for super-abrasive wheels can be roughly classified in Table 4.1. As seen here, the dressing methods for super-abrasive wheels are in contrast to the ones for conventional wheels, where diamond dressers are exclusively used. That is to say, the single point diamond dressers are not suitable for super-abrasive wheels from their impact mechanism between the hard particles in the dressing action, though there are some exceptions. So, the principles of the dressing for the super-abrasive wheels must depend mainly on the bond removal mechanism, which are induced mechanically by a scratching, plowing, or scooping action by non-diamond abrasives. In some cases, chemical or thermal actions are also applied.

Table 4.1: Classification of Dressing Methods

Mechanism	Action	Tool
Mechanical	Scratching, Impact	Single Point Diamond Impregnated Diamond
	Crashing, Scooping	Abrasive Bonded Products Soft Steel block
	Blasting	Free Abrasives Glass Beads
Electrical · Thermal	Electrical-discharge	Electrode
	Electrolytic	Electrode
Chemical	Etching	Chemicals

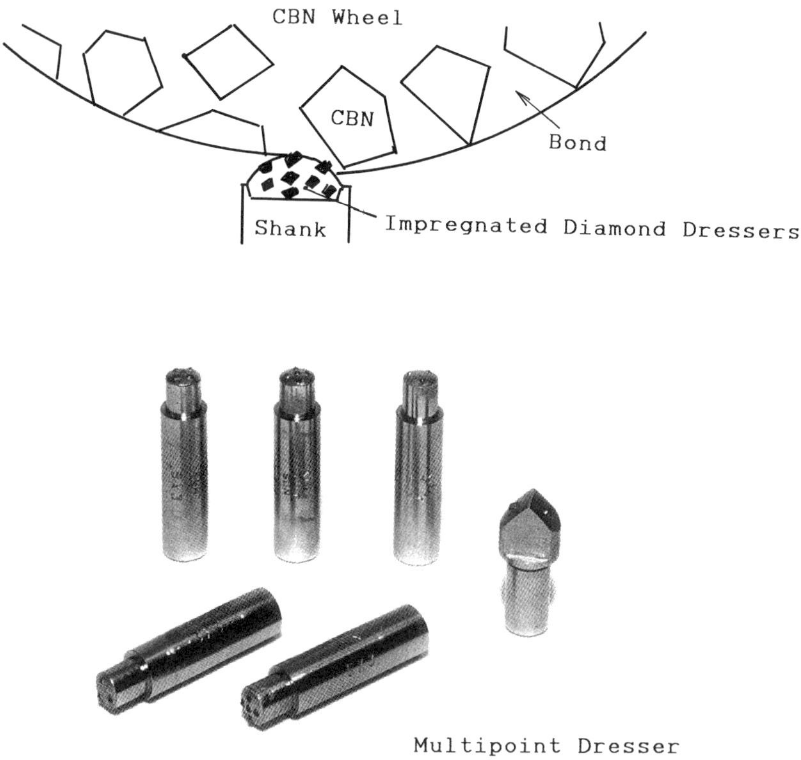

Figure 4.27: Dressing by Shank-type Diamond Tool.

Dressing by Diamond Dresser. The typical shank type dressing tools are single point diamond dressers and impregnated dressers (cluster and muti-point diamond nibs). As shown in Figure 4.27, these dressing tools cut the wheel face at a high turning speed with a dressing depth and a dressing lead. They not only remove the bonding material but also pull out or fracture grain particles. Therefore, some grain particles are irregularly fractured and sharp cutting edges can be expected to be generated on the wheel surface. For the dressing of conventional grinding wheels, the single point diamond dressers have been utilized so far because they are simple and easy to handle, and a desired distribution pattern of the cutting edges can be obtained relatively well by the principle of the threading pattern formation on the face of the wheel. Though the shank type dressing tools are superior in geometrical accuracy and in the generation of their sharp cutting edges, they are not often used as the dressing tool for the super-abrasive wheels. This is the reason why its operation efficiency is extremely low and why the dresser wears significantly on the super-abrasive wheels. There are only some exceptions for resin and vitrified-bond CBN wheels. To say the least, they can never be applied to the dressing procedure of diamond wheels due to the accelerated dresser's wear and accompanied damages caused on the wheel surface. Consequently, some types of tools equipped with a wear compensation mechanism such as the prismatic diamond dresser and the rotation-type single point diamond dresser have been developed. In special cases, the end face of an electro-plated diamond wheel could be used effectively for the dressing and trueing of a formed super-abrasive wheel named EF dressing (Figure 4.28). Supersonic dressing is a new procedure for improving the dressing efficiency.

Dressing by Non-Diamond Grains. As often pointed out, the principal purpose of the dressing for super- abrasive wheels should be to expose cutting edges by relieving the bonding matrix below the tip of the abrasive particles. The large height of thc protrusion brings about such profit in regards to high grinding ability and low specific grinding energy. Contrary to this, however, it tends to promote the shedding of the particles and generates a rougher surface finish. Therefore, practically, it is important to provide the protrusion of the abrasives properly from an economical viewpoint, considering the grinding ratio of the expensive super-abrasive wheels. Generally speaking, the grain cutting edges are believed to be desirable when exposed by 25~ 45% of mean grain diameter. Once a dial indicator or an electric micrometer has been

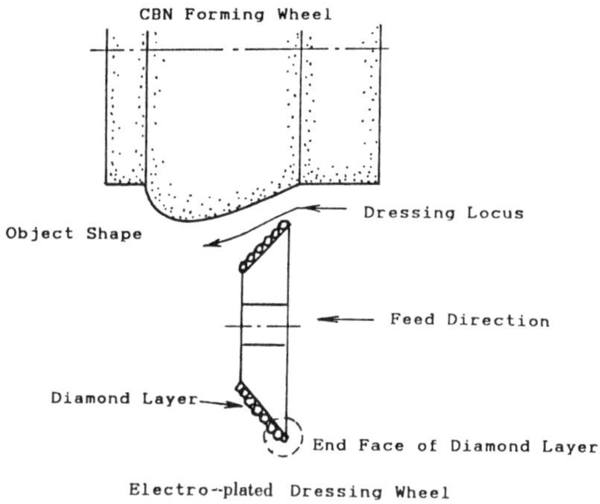

Figure 4.28: Dressing of a Formed Diamond Wheel

applied in trial for measuring the protrusion height of abrasives above the bond, it is practically estimated, from its observation by an optical microscope, by measuring the ground surface roughness of soft steel blocks or by the sense of its worker's finger touch.

Part A. Dressing Procedure by Grinding Conventional Abrasive Bonded Products or Soft Steel Blocks. The free grinding method with abrasive sticks (GC, WA; ~#1000), which is pressed manually or with a mounting apparatus against the wheel face, is one of the convenient and practical methods. Therefore, the dressing can be done economically under a wide variety of dressing conditions (selection of grain size, grade, kind of bond, etc.). The principle which is illustrated conceptually in Figure 4.29 suggests that grain debris fractured and shed off from the abrasive sticks crash and scoop out the bonding matrix at the front faces of the grains to expose them above the bond. Simultaneously, the cleared-away bond matrix allows chip clearances (pockets). At the backside of the grains, there is a bond face which remains unaffected in a tail-like shape (bond tail) which plays an important role for supporting the grain from the shedding process (Figure 4.30). These dressing effects are believed to be remarkable under relatively low dressing speed with friable SiC particles.

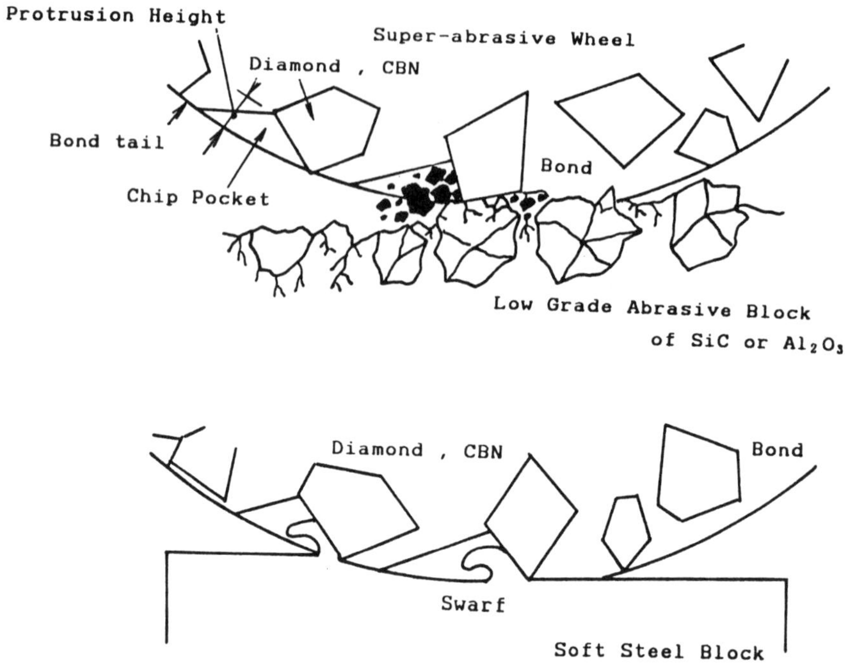

Figure 4.29: Dressing by Crashing and Scooping out of Bond Matrix

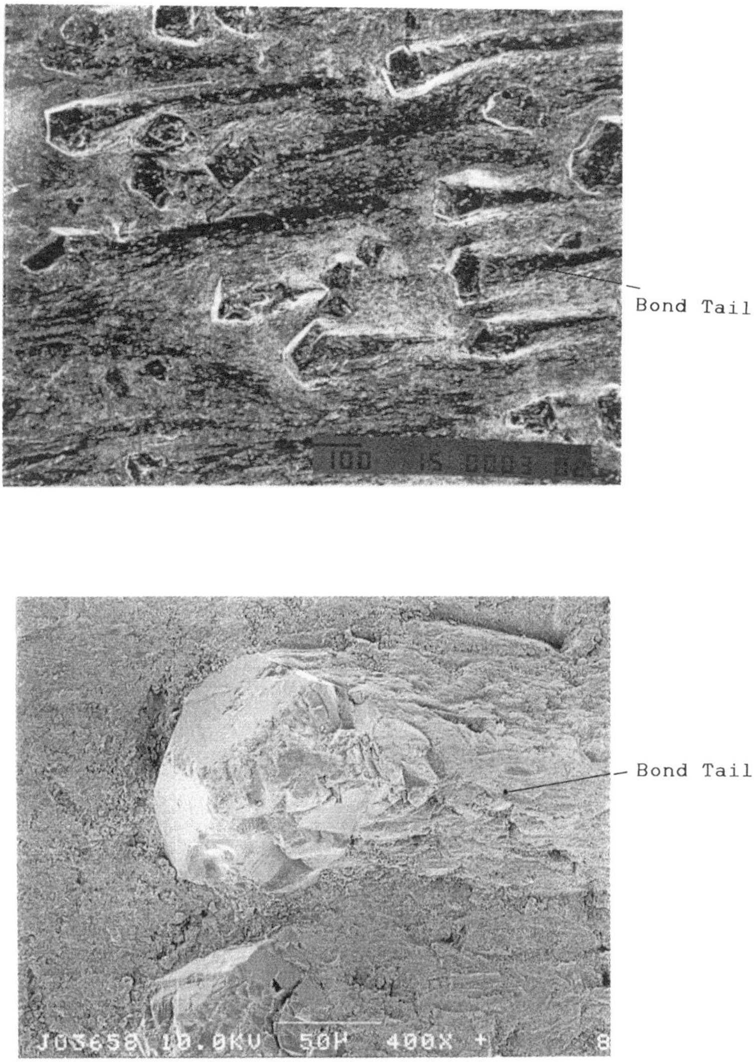

Figure 4.30: Wheel Surface Where Bond Tail Can Be Seen. (Courtesy of Dr. Setsuo Suzuki, the Tokyo Metropolitan Industrial Technology Center)

A recommended example in the bond-crash-type dressing methods is surface grinding of a soft steel block or a low grade abrasive block of SiC or Al_2O_3 (Figure 4.31). In this procedure, as in the free grinding method with abrasive sticks, grain debris, or grinding swarfs, which are removed from an inexpensive abrasive works or soft steel blocks in their wet grinding, can crash or scoop out the bond matrix, expose grains, and form chip clearances. In some cases, the trueing action referred to in the latter can be simultaneously achieved here. From the viewpoint of the operation efficiency, however, this procedure is poor despite its excellent ability to generate high quality in the wheel cutting face.

Some elaborate units which realize the dressing principle effectively have been developed. They are brake-type and driving-type units which are now available commercially. The former has a straight-type or cup-type conventional abrasive wheel (Al_20_3, SiC) mounted on a spindle. As the dressing wheel rotates freely without any driving mechanism, the wheel is driven by frictional drag with a rotating dressed wheel face. The latter is an improved

Figure 4.31: Dressing by Surface Grinding of an Abrasive Block.

Figure 4.32: Driving-type Dressing Unit Equipped with A Driving Mechanism.

one equipped with a driving mechanism of the dressing wheel so as to yield the sliding speed between the dressing wheel and the dressed wheel arbitrarily (Figure 4.32). Here, a conventional wheel as the dressing tool is driven by a motor at between 100 ~300 rpm. In order to induce the velocity difference more effectively between the two wheel faces of the dressing unit, from a practical viewpoint, it is desirable for the two wheels to be set in contact with a certain angle of 15°~ 20° to each other. The dressing unit needs a somewhat sophisticated design and is a little expensive, but stable dressing performances can be expected. As a whole, these dressing procedures mentioned above tend to create a rougher wheel face due to the macroscopic bond fracture mechanism. If the delicate control of grain cutting edges is needed, it would be necessary to select a grain size, a grade, and a rotation number of a dressing wheel corresponding to the grinding purpose. These dressing procedures are also recognized not to be satisfactory in keeping their profile for formed wheels with respect to the form collapse caused by the wear.

Part B. Dressing by Free Abrasives. Free abrasive methods are also applicable as one of the dressing procedures, where crashing of the bond materials is the main action in their dressing mechanism. For example, there is the abrasive jet dressing, where water mixed abrasives or glass beads is spouted at a high pressure from a nozzle to the wheel face and the dressing action is achieved efficiently (only within several seconds). That is to say, it is a type of blasting process. In addition, though it is not so popular, the abrasive slurry method also belongs to this procedure. Here, its dressing manner is as follows: abrasive slurry of SiC or $A1_20_3$ poured between the dressing and the dressed wheels, or between the dressed wheel and a steel block or a steel roll, which are set in close contact with each other and rotated at a low speed and in the same direction.

Cup-wheel can be dressed and trued at the same time manually, by hand lapping with a lapping compound on a dressing plate (glass or cast iron). Grain size of SiC applied (usually, #150, #200, #400) should be selected according to the grain size of the dressed diamond wheel. It is a serious disadvantage for the grinding operation that we must remove the wheel from the grinding machine for the dressing operation, because it interrupts the grinding operation.

Finally, from the above-mentioned viewpoint, it could be concluded that the resin bond and the metal bond wheels of poreless bond structure should be most suitable for the crash-type dressing.

Electrical-, Chemical-, and Thermal-type Dressing. For the dressing of conductive metal bond wheels, the electrical discharge processing is effective (Figure 4.33). Various shapes of electrodes like fixed type, rotational type, brush type, and wire type are utilized. In these dressing principles, the top part of the particles are completely exposed (Figure 4.34).

The twin electrode apparatus for the electrical discharge dressing is quite useful due to its easy handling: one does not need to reconstruct the grinding machine in order to provide electricity to the electrode. On the other hand, ELID (Electrolytic In Process Dressing) has attracted special interest recently. In ELID, a wheel face of a positive electrode and a negative electrode mounted against it are brought close together in a gap of around 0.1 mm. Entering grinding fluid fills an electrolyte fluid and an electric-chemical action

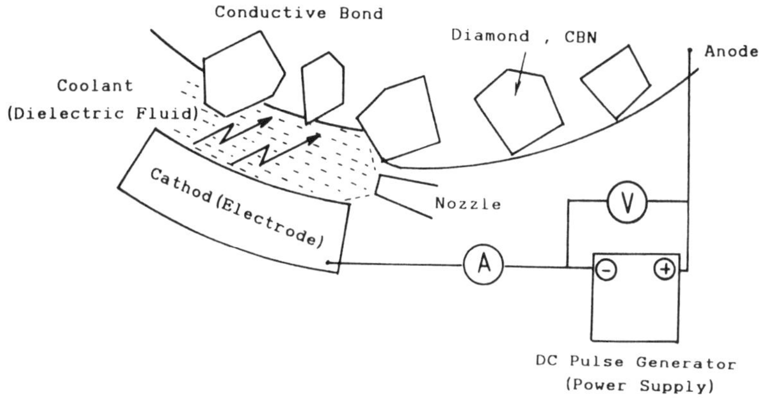

Figure 4.33: Dressing by Electrical Discharge Action.

takes place while the direct-current (DC) pulse ionizes and removes the conductive bond materials. The surface layer coated with a nonconductor like iron oxide or hydroxide produced by this process restrains its excessive action to properly ensure the protrusion of grains. By means of this dressing principle, a very small amount of the dissolution can be controlled. So, ELID is effectively applicable to fine grain size wheels. On-line dressing also becomes possible as the practical application of this principle.

Technically, chemical dressing would also be possible, where a grinding wheel is soaked in a chemical liquid which dissolves the bond material. For instance, for a Cu and Sn bond wheel, the next chemical reaction will form an insoluble precipitate to remove the bond material:

$$Cu + H_2O + H_2S0_4 + 3H_20 \rightarrow CuS0_4\downarrow + 5H_2O \quad (4.26)$$

This might be an ideal dressing procedure because of a short dressing time as well as no physical and mechanical damage to the dressed grains. In this procedure, however, we must be aware of inducing corrosion to the grinding machine.

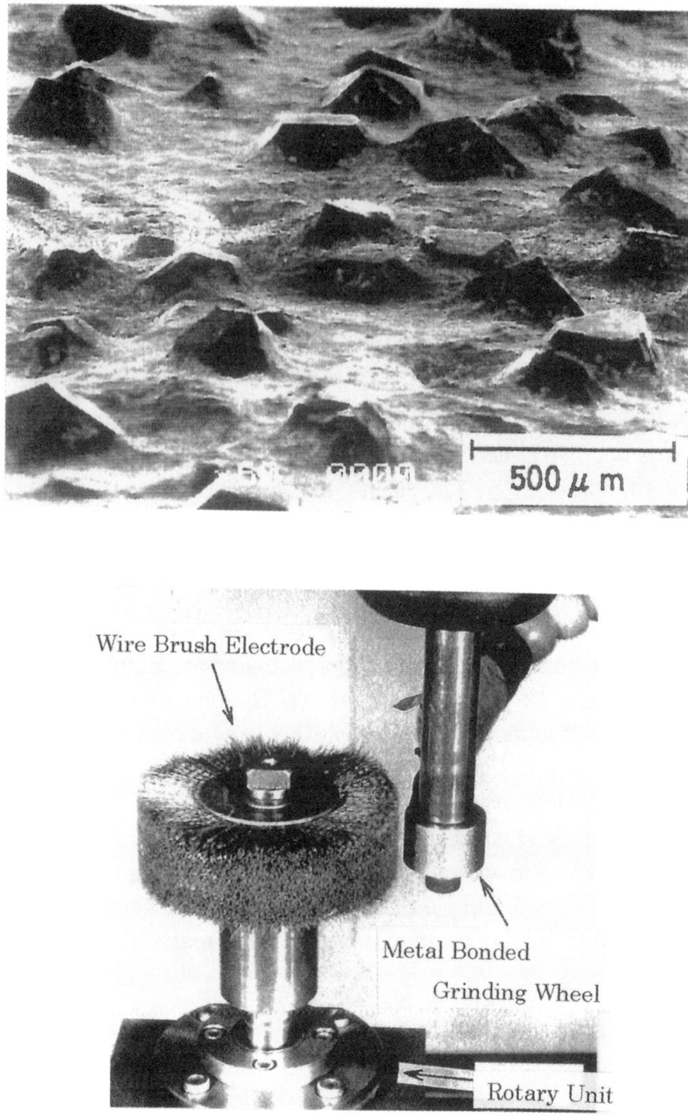

Figure 4.34: Abrasive Particles Exposed Completely above Bond Face by Electric-discharge Dressing Method with Wire Brush Electrode. (Courtesy of Dr. Kiyosi Suzuki, Nippon Institute of Technology)

Trueing of the Wheel

Purpose and Significance of Trueing

In the trueing operation, the excessive outer surface of the wheel is removed mechanically. The tool used for this purpose is sometimes called a truer or a cutter (usually, a grinding wheel is used). There are two types pf trueing as follows:

- Deflection-trueing: to make the wheel round and concentric with its axis of rotation, and at the same time to expect correction of the poor wheel balance which causes bouncing of the wheel on the work.
- Form-trueing: to impart the desired geometry to the wheel face such as flatness, straightness, roundness and formed profile, or to regenerate them when they are deteriorated due to the wheel wear.

With the trueing operation, similarly to the dressing, the work flow must be interrupted in the production line, causing the precious super-abrasives to be lost. This is not favorable, although trueing is very important in securing the product's quality and the safety of the operations through the correction of the wheel deflection and the wheel balance, especially in today's prolific trends for high speed grinding (extremely high revolution of the wheel).

Trueing of conventional grinding wheels can be achieved simultaneously with the dressing process. The trueing of super-abrasive wheels, however, must usually be done as an independent process from the dressing operation. This means that the conditioning operation is so worrisome that after the wheel geometry is generated or regenerated by trueing, the dressing must follow it to make cutting edges desirable. The reverse would also be possible. The dressing procedure by grinding a soft steel block or an abrasive stick (Table 4.1) sometimes has the advantage of achieving the trueing effect simultaneously.

Types of Trueing and Their Principles

One of the trueing methods is the utilization of a SiC grinding wheel. Its principle is basically the same as the one for the dressing. For example, a brake-controlled trueing device is chosen and mentioned.

After the trueing device is fixed on the table of a surface grinder, the SiC vitrified trueing wheel is advanced in a plunge grinding manner toward a trued super-abrasive grinding wheel, which is rotating at a usual speed. In many cases, the trueing device is traversed. With the contact of these two wheels, the trueing wheel starts running in the same direction as the trued wheels rotate. The rotation speed of the trueing wheel is kept at a certain low speed by the braking action of the trueing device. The process creates a rubbing action between the trueing and the trued wheels. As a result, the trueing action of the super-abrasive wheel is accomplished by fractured or extracted debris of the abrasive particles. These are created on the trueing wheel face by the frictional drag between the trueing and the trued wheel. A brake-controlled trueing device with a cup-type trueing wheel (cup truer) can generate a straight wheel face accurately and effectively without giving any cross-feeding to the truer; this action can produce a dressing action together.

Generally speaking, the trueing efficiency by Al_20_3 wheels is inferior to the one by SiC wheels. This derives from the fact that SiC abrasives are more friable than Al_20_3. Consequently, Al_20_3 wheels are recommended only for the dressing, not for the trueing.

Diamond blades used as slicing wheels would be trued by a motor-powered trueing device more effectively than by brake-controlled ones. In the trueing for cylindrical and internal grinders, a SiC trueing wheel is attached in place of a work and ground by the super-abrasive grinding wheel which is to be trued.

In the form-grinding of threads, dies, etc., the negativc of thc cutter shape (the wheel profile) is imparted to the work. Conventionally, most of formed-wheels are electro-coated wheels, which, other than special cases, need not to be trued. Recently, however, new formed-wheels capable of being trued have been developed. For form-grinding wheels, not only the trueing ability of the tight tolerances for intricate forms, but also its durability have also been expected. In response to the demand, form-grinding wheels made from various kinds of bond materials like soft resin bond, porous vitrified-bond, and specially developed metal-bond are now available commercially. For their form-trueing, especially in the case of intricate forms, crush rolls or crushing truers (Figure 4.35) are recommended because of their superior trueing efficiency. For the form-trueing of a simple profile (V-shape, R-shape), the rotary-type-formed diamond dressers are utilized effectively for the benefit of a short trueing time, high accuracy, and a long trueing interval. Crushing block dressers (electro-coated, impregnated, metal-bonded, and electro-formed) are also applicable for this purpose with the advantage of the tool cost, though they are believed to be inferior in their trueing accuracy and durability (Figure 4.36).

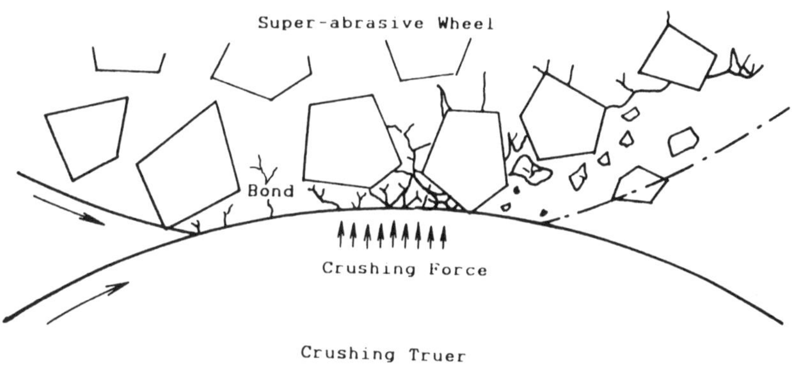

Figure 4.35: Trueing Model by A Crushing Truer.

Figure 4.36: Crushing Block Dresser.

4.0 WEAR MECHANISM

Junji Shibata

Wear of Abrasive Cutting Edges

The wear mechanism of abrasive particles are classified into the following 3 patterns:

- Attrition Wear,
- Fracture, and
- Shedding (Pullout).

As long as abrasive particles take part in cutting actions, their wear is unavoidable with the grinding time, even though they are super abrasives. So, the important requirement for grinding operations is how to keep grain cutting edges sharp and promote their self-sharpening action in the suitable wear behavior.

Attrition Wear

When super-abrasives of high impact strength are held firmly by a tough bond material like a metal, abrasive particles are abraded gradually from their tips to generate wear flats, the so called "dull."

Generally, the diamond might be considered to have high impact strength, chemical inertness, and thermal stability of physical properties. However, diamond is composed of a carbon atom structure and is susceptible to heat; that is to say, it will oxidize around 760°C (Celsius degree) to form oxides. Presumably, under the grinding action of such a high temperature, diamond abrasives react chemically not only on Fe but also on the elements like C, W, Co, and V to easily form carbides and promote attrition wear. This is why diamond wheels are unsuitable for the grinding of ferrous materials such as low carbon steel, tool steel, nickel and cobalt-based alloys etc., which contain some of these elements in small (yet noticeable) quantities. Today, nickel-coated diamond abrasives have been developed and used effectively with the choice of the proper grinding fluid and considerations of its delivery method, (e.g., in flood coolant) for these work materials.

On the other hand, CBN (Cubic Boron Nitride) is synthesized at a high temperature and pressure in the same process as synthetic diamond. So, CBN is also the same in its crystalline form as is diamond. The only exception is that it does not contain the carbon element in its composition. This is why it can avoid the carbon affinity problem in the grinding of ferrous materials and hold the chemical inertness as high as around 1,400°C. Therefore, it would show remarkably superior grinding performances when it grinds the above mentioned ferrous materials. However, in high temperature circum-stances above approximately 1,000°C, it reacts with steam molecules (H_2O) and is decomposed. The chemical compound is as follows:

$$BN + 3H_2O = H_3BO_3 + NH_3 \tag{4.27}$$

CBN wheels can fully show their excellent grinding ability with straight oil type grinding fluids. However, an expensive apparatus for collecting the oil mist is necessary to maintain a clean working environment. As a conclusion, in the case of the wet grinding, the emulsion type grinding fluid of high concentration has been satisfactory for the CBN wheels. In addition, when using CBN wheels, other careful considerations must also be made, e.g., the best choice of fluids such as the adoption of extreme pressure agents like sulfur-chlorinated oil. The fluid delivery system is very important to achieve satisfactory economics of CBN wheels. Effective fluid delivery decreases frictional heat generated in the grinding zone and reduces bond/abrasive wear.

Fracture

From a viewpoint of crystalline structure, CBN should be cleaved more easily than diamonds. Actually, however, it has been recognized that properties of the super abrasives are significantly affected by the synthesized process in their production. For example, regarding synthetic methods, the type of catalyst (initiator) used influences its toughness more remarkably.

Shedding (Pullout)

Shedding is a loss of the super abrasives due to extraction from the bond. Holding strength of the abrasives by the bond matrix against the shedding is (in order) as follows:

Electro-coating > Metal-bonding > Vitrified-bonding > Resin bonding

The phenol bond material, which is representative of the resin bond materials, is very weak in holding grains as it begins to soften at around 100°C. Some geometrical properties of grain particles are also very important factors for the shedding wear, because the irregularity in their shape and surface texture have an effect on physical bond anchoring. As an example, Ni-coated grains whose surfaces show microirregularity are expected to be improved in holding the abrasives firmly in their bond matrix against shedding wear. In order to resist the shedding wear, the effect of protrusion height of each grain should not be ignored. Overdressing will cause premature grain particle loss.

Wear of Grinding Wheel

The wear of super-abrasive wheels is extremely small. Specifically, the reduction is a fraction of tenths to a fraction of hundredths compared to the wear of conventional grinding wheels. In another measure of the grinding ratio, the value of the super-abrasive wheels is so large that it is increased by a hundred fold to ten thousand fold. Although thermal as well as chemical factors are the main source for wear of super abrasives, a mechanical factor should not be overlooked.

If its grinding ratio is considered for the grinding of ceramics with a diamond wheel, its value tends to become larger with the increase of the work's fracture toughness to the impact. However, extremely high or low impact resistance of the grains may be unfavorable to the wheel life, such that one should choose appropriate intermediate values. One should also remember that wheel wear is influenced not only by the quality of super abrasives but also by the grade of bond matrix, the concentration of grain particles, the filler, and various other grinding conditions such as work materials, machine properties, the work holding system, etc.. It is a basic perspective that grains of low friability are held in the resin bond matrix of the softer grade. Grains of blocky shape and high impact resistance should be held by metal or an electric-coated bond which can anchor the particles more firmly to resist the impact load. CBN is only half as hard as diamond. Therefore, CBN cannot be applied for grinding works of high hardness such as ceramics and Tungsten Carbide (WC) because of the excessive wheel wear.

Concerning profile deterioration, metal bonds hold the wheel shape best because they have the strongest and most durable bond system; however, they require more time for the trueing and the dressing.

5.0 COOL LUBRICATION

Gunther Spur

Cooling Lubrication Systems

The task of cooling lubrication during grinding is to minimize the mechanical, thermal, and chemical stresses arising in the active zone between the tool and workpiece. On the one hand, this cooling lubrication lowers the friction between the grinding grain and workpiece and thus, reduces the development of heat in the active zone; on the other hand, the workpiece, tool, and grinding machine are cooled or thermally stabilized by absorption and transportation of the developed heat. In addition, the process of cooling lubrication is aimed at cleaning the grinding wheel from chip products and at rinsing out chips from the contact zone or the working room of the machine.[20][21][22][23][24][25][26][27][28]

Industrial cooling lubrication systems can be produced in individual or central design. They consist of numerous components which include in addition to the cooling lubricant itself, the circulation system with feed and cleaning devices, the storage bin, and accessories. In order to safely fulfill the tasks of cooling lubrication, all elements of the cooling lubrication system are designed appropriately to meet the demands and are coordinated with each other.

Cooling Lubricant

The cooling lubricant is of central importance to the establishing of favorable cooling lubrication conditions. With its properties which are determined by type, base oil, type of additive, concentration, and state, the cooling lubricant significantly defines the efficiency of the cooling lubrication. In addition to the ability to perform the requested functions, its action is also characteristic for a cooling lubricant, (Table 4.2).[23]

Table 4.2: Important Usage Characteristics of Cooling Lubricants

Usage Characteristics	
Fulfilment of Functions	Action
Lubricating effect (pressure capacity)	Human and ecological compatibility (toxicity, smell, dermatological compatibility)
Cooling Effect	Age and bacterial resistance (stability)
Rinsing effect (purification, chip transportation)	Ability to filtration and recycling, mixing behavior, emulsifiability
Corrosion prevention	Washability, residual behavior, solid carrying capacity
	Foaming and fog behavior, flammability
	Material compatibility

Non-water-mixable and water-mixed cooling lubricants are available for application to grinding machines.[27][29] The term "non-water-mixable cooling lubricants" describes cooling lubricants such as mineral oils and esters which do not need to be mixed with water in order to be applied. Additives are added to these cooling lubricants in order to improve their lubricating effect, age resistance, and corrosion prevention as well as to avoid fog formation (Figure 4.37).

Ready-made mixtures from concentrate and water, mostly solutions or oil-water-emulsions, are defined as "water-mixed cooling lubricants." Concentrations mainly lie in the range between 2 and 6%: in some cases, they reach up to 20%. Here, additives with the functions mentioned above are also applied. Moreover, the wetting power of the medium must also be regarded in this case.[23]

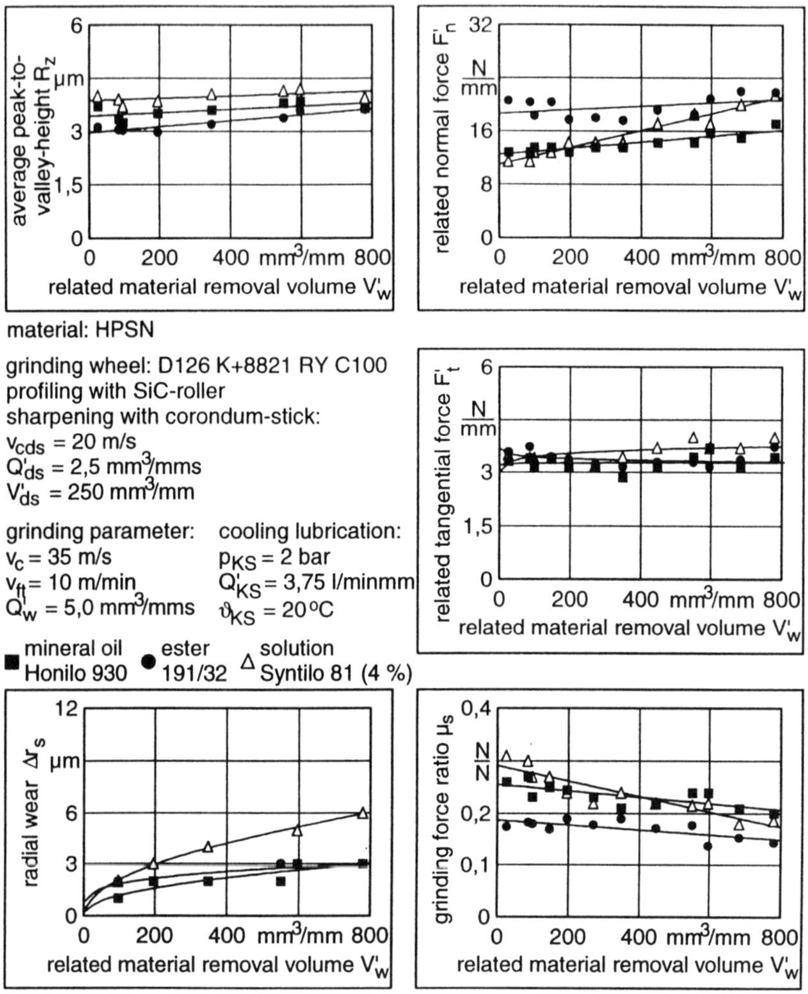

Figure 4.37: Related Grinding Forces, Force Ratio, Radial Wear, and Mean Total Height in Relation to the Related Material Removal Rate for Water-mixed and Non-water-mixable Cooling Lubricants[41]

Supply, Circulation, and Purifying Systems

The cooling lubrication system must lead the required cooling lubricant directly into the active zone with the help of supply devices. Hereby, the efficiency of the cooling lubricant feed is strongly dependent on the design and arrangement of the feed jet as well as on the selection of feed pressure and volume flow. The circulation system has the objective to transport the cooling lubricant in a defined state to the supply devices in the working room of the machine and to carry it off again. For this purpose, storage and conveying elements such as containers, pipelines, channels, and pumps as well as measuring and regulating devices are necessary. Other parts of the circulation system serve to maintain the usage characteristics of the cooling lubricant. These include monitoring systems, which indicate the current state of the cooling lubricant using suitable characteristic quantities such as the pH-value and electric conductivity, as well as continuously or discontinuously operating maintenance devices. Systems of this kind are ventilating, attemperation, and purifying devices which can be led into the cooling lubricant circulation through a main or secondary stream.[30][31]

Considering the aspect of fulfilment of functions and durability of the cooling lubricant, special attention must be directed to its purification. The technologies of flotation, filtration, and centrifugation are available for the machining of ceramics. Combinations of these technologies in series or parallel connection are possible.[23]

Magnetic grate separations, however, cannot be used for the separation of ceramic particles. Sedimentation technologies, too, can only be used as a coarse purification phase for particle diameters of sometimes $dp < 3$ µm created during machining of ceramics.[23] Due to the Brownian movement, particles of this size are kept hovering in an increasing manner.

Filtration is the most common purification technology for cooling lubricants. The most frequent design of filters are band-pass and candle filters. Band-pass filters are used in simple plants as pure gravity filters. An increase in the amount of throughputs is made possible by vacuum filters, whereas an improvement of filter fineness is reached with candle filters and filter accessories. Investigations prove that with regard to the minimization of grinding wheel wear, a medium particle size of $d_{pm} = 3$ µm and a solid concentration of $K = 0.1$ g/l comprise the upper contamination limit.[32] Brücher was able to demonstrate the potential of new purification devices such as alluvial, gravel, and full-jacket centrifuges for the machining of ceramics.[23]

Auxiliary Attachments

Supplementary components such as suction plants, metal clads, mixing devices, and chip or mud concentrators, as well as emulsion separate units count among the accessory devices.

However, literature on the machining of ceramics does not report of a purification of the grinding wheel by means of purification jets, which is often recommended in metal machining. It is especially unnecessary when a permanent influence on the chip area is achieved by the technology of continuous in-process-sharpening.[33][34][35]

Impact of Cooling Lubricants and Feed Conditions on the Process

Influence of Cooling Lubricants

Investigations of pendulum grinding of ceramic materials with diamond grinding wheels revealed advantages regarding the surface quality and process conduct during application of grinding oil compared to water-mixed cooling lubricants. While a strong increase in the normal force is characteristic for the grinding of $A1_2O_3$ and HPSN during usage of water-mixed cooling lubricants, a process conduct characterized by low and almost constant grinding normal forces could be observed up to a related material removal rate of $V'_w = 780$ mm^3/mm during cooling lubrication with grinding oil. In Figure 4.38, the course of characteristic quantities in tests of HPSN is described as an example.[41] Moreover, a lower radial wear of the grinding wheel occurs during grinding with grinding oil.[23][32][36][37] Significant differences in the topography of $A1_20_3$-surfaces ground with grinding oil and emulsion hereby indicate a significant influence of the cooling lubricant on chip-formation mechanisms. In contrast to surfaces that were machined with

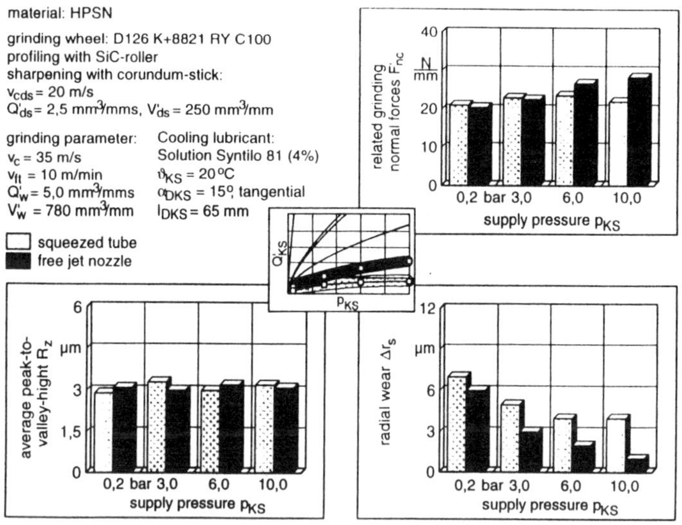

Figure 4.38: Related Normal Force, Radial Wear, and Surface Quality in Relation to Jet Quality for Various Feed Pressures During Grinding[23]

grinding oil, directional grinding traces barely occur on surfaces machined with water-mixed cooling lubricants.[23][37][38][39][40]

These insights are also confirmed for face-side-grinding of an alumina reinforced with 10% ZrO_2, for which different surface structures are also generated in relation to the cooling lubricant, and for which almost constant normal and tangential forces as well as a lower grinding wheel wear occurred during the application of grinding oil.[21][39][40][42] Compared to this, other studies on flat grinding of various oxide-ceramic materials showed lower normal forces during cooling lubrication with petroleum or petroleum fog than with emulsion, emulsion fog, or compressed air. Yet, the lowest tangential force as well as the lowest wear were measured during grinding with emulsion or under emulsion fog. The attainable surface quality was almost independent of the cooling lubrication in these investigations. Deviating results were also revealed during side grinding of HPSN and Al_2O_3/TiC. In these cases, higher normal forces were measured with grinding oil than with water-mixed cooling lubricants, which are referred to as increased thermal stress on the grinding wheel. The grinding wheel wear, however, was also lower for these materials during application of non-water-mixable cooling lubricants.[21][39][40][42]

Discrepancies are rarely observable between water-mixed cooling lubricants of various compositions with regard to grinding forces during face-side-grinding of Al_2O_3 + 10% ZrO_2, HPSN, and Al_2O_3/TiC. Yet, the grinding wheel wear was, irrespective of the material, slightly lower with an emulsion than during grinding with a solution.[23][40]

Influence of the Feed of Cooling Lubricants

The influence of altered conditions for the feed of cooling lubricants on the process and result values was investigated for pendulum-face-grinding, deep grinding, and cylindrical grinding.[23][35] A free jet nozzle or an enclosing nozzle respectively, with a width matched to the width of the grinding wheel, as well as a squeezed tube with identical exit section were used. Due to the lower jet quality of the squeezed tube, which causes heavy air intake, the reached cooling effect is strongly diminished for this nozzle, so that a high yielding of the bond occurs, which then results in increased grain chippings. This in turn leads to a significantly higher grinding wheel wear, lower surface quality, and lower normal force than is the case for the improved nozzle, (Figure 4.38).[23][35][41] The technological disadvantages of the squeezed tube, particularly with respect to the grinding wheel wear, can also not be compensated by an increase in feed pressure. If the cooling lubricant pressure, however, is increased from p_{KS} = 0.2 to 10 bar at simultaneous alteration of the volume flow for the free jet nozzle, a decrease in grinding wheel wear can be observed. The total grinding force, which consists of cutting forces and cooling lubricant forces which rise with pressure, increase strongly with an increase in feed pressure. If, however, only the pure grinding forces are observed, only a slight increase in forces is recorded, which is explained by the fact that diamond grains are kept longer in the bond at lower surface temperatures of the grinding wheel, and that they pass longer friction ways under formation of stronger grain flattenings.[23] The wear then increases slightly at even higher pressures of p_{KS} = 15 bar, because the cooling lubricant supply of the contact zone deteriorates as a result of an increase in turbulence at very high flow speeds. The 4-point-bending strength decreases slightly with increasing feed pressure, which is traced back to crack-stop-effects at lower cooling lubricant pressures.[23][35][36][41][43][44]

During high-speed-plain grinding of Si-SiC with a cutting speed of v_c = 100 m/s, significant tool wear was determined at insufficient heat removal due to a low cooling lubricant volume flow of Q'_{KS} = 1.01/min mm in connection to a free jet nozzle, because the synthetic resin bond was stressed above its permissible temperature limit. An overly proportional wear can be avoided by an increase in cooling lubricant volume flow to Q'_{ks} = 10.0 l/min mm, because higher amounts of heat are removed from the contact zone. Similar to the case of deep grinding, the positive effect of an increased volume flow is overly compensated by rising jet turbulences at increased flow speeds, so that radial wear increases again. The usage of an enclosing nozzle, on the contrary, allows one to decrease the grinding wheel wear already at a related cooling lubricant volume flow of Q'_{KS} = 5.0 l/min mm below the level which is reached at double flow capacity with the free jet nozzle. Obviously, the interruption of the air buffer caused by this nozzle, as well as the related toothing of the cooling lubricant with the grinding wheel, lead to a more effective heat removal from the contact zone. In contrast to the free jet nozzle, a further volume flow rise does not result in an increase in wear, but in another, minor decrease, because turbulence effects are less important due to the good adherence of the cooling lubricant to the grinding wheel. Irrespective of the used nozzle systems, the total normal forces increase as a result of increasing liquid pressure forces with increasing volume flows (Table 4.3).

Table 4.3: Pressure in Contact Zone for Different Feed Conditions[23]

Pressure in Contact Zone p_{KSK}	Feed Pressure			
	p_{KS}=0.2 bar	p_{KSK}=3.0 bar	p_{KSK}=6.0 bar	p_{KS}=0.2 bar
squeezed tube (A_{PKS} = 31.5 mm^2)	0.53 bar	0.85 bar	0.96 bar	1.14 bar
free jet nozzle (A_{PKS} = 22.5 mm^2)	0.55 bar	1.40 bar	2.53 bar	3.92 bar

Ecological and Physiological Aspects During Cooling Lubrication

Health Hazards

The risk of a disease of the skin, respiratory tracts, or cancer is given by the contact of staff with cooling lubricants. This can be caused by direct skin contact, swallowing, inhaling, or by contact with fogs and vapors.[45][46][47][48] Physiologically relevant substances are: emulsifying agents, which at times have a degreasing effect on the skin;[49] corrosion inhibitors, of which some are toxic and therefore should no longer be used;[49] preservatives such as formaldehyde;[50] and heavy metals. Above all, chlorine leads to chlorine acne in addition to creating problems of disposal.[47][49] Further health hazards are found in reaction products such as nitrosamines; in microorganisms and their metabolic products;[22][51] in solid materials which can cause micro-damages; in dissolved, toxic material elements such as nickel or cobalt; and in personal hygiene.[49][52] During application of water-mixed cooling lubricants, the risk of skin irritation generally increases with rising pH-value and concentration.

Ecological Damage

The use of cooling lubricants can lead to ground, air, and water contamination. Emulsions and solutions are hereby most problematic with regard to their disposal and water damage. In connection with their relatively low service lives, which are mostly limited by bacterial infestation, a relatively high amount accumulates for disposal. The life of alloyed mineral oils is longer and is characterized by contaminations.

The least risk arises from biologically-decomposable cooling lubricants such as esters. In their initial stage, they can even be described as ecologically neutral. The lower disposal expenditures, however, must be compared during cost calculation to the often higher cost-values. Hence, an efficiency analysis is recommended for the selection of an optimal cooling lubricant under consideration of all limiting quantities, with the single criteria of the objectives being mathematically combined with the capacity to fulfill these objectives.[23][43] An exemplary application supported the recommendation of an ester as a cooling lubricant based on its good ecological behavior.[43]

Approach to the Reduction of Cooling Lubricant Volume

The most strict step to solve the ecological, physiological, and economic problems related to the application of cooling lubricants is made by the complete avoidance of cooling lubricants by dry machining. Major advantages of dry machining are the omission of health hazards and ecological damages, of purifying, drying, or disposing processes for workpieces, chips, or the cooling lubricant itself, and correspondingly unnecessary costs for cooling lubricants and purification. These result in a higher process flexibility and better utilization of cutting products.[22][45][46][47][53] However, these advantages are opposed by problems which result from the fact that the primary functions of cooling lubricants such as cooling, lubricating, and rinsing are neglected. Cases of application of dry grinding are so far only mentioned in the literature.[48][54] Yet, a reduction in the required circulation amount can be realized by optimization of the feed process in the active zone.[22][45][46][55][56] Improved nozzles and nozzle-adjustments create the prerequisites for this. Another approach suggests taking the cooling lubricant directly into the active zone by means of tools with internal cooling.[47] In an extreme case, the cooling lubricant is measured so precisely in minimal-amount-cooling-lubrication systems that it is completely used up through vaporization. A major advantage of this cooling lubrication technique lies in the low amount of cooling lubricants required, as well as in the fact that the cooling lubricant is only once sprayed and not operated in a circulation. Thus, high expenditures for the control, maintenance, and disposal of cooling lubricants become unnecessary. Insight into the effects of cooling-lubricant-fog released into the surrounding air as well as corresponding counter-measures are not yet available.[45][54][57][58][59][60[61] A discontinuous feed of the cooling lubricant or delivery pumps regulated by machine regulation facilitate amounts of cooling lubricants which are exactly adjusted in time and place to the process. This again allows a reduction of energy requirement in addition to a decrease in the total amount of cooling lubricants.[47]

6.0 ENVIRONMENTAL ASPECTS

Hans K. Tonshoff

Production science has developed four vital criteria to evaluate manufacturing processes:[62][63]

- producible dimensions and shapes (principal technology),
- dimensional and surface accuracy (quality technology),
- economic efficiency (productivity), and
- man-environment technology.

The manufacturing industry has only lately followed the man-environment technology criterion. This is because consumers' and investors' awareness of environmental aspects were undeveloped, and the market was not prepared to pay an appropriate price for positive environmental properties. This has changed in the past decade as a result of public opinion and enforcement by governmental regulations. Public opinion, the market, laws, and regulations impacted industry to take environmental risks of manufacturing into account.[64]

As a result, waste disposal costs have increased as manufacturers assume liability for the effect of wastes upon the environment and for the impact of manufacturing processes upon workers' health. Concerning the machining of ceramic parts, environmental aspects concentrate on the usage and handling of grinding fluids.[65] In this case, emphasis is placed on fluid maintenance in order to minimize the amount of waste generated, to extend the life of the fluid, and to minimize negative effects on health.

Health Hazards

Grinding fluids within the factory environment may have negative health effects for workers. The common health effects appear to be dermatological, but respiratory and pulmonary effects on operators may also result from exposure to grinding fluids. Studies report that of all occupational health hazards, skin problems are the most common. This is the case for the metal- or ceramic-working industry. Of all reported occupational diseases, those related to the skin have been said to account for 40 to 60%.[66][67]

Occupational dermatitis takes two forms: irritant contact dermatitis (or skin irritation), which accounts for 50 to 80% of all cases, and allergic contact dermatitis (or skin sensitization), which accounts for 20 to 50% of all cases.[67][68] Irritant contact dermatitis is caused by exposure to substances that can damage the skin. Such dermatitis is curable and should not pose a permanent threat to the worker. Allergic contact dermatitis results from a worker's allergic intolerance to a chemical or chemicals and is generally non-curable.[68][69] Moreover, as a result of repeated exposure, a worker's reaction to a chemical may become more acute. Once diagnosed as an allergic reaction, the worker should avoid all contact with the offending chemical and should possibly take a different working place where there would be no such exposure.[66]

Actually pinpointing the cause of skin problems of workers is difficult, however, because causes are often multiple rather than single. A worker who develops a rash might indeed be suffering the negative effects of contact with a grinding fluid but may just as easily be reacting to chemicals found within toiletries or other consumer products.[66] Researchers have, however, identified factors such as worker hygiene, plant cleanliness, air quality, and time of the year as contributing to the likelihood of the onset of dermatitis among workers.[66][67][68] Factors such as psychological stability, age, and ethnic background can also influence a worker's susceptibility to dermatitis.[66][67][68]

According to another source, several factors leading to skin problems can be identified: the pH of grinding fluids is approximately 8.5, whereas the human skin has a pH value of 5.5 to 6; the use of surfactants in grinding fluids act to remove fats and oil from the skin. Certain metals, nickel, chromium, and cobalt for example, have also been identified as factors for skin irritation. It has also been suggested that microbial contamination of grinding fluids might contribute to the onset of dermatitis.[70]

Researchers investigating respiratory effects of grinding fluids on exposed workers have generally failed to establish definite causal relationships. It has been reported that exposure to mists caused by application of grinding fluids has not been definitively linked to respiratory and pulmonary problems, but it has been suggested that such exposure may well increase a worker's susceptibility to respiratory problems. This susceptibility can be linked to the level of chemicals and particles contained in mists generated by grinding. One study reports that oil mist exposure may lead to increased incidence of gastrointestinal cancer in exposed workers.[71] Up to now, inadequate information is available to fully determine the extent of respiratory problems from exposure to cutting fluids.

Managing of Grinding Fluids to Remove Potential Hazards

Environmental regulations and the increasing demand for high engineering performance have led to the development of the concept of fluid management. In short, fluid management refers to the full integration, in all stages of production, of considerations related to grinding fluids. This idea might be extended to considering the choice of grinding fluid very early in the machining of a ceramic part, and consideration of the environmental impact would become a matter of concurrent engineering.[72][73]

Deterioration of the grinding fluid through use results in many adverse effects upon its performance. For example, it is inevitable that small metal particles are formed during grinding operations; recirculating such particles will increase wheel wear and will cause the workpiece surface finish to deteriorate. Moreover, contaminants will hasten fluid degradation.[74][75] Problems associated with bacterial growth, additive absorption, demulsification, and catalytic oxidation, are all exacerbated. Another source of fluid contamination is tramp oil lubrication and hydraulic oils that leak into the cooling system. Oil contaminants react chemically with metallics of the machine and provide breeding grounds for micro-organisms. Depleted or improperly formulated grinding fluids can have other effects upon process and machinery operations; they can cause corrosion of machine components, paint stripping, foaming, and other process complications that adversely affect productivity and incur costs.

The first critical step in a fluid management program is the selection of a fluid for its appropriateness to the grinding operation and its durability. High quality fluids are more easily maintained and have a longer life than lesser quality fluids. Cleaning systems are designed to remove liquid and solid contaminants from the grinding fluid. To accomplish this, mechanical separation and filtration processes are generally used. Figure 4.39 shows a summary of different separating processes which are able to remove conta-

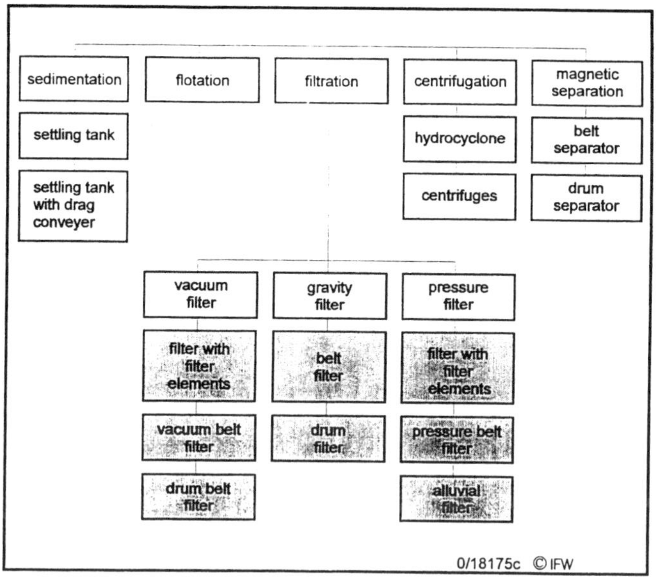

Figure 4.39: Mechanical Separating Processes

minants from fluids.[74][76][77] Concerning the machining of ceramic parts, only two mechanical separation processes are important.

Filtration systems are the most commonly used group of mechanical separation equipment. In filtration, a mixture passes through a barrier, the filter media, which catches particulate matter. Filters are available in a broad range of shapes, sizes, and throughput. Permanent media filters are made of metal meshes, belts, and screens; disposable media filters are made of paper, cotton, synthetic fibers, and other materials. In recent years, inorganic mineral filter media, such as diatomaceous earth,[78] have become more important. These kinds of filters, which are occasionally used as a filter medium in specialized oil systems, are extremely efficient at removing contaminants and very small particles, but they can also remove desirable components of the fluid such as extreme pressure additives and emulsifiers.

Papers or gauze elements are commonly used but must be checked or changed regularly. Halting production to change filter elements or media may be avoided by self-cleaning systems. In belt filter clarifiers, for example, fluid is fed onto a bed of filter paper which is carried on a pool-shaped conveyor.

Liquid passes through the filter medium and is pumped back to the work area. When the filter becomes blocked with solids, the conveyor moves forward, discharging the sludge and the used filter medium into a bin (Figure 4.40a). The only maintenance required is periodic emptying of the sludge box and replacement of the roll of filter paper.[74][75][76][77]

Because of growing disposal costs and increased environmental sensitivity to industrial waste, machine tool users are viewing reusable filters as a promising alternative to common throw-away filters. Clearable plastic filter elements and belts have a working life of about one year, dependent on the kind and quantity of swarf, and generate much less waste. The cleaning of plastic filters can be integrated into modern recycling systems by using compressed air blowers or scrapers.[79] Permanent media filters can also be cleaned by backwashing.

Filters can be alternately classified as gravity filters, pressure filters, and vacuum filters. As the name implies, gravity filters rely upon gravity to force fluid through the filter media by a pump. Vacuum filters rely on suction to bring fluid through the filter media.[80][81]

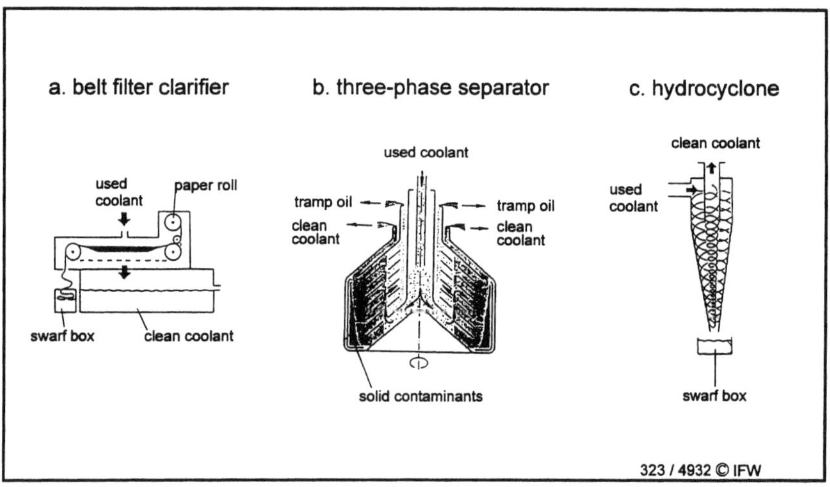

Figure 4.40: Examples for Separating Equipment

Centrifuges occupy less space than other cleaning methods because the centrifugal force can be much higher than gravitational force. Using this method, small particles can be removed and a higher proportion of the cutting fluid recovered.[78] Centrifugal separators can remove liquid as well as contaminants. Figure 4.40b shows a three-phase centrifugal separator which processes the dirty fluid to yield clean fluid, tramp oil, and solid particles.

With the hydrocyclone (Figure 4.40c), used fluid is pumped tangentially into a conical vessel, shaped to impart a vortex. Solids are thrown to the walls of the vessel and are discharged by gravity from the bottom. The vortex creates a vacuum at the axis of the vessel so that the cleaned fluid is removed from the top.[75][76][77]

A problem associated with the use of water-based grinding fluids is that they are subject to microbial attack. Micro-organisms (bacteria and fungi) can colonize in the free water phase of the grinding fluid and on the solid surfaces of the machines. The growth of micro-organisms is a predominant cause of fluid failure.[82] Microbial activity can result in foul odors, rust problems, foaming, reduced tool life, poor product finish, clogged lines, and increased disposal costs. In addition, relationships may exist between microbial contamination of fluids and human health hazards.[70]

To control the growth of micro-organisms, sources of contamination, such as swarf and organic matter, to the fluid system should be avoided. Thorough cleaning of machine pumps and pipes after disposal of used fluid and mixing with de-ionized water help to control microbial growth.

Conclusions

From the survey of literature and opinion summarized in this chapter, the following conclusions can be drawn in the case of grinding fluid and disposal:

- Responsible handling of grinding fluids and reduction of waste are necessary to secure a safe environment for the future.
- Treating the environmental problems posed by grinding fluids will incur cost, but the resulting improved control and maintenance of fluids should result in a more consistent part quality.

7.0 PROPERTIES OF GROUND SURFACES

Hans G. Wobker

The quality of high-performance ceramic components is mainly characterized through the steps of the production process and of machining. The technical strength of ceramics differs considerably from theoretical, molecular strength. Mechanical strength concentrations, due to flaws and cracks, cannot be reduced by plastic deformations as known for metals. The failure of a ceramic due to strength concentration is characterized by randomly distributed flaws. Therefore, with the strength data variate high, standard variations can reach up to 20 to 25%.

For an optimization of both efficiency of machining and strength of machined ceramic parts, the complex relationship between machining parameters, elementary mechanisms of material removal, and strength-controlling surface effects - surface integrity - must be understood. The aim of this chapter is to clarify how surface integrity influences the strength of machined ceramics.

Strength Testing Methods

The most common strength testing methods are three- or four-point-bending methods because of the simple specimen geometries and experimental arrangement. Other methods for static strength testing are the tensile test, C-ring test, and double ring test. Almost uncommon for testing ceramics are dynamic testings because of ceramics brittleness, except for high temperature fatigue testing.

Double Ring Test

The double ring test[83] uses circular or square ceramic plates that are forced by two different sized concentricity arranged rings (Figure 4.41). The nature of load is bending. The main advantage of this test is that chamfer effects do not influence the strength data. The linear theory of plates evalu-

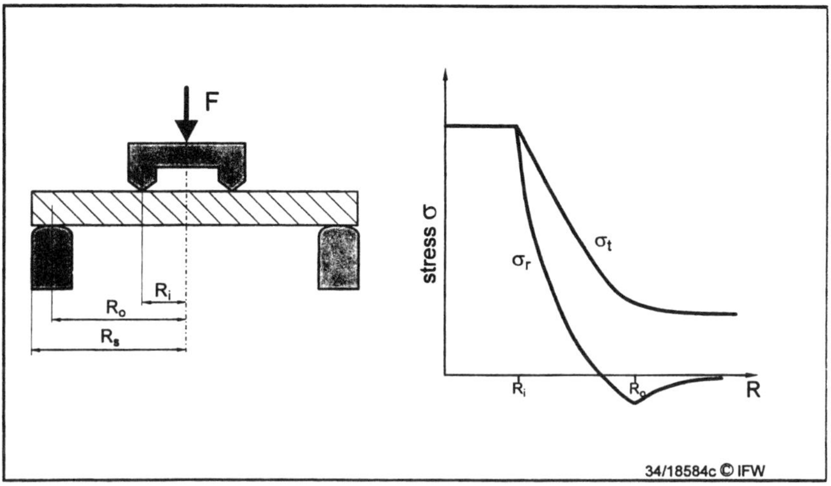

Figure 4.41: Double Ring Test

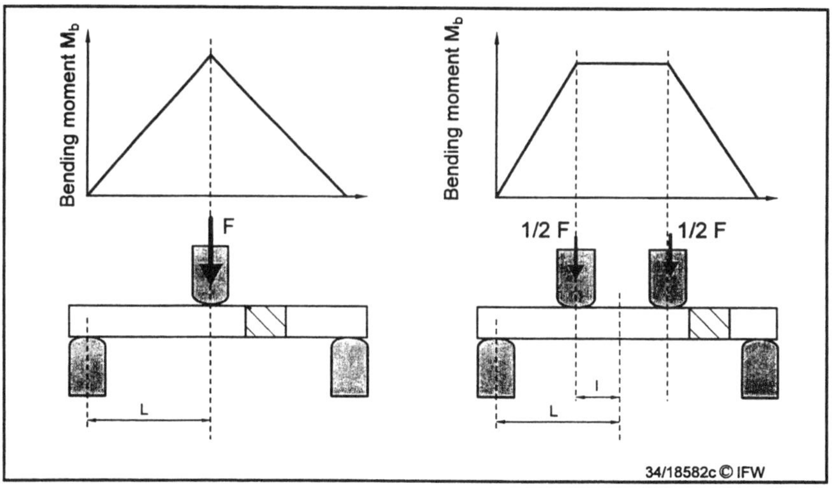

Figure 4.42: Three- and Four-Point-Bending Tests

ates equal radial and tangential tensile strength in the inner region of the specimen:

$$\sigma_r = \sigma_t = \frac{3(1+\upsilon)\cdot F_{max}}{4\pi\cdot d^2} \cdot \left(2\ln\frac{R_0}{R_i} + \frac{1-\nu}{1+\nu}\cdot\frac{R_0^2 - R_i^2}{R_s^2}\right) \tag{4.28}$$

R_o: radius of outer ring, R_i: radius of inner ring, R_s: radius of specimen, d: thickness of specimen

Bending Tests

The most common strength testing methods for ceramics are bending tests. Figure 4.42 shows the principles and the bending moment lines of both testing methods. In 4-point-bending tests, the specimen are loaded at four bearing areas. In contrast to the 3-point-bending tests, the bending moment in 4-point-bending tests is constant without traverse strain between the bearings. Therefore, the material properties will be analyzed better because ceramics bending strengths depend on volumetric flaws, surface cracks, and residual stresses. The moment stressed region in the 4-point-bending specimen is greater than in the 3-point-bending specimen. Therefore, bending strength determined by 3-point-bending tests can be higher than in 4-point-bending tests.

The geometry and measure of the specimen are standardized.[84] The sticks with rectangular cross-section are dimensioned as follows:

- length ($l \geq 45$ mm),
- width ($b = (4 \pm 0.1)$mm),
- thickness ($d = (3 \pm 0.1$ mm), and
- plane-parallelism (≤ 0.05 mm).

The edges should be rounded off or chamfered.

The used specimen can be manufactured separately or from a big part. The minimum number of each kind of specimen is 30. The 4-point-bending strength σ_{4B} is calculated by using the measured force F_{max}:

$$\sigma_{4B} = \frac{3 \cdot F_{max}(L - l)}{2 \cdot b \cdot d^2} \tag{4.29}$$

Statistical Evaluation. The Weibull-distribution of the bending test sampling can be fitted by the parameters m and σ_0. To evaluate these parameters, the Maximum-Likelihood-Method can be used.[85]

Influence of Grinding Conditions on Ceramics Strength Behavior

Grinding Direction

Many components can be ground parallel or perpendicular to the stress direction. Fracture of ceramics originates from machining flaws as well as from material intrinsic failures. The strength degradation effect of microcracks in the surface of ceramics can easily be understood by equations of fracture mechanics. The stress necessary to propagate a microcrack is given by the following equation:

$$\sigma_B \approx K_{Ic} \cdot Y \cdot \frac{1}{\sqrt{c}} \tag{4.30}$$

While K_{IC} is the fracture toughness of the material, Y refers to the crack geometry and c gives the crack length.

In view of their strength, the grinding direction should be parallel to the stress direction because perpendicular ground stressed components can be strength-reduced up to 30 to 40% (Figure 4.43). This strength degradation can be explained by the different geometric crack distributions. The large surface longitudinal cracks lead to lower fracture strength than the smaller surface radial cracks.

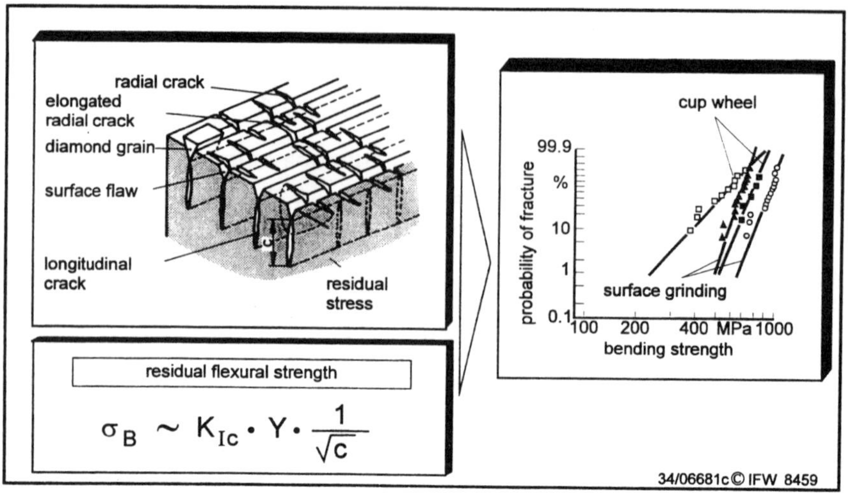

Figure 4.43: Influence of Machining Direction on the Strength of Brittle Materials

	method	active means	setting points	trueing
A	two sides plane-parallel lapping	lapping abrasive boron carbide+ water+additives, F800	p_L = 0.01 N/mm² v_L = 8.00 m/min	———
B	pendulum surface grinding -singlepass-	diamond wheel: D126K+8821 JY C50	a_e = 30 µm v_{ft} = 10 m/min v_c = 35 m/s	throughfeed trueing with corundum block a_{eds} = 2.5 mm v_{cds} = 20.0 m/s Q'_{ds} = 2.5 mm³/mms
C	pendulum grinding -multipass-		a_e = 30 ;5;2;0 µm v_{ft} = 10 m/min v_c = 35 m/s	
D	deep grinding -singlepass-		a_e = 1.0 µm v_{ft} = 0.3 m/min v_c = 35.0 m/s down-grinding	

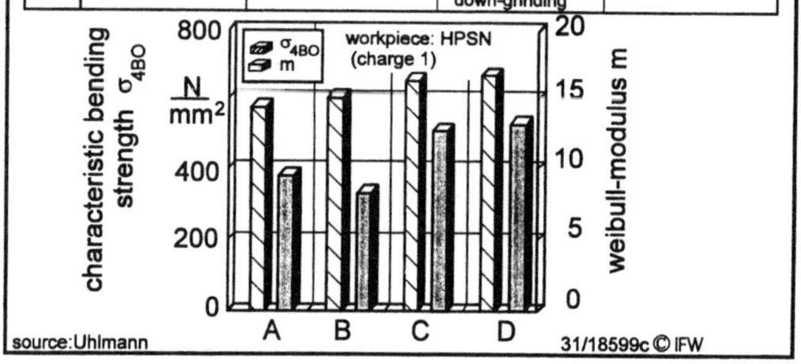

Figure 4.44: Influence of the Machining Process on Fracture Strength

Influence on Grinding Condition Compared to Lapping

Figure 4.44 shows the strength of HPSN for different machining conditions. Besides conventional grinding, creep-feed and multistage grinding as well as lapping were used. Compared to creep-feed grinding, the other machining processes lead to lower bending strengths and Weibull-modules because of the process-induced crack population.[86]

Process Model and Strength

The setting and system quantities of the grinding process influence the strength of machined ceramics. Most of the investigations found that the

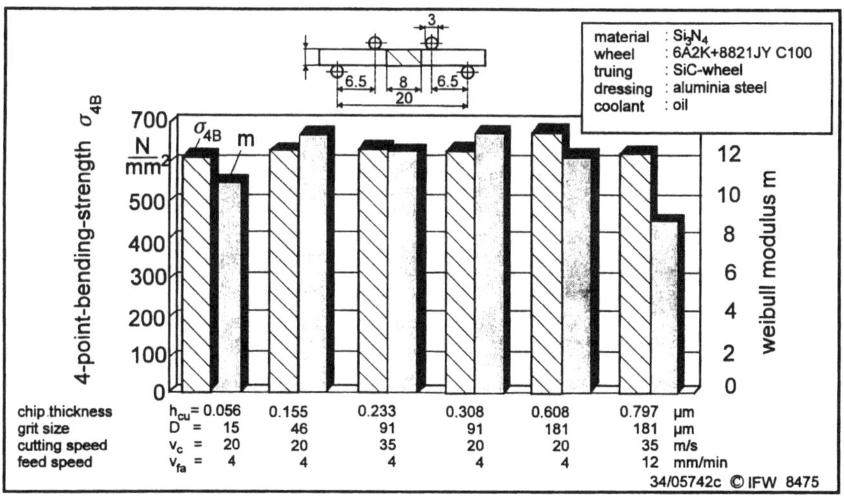

Figure 4.45: Influence of Grinding Conditions on Bending Strength and Fracture Probability of Silicon Nitride

grit depth of cut has a very important role in grinding[87]: there exists a correlation between the grit depth of cut and the strength degradation in cup-grinding of silicon nitride.

The variation of the grit depth of cut by varying the grit size of the diamond wheels, the cutting speed, and the feed speed induced different stages of surface integrity. The influence of grinding conditions on bending strength and Weibull's modules of machined alumina ceramic is shown in Figure 4.45.

8.0 GRINDING MACHINES

Hans K. Tonshoff

Ceramic parts require narrow tolerances for dimensional, form and position deviations, and specific microgeometric and physical surface states. These requirements are due to the typical application fields and the material properties. Very often, ceramic parts are chosen to meet high mechanical, thermal, chemical, and tribological loads. On the other hand, the specific sensitivity of many ceramics against tensile stresses and spot-like loading necessitates the control of dimensional accuracy to a higher level than by using ductile or tensionless endangered materials. These dimensional accuracy and surface integrity requirements must be considered when grinding machines are designed or installed. Up to now, grinding machines applied to ceramic parts have mostly been designed for machining of ductile metals. The adaptation of machine tools to the requirements of ceramics calls for the optimization of several functional units of a machining system. These are mainly the:

- machine structure,
- guiding system, and
- spindle system.

The drives and control units require some modifications but are in general compatible to the modern type of design. The machine periphery like the clamping devices, the measuring and sensoring system, the balancing device, and conditioning system for the tools must be adapted to the machining of ceramics.

Machine Structure

The structure of a machine tool, comprising the body, the slides, and tables has two main functions:[88]

- to ensure the geometric position of the accuracy-determining components of the machine, and
- to support against forces and moments.

This functional system is influenced by some disturbances. They lead to deviations between tool and workpiece. These influences can be summarized into four groups (Figure 4.46).

One of the main disturbing effects is caused by **thermal induced deformations**. This so-called "thermal stability" is of special importance in automatic fine finishing because the position of the workpiece to the grinding wheel normally changes by the thermal deformation of the machine structure.

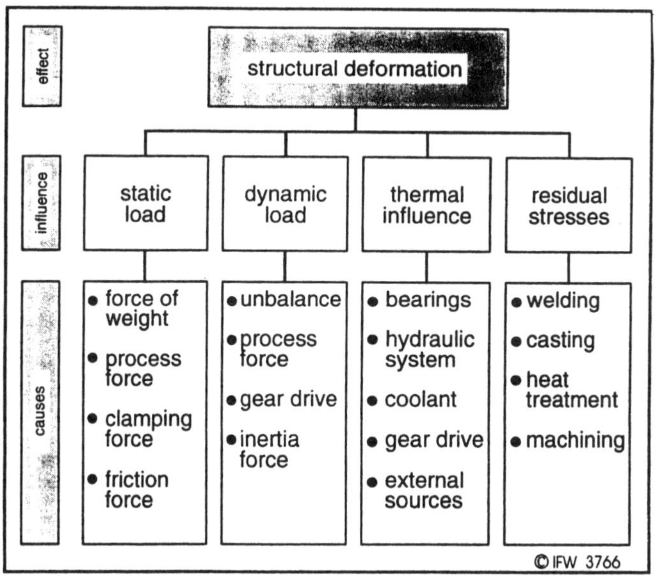

Figure 4.46: Structural Deformations and their Causes

Thermal deformations Δl result in the simple case of uniform temperature distribution $\Delta\vartheta$ proportional to the length of a component according to

$$\Delta l = \alpha l \Delta\vartheta \tag{4.31}$$

the linear coefficient of elongation α. In non-uniform temperature fields, even though additional bending and tilting takes place, the effect is normally much more severe. Figure 4.47 shows the deformation of a surface grinder resulting from an instationary temperature field. This may be caused by the front bearing of the spindle. First, it warms up the spindle stock which moves in the negative Y-direction because of the clamping at the upper part of the spindle stock. Later, the column is also warmed up but only partly at the front side. This generates a bending deformation of the column. Thus, the typical movement of the wheel against the workpiece over time takes place as shown in Figure 4.47.

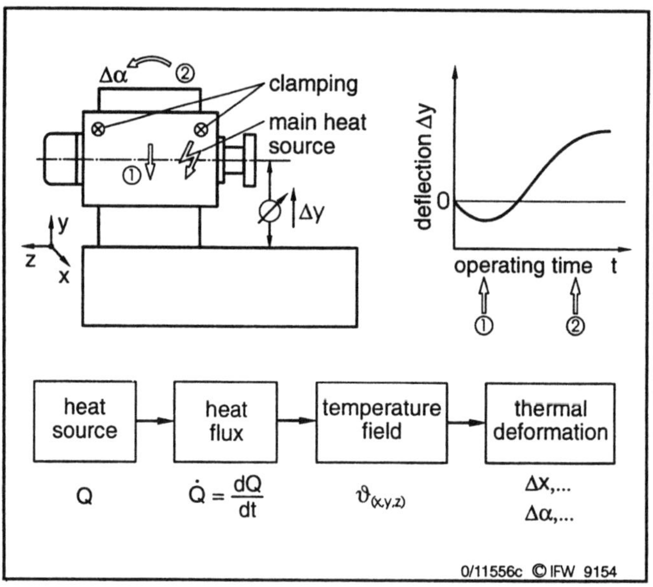

Figure 4.47: Typical Deformation Trend and Causal Connection

The main heat sources in grinding machines are the:

- abrasive process itself,
- electric motors,
- spindle bearings,
- hydraulic system, and
- cooling system.

To minimize the effect of temperature fields in the machine structure, passive or active actions can be taken (Figure 4.48). Passive measures must be introduced when the machine is designed, for instance by removing heat sources from critical areas of the structure or by providing compensating elements in the structure. Active actions influence the energy flux in the body by cooling, heating, or compensating deformations via the numerical control of the machine.

Another disturbing influence results from static loads. As mentioned before, the ratio between normal and tangential force, or the thrust and cutting force respectively, in grinding is much higher than in cutting, i.e., in turning or milling (Figure 4.49). This is due to the highly negative rake angles of an

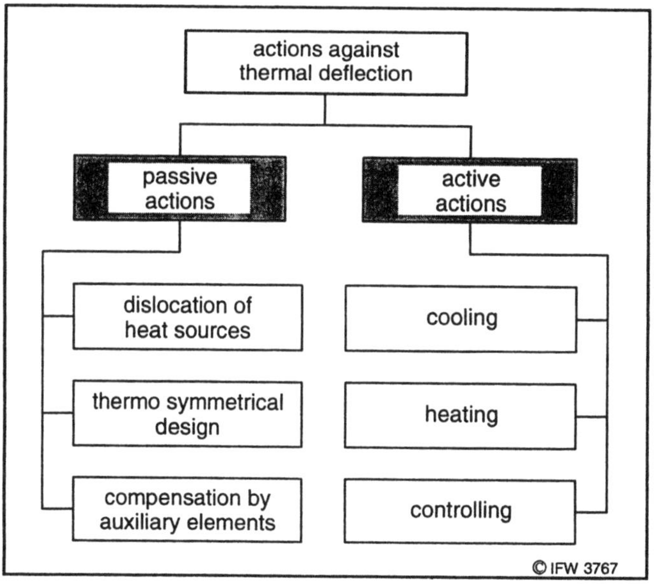

Figure 4.48: Actions against the Thermal Deflection.

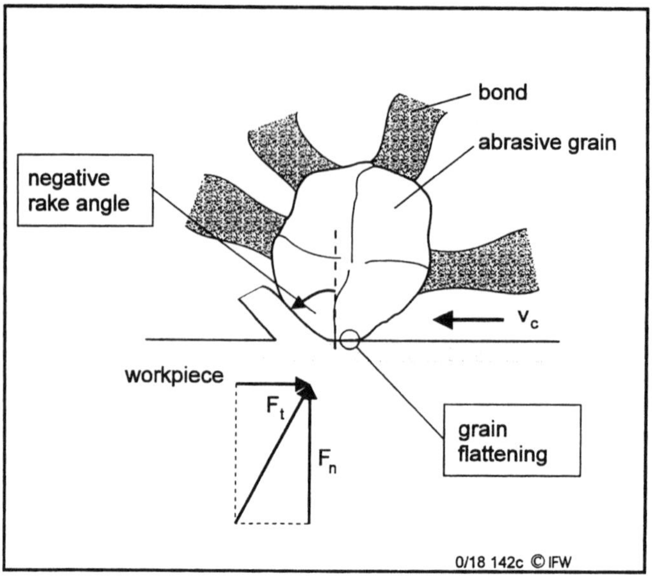

Figure 4.49: Force Ratio in Abrasive Machining.

average abrasive grain which is relatively worn and thus shows a flat shape. The chip generation in machining of ceramics leads, furthermore, to a substantial increase of the force ratio from about 4:1, in grinding ductile metals, to 10:1. The high normal forces bring high thrust deformations which directly influence the accuracy of the machine. Therefore, the static stiffness normal to the active surface of the grinding tool is of great importance for grinding machines applied to ceramics.

Besides the accuracy, the productivity of a machine is of concern. The high thrust forces bring deviations of the effective position against the nominal position of the tool. These deviations are normally compensated by spark-out dwelling (Figure 4.50). The spark-out time depends on the thrust force and on the stiffness of the system. The stiffer the machine, the shorter may be the spark-out period which can be followed from the time constant T.[89] The static stiffness of the structure is dependent on its material and the design of the components in the flow of force.

Figure 4.51 gives some material data that is of interest for static loading of a structure.[E: Young's modulus, v: Poisson ratio, G: Shear modulus,

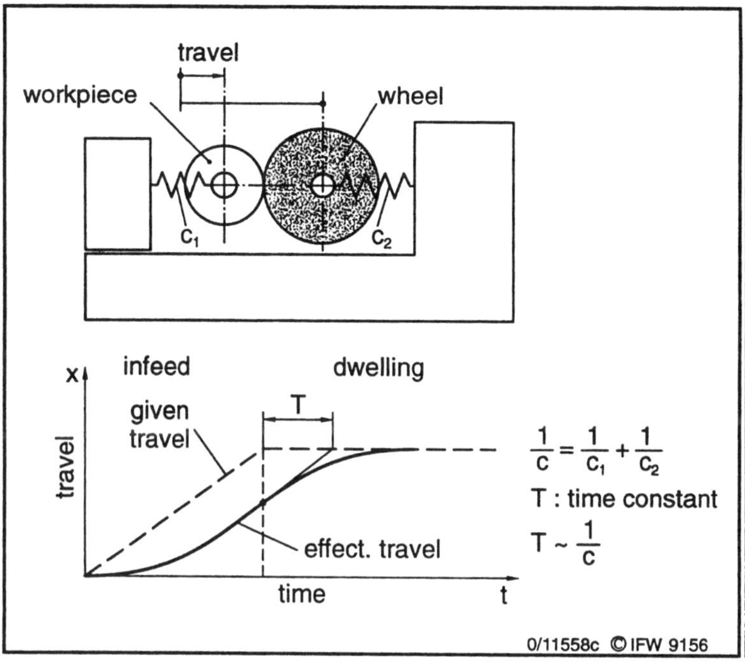

Figure 4.50: Elastic Deformation of a Grinding Machine

Material	E [kN/mm²]	ν	G [kN/mm²]	R_m [N/mm²]	R_{eH} [N/mm²]	ρ [g/cm³]
Cast iron	110	0.25	45	250	140	7.35
Steel	210	0.3	80	400	235	7.8
Tungsten carbide	400	0.25	.170	1000	700	13.2
Aluminum Al 99.5	72	0.34	27	100	70	2.7
Silicon nitride	315	0.28	123	300	-	3.19

Figure 4.51: Properties of Materials

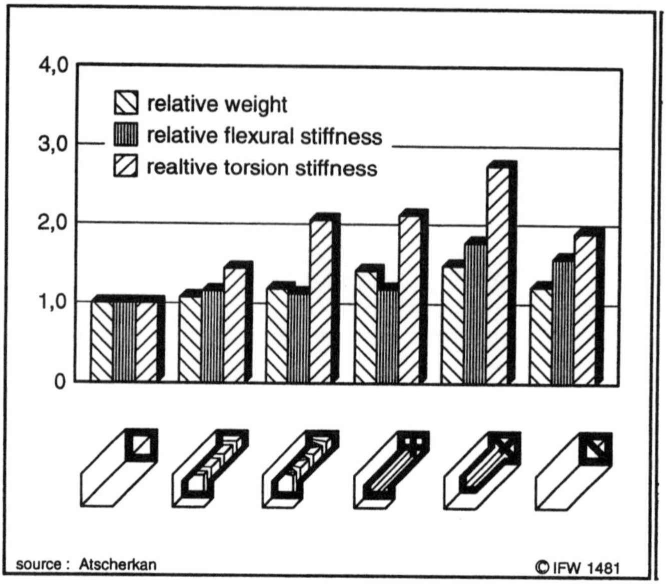

Figure 4.52: Stiffness of Structure Elements.

R_m: Tensile strength, R_{eH}: Apparent yielding point, ρ: Density] Because of economic reasons, the choice of materials is limited. The stiffness therefore has to be reached by design measures. For given material properties, the lowest compliance show components that are loaded by pure tension or compression only, because the material cross-section receives a uniform stress distribution and is thus used most effectively. In machine tool structures, bending and torsional loading situations cannot be avoided. Figure 4.52 shows that the form features of components open an additional degree of freedom to influence the stiffness. The figure gives relative weights, bending, and torsional stiffness values in comparison to a square profiled beam. In the case of bending, the axial momentum of area is to be maximized. The distance of the material from the bending axes must be as large as possible. The double T cross-section is most favorable. In the case of torsional loading, the polar momentum of area is important. Additional effects can be reached by three dimensional design components. The cross-sectional profiles with diagonal ribs or closed cross-sections with maximal diameters yield the highest stiffness.

In general, the structural elements of a machine body are not prismatic beams but are cross-sectional changes. Here, it is important that the flow of forces is smooth, because the introduction of transversal forces into flexible walls has a significant weakening effect. The same is true for joints in the flow of forces. They can be stiffened by pre-stressing.

Accuracy and productivity are also influenced by the dynamic behavior of a machine structure. It has to be distinguished between enforced or externally-excited and self-excited vibrations (Figure 4.53). Enforced vibrations follow the excitation force or an exciting movement in their frequency. Excitations may be generated from unbalance, from periodic forces of the grinding process, from hydraulic pressure or volume oscillations, and from drives.

Grinding wheels rotate with high speed. In general, their mass is not exactly concentric to the axis of rotation. There is a small eccentricity e_μ, or a small angle between the principal axis of inertia and the axis of rotation, or centrifugal forces generated by the imbalance act on the bearing system. If the rotor is disk-like, it may be balanced statically. This means that the axis of inertia and the axis of rotation can be brought together by applying one comp-

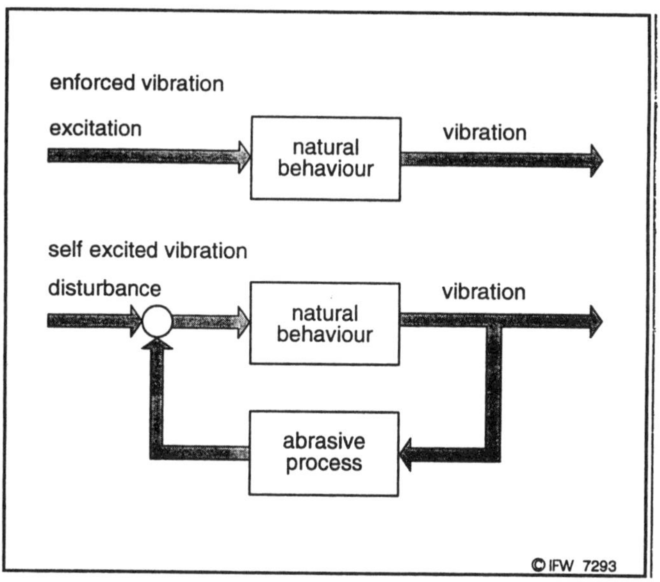

Figure 4.53: Vibrations in Machines

ensation mass. If the rotation system has a considerable length, as, for instance, a broad wheel for centerless grinders or if the disk-like wheel tumbles, the rotor must be balanced at least in two planes in the so-called dynamical balancing mode.[88]

For static balancing, the rotating mass m with the eccentricity e_μ, must be compensated by a balancing mass u on the radius r_μ,

$$e_\mu \cdot m = u \cdot r_\mu, \tag{4.32}$$

The allowable eccentricity $e_{u\,max}$ is defined for rotors of the same type. It is inversely proportional to the angular frequency.

$$e_{umax} \cdot \Omega = \text{const.} \tag{4.33}$$

The following quality classes are defined :

G 2.5: $e_{u\,max} \cdot \Omega \leq 2.5$ mm/s for machine tools in general
G1: $e_{u\,max} \cdot \Omega \leq 1.0$ mm/s for grinding machines
G 0.4 $e_{u\,max} \cdot \Omega \leq 0.4$ mm/s for high precision grinders

In order to compensate in imbalances, balancing devices are applied. The different operational methods can be classified by three functions. For each function, there are different solutions (Figure 4.54). As an example, type 2.2.4 is used for automatic balancing. An accelerometer measures the vibration in amplitude and phase. These signals are used to actuate four valves which lead fluid, for instance coolant, into four chambers. As a result, a compensating mass is generated.

The other balancing devices, mostly using the centrifugal action to detect the imbalance (detection mode 2) and an active compensation mode (mode 2), are applied when high surface quality (especially in grinding of cermics) is required.

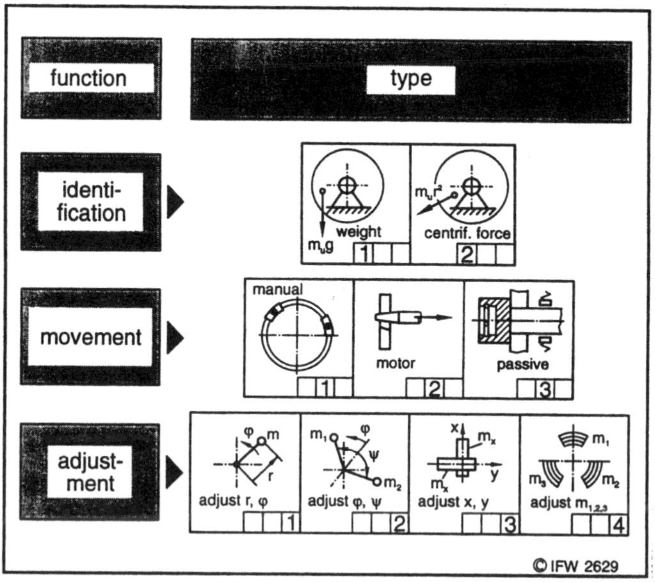

Figure 4.54: Balancing Methods for Grinding Wheels

External excitation can also be caused by gears. This is why the main spindle of a grinding machine is driven via belts or directly by a motor whose rotor is fixed on the spindle. Besides the mechanical uniformity of the speed, magnetic constancy of the electric motor is necessary to avoid excitations.

Because enforced vibrations can principally be influenced from the outside, they are not considered as critical as self-excited vibrations. Especially in grinding of ceramics, the danger of chatter is high because of the heavy normal forces. The excitation is determined by the dynamic system itself, i.e, the machine structure, the tool, and the workpiece. The system vibrates with one or several natural frequencies. This kind of vibration can be suppressed by influencing the vibrating system by increasing the damping. This is not achieved so much by the material of the machine body because material damping only contributes to a minor extent to the total damping. The influence of joints and guideways is two or three orders of magnitude higher. A special effect is possible by using auxiliary mass dampers (Figure 4.55). The system must be tuned on the critical frequency. The effective band width is limited. In the case of enforced vibrations, the minimal compliance is determining the

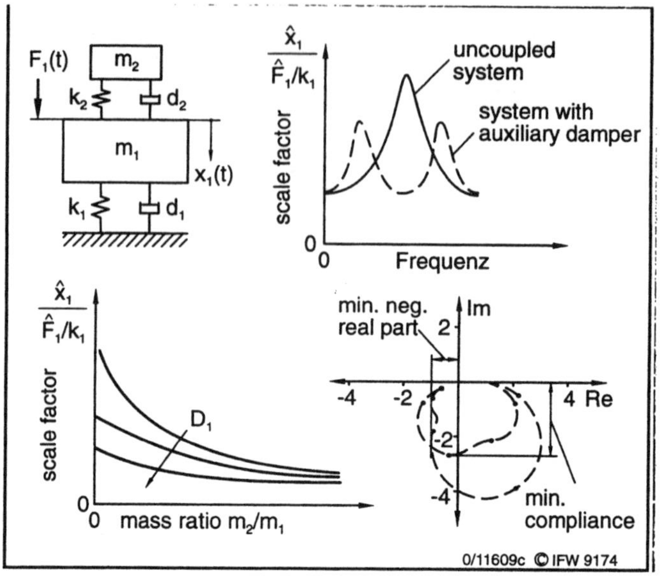

Figure 4.55: Effect of an Auxiliary Damper

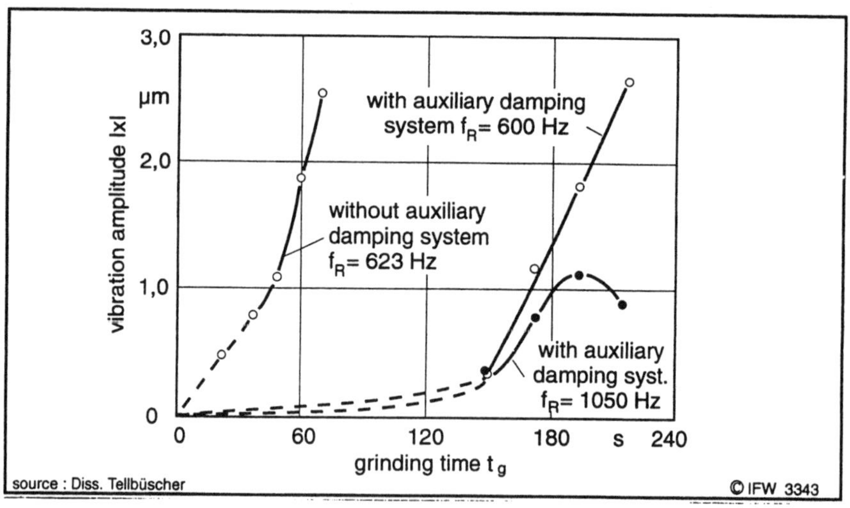

Figure 4.56: Influence of Auxiliary Damping Systems in Grinding

dynamic stiffness: for self-excited vibrations, the minimal negative real part of the compliance is critical.

Figure 4.56 shows the influence of a damper fixed to the workpiece support. It consists of an auxiliary mass, a bending beam as a spring, and a high absorption material which is loaded in shear. This measure doubled the dynamic stiffness at the working spot. As can be seen in the figure, the development of amplitudes at the workpiece could not be suppressed but could be retarded considerably by a factor of three.[90] The experiments are well reproducible; they were repeated four times. The scatter of the results was 12% measured in the time scale. The effect can be further increased by directly applying the damper to the workpiece. This is not always possible but in many applications when using an arbor, the damper can be built in. Another typical measure to pass chatter problems in the grinding of ceramics is to use grinding wheels with high material damping.

Guideways

Guideways determine the linear movement of a machine component like a table or slide in one direction and hinder movements in the other five degrees of freedom. The guiding function is disposed to several disturbing influences from which the requirements for guideways can be derived (Figure 4.57).

Guideways should be protected against swarf of the workpiece and particles from the tool. This is especially important for the grinding of ceramics because the work particles have diameters down to micrometers or less and are highly abrasive. Sliding guideways may be worn out quickly. Antifriction guideways can also be disturbed by rolling the abraded particles into the guiding surfaces. Even for hydrostatic bearings, close protection is necessary to avoid pollution of the oil. There are several means to protect guideways (Figure 4.58). They may act as scimmers or as covers. The accuracy and lifetime are very much influenced by these protective devices.

Guideways can be distinguished according to the kind of force transmission between the moving partners as follows:

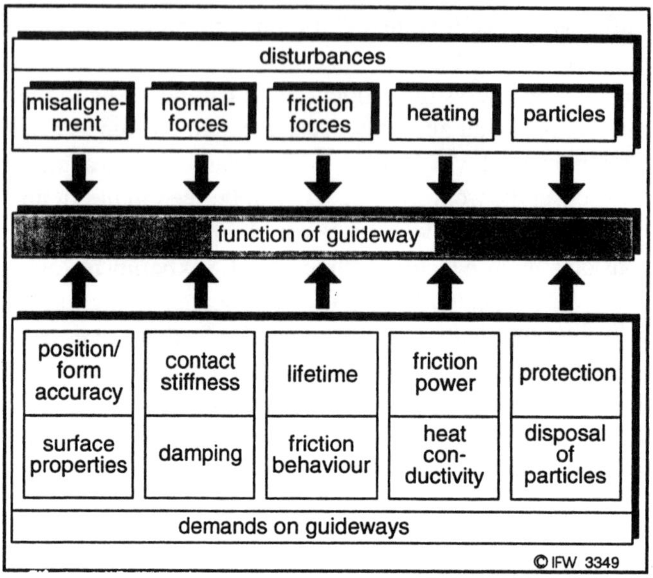

Figure 4.57: Disturbing Influences and Demands On Guideways.

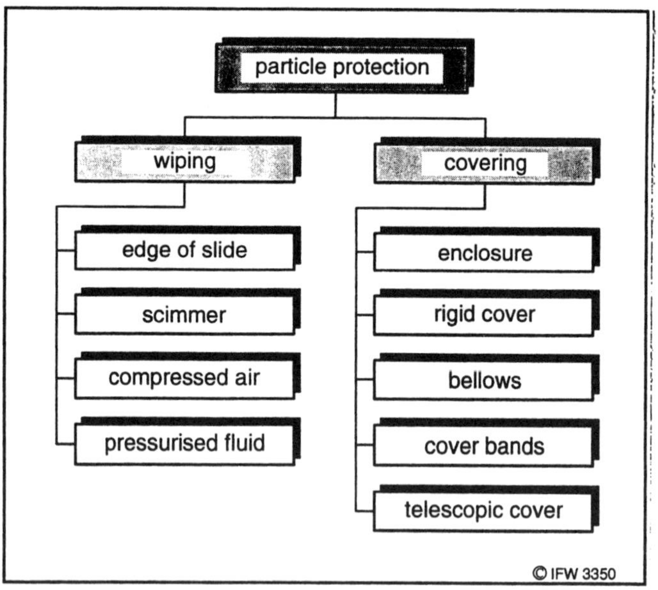

Figure 4.58: Particle Protection Devices.

- sliding guideways (slideways),
- roller guideways (rollerways), and
- hydrostatic guideways.

Slideways work in the mixed friction domain: partly solid friction and partly fluid friction takes place. Normally, the traveling speed is not so high that a pure hydrodynamic state can be reached. For ceramic grinding machines, it is important that the travel of the wheel against the workpiece is exactly performed as dicated by the control and the drive of the machine. This means that the guiding system must not show stick-slip behavior (Figure 4.59) otherwise small engagements will not be brought through from the driving system to the working point. The precondition to avoid stick-slip is that the friction coefficients of movement and standstill are equal. This can be achieved by polymer sliders, for instance reinforced polytetrafluorethylene (PTFE). These layers on one side of the guideways solve the stick-slip problem for slideways. Because of the continual low pressure between the sliding surfaces - the weight of the workpiece and the process loads are normally of minor importance - guideways are often designed according to the sliding pri-

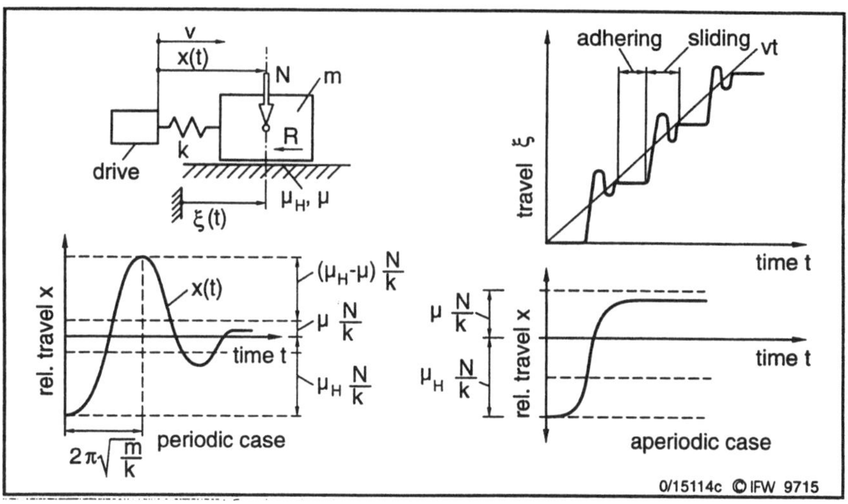

Figure 4.59: Stick-slip Model.

nciple. This is also a fairly inexpensive solution. They may be built in an open form as prism-flat or prism-prism versions (Figure 4.60). The prism-flat design is considered as statically determined, though slideways consist of larger matching planes which are principally overdetermined. The friction characteristics of the two guideways are different because of the wedge behavior of the prism. This is avoided by two prisms, although the design is statically overdetermined. This can be overcome by giving a floating freedom to one of the prisms.

A sophisticated technical solution for linear guiding systems are hydrostatic guideways. Oil is pressed by an external source into the gap between the moving partners. The guiding surface is divided into several pockets which must be hydraulically decoupled enabling the system to carry eccentric loads. Very often, capillary hydraulic throttles are used to obtain this decoupling effect. Stiffness and transversal damping abilities of hydrostatic guideways may be extremely high. The fluid friction force is almost negligible. The damping in the moving direction, therefore, may be critical when using path or position control loops. The systems are very sensitive to pollution of the hydraulic medium. Hydrostatic guideways must be thoroughly protected

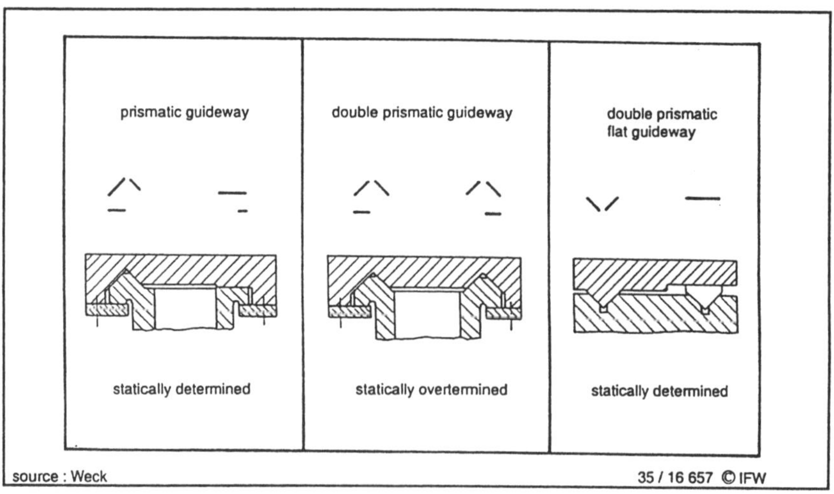

Figure 4.60: Prismatic-flat Slideways

against swarf and particles, especially when grinding ceramics. The system is dependent on the reliability of the hydraulic oil supply and is expensive compared to slideways or rollerways.

Rollerways contain balls or rolls as rolling elements. Their systems return the elements like roller chain, roller shoes, and others in which the rolling elements move against the guided table or slide. An important advantage of rollerways is that they are standardized and can be purchased from special suppliers. The stiffness is dependent on the number and diameter of rolling elements and especially of the preload. Pre-stressed guideways gain considerably in stiffness but also show higher friction. The damping ability is low compared to slideways or hydrostatic guiding systems. As mentioned above, the protection against swarf and particles is critical, especially in grinding ceramics.

Spindles

Generally, spindles are the most important components of a machine tool. They are located in the flow of force and determine the static and dynamic stiffness; thus they determine the accuracy and productivity of a machine tool to a great extent.

In grinding machines for ceramics, the design of the wheel carrying spindle is of special interest. The high spindle speeds, high radial forces, and the required extremely-low, radial run-out set specific demands. High dynamic stiffness is necessary to suppress self-excited vibrations as well as to mini-mize the vibration amplitudes in the case of enforced vibrations. Besides a high static stiffness, high damping values are necessary. These properties are mainly dependent on the bearing principles which are hydrodynamic, hydrostatic, aerostatic, roller or magnetic bearings (Figure 4.61).

Hydrodynamic bearings are fed by non-pressurized oil. The pressure distribution and its resultant force direction is dependent on the spindle speed. Thus, the bearing capacity, the stiffness, and the position of the spindle axis change with the rotational speed. For stationary conditions, the bearing shows a favorable behavior.

Hydrodynamic radial or axial bearings are normally designed with at least three pockets on the circumference to develop the necessary pressure

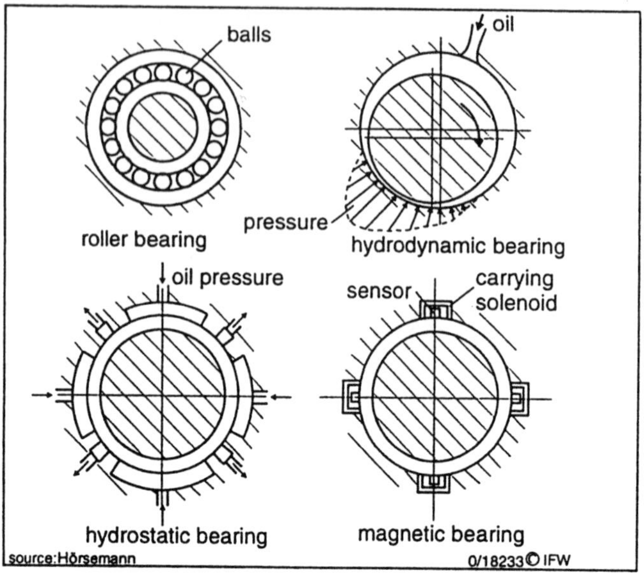

Figure 4.61: Spindle-bearing Principles.

fields. In the past, this type of bearing was typical for grinding machines because the spindle speed was mainly kept constant. Today, hydrodynamic bearings are still in use, especially for large spindle diameters. They have design features which enable pure fluid friction with low speeds.

A fluid layer of oil or air is generated between the rotating spindle and an outer ring. The behavior of these bearings is not dependent on the speed. In hydrostatic bearings, the oil temperature is critical for the performance because of the high shearing velocity in the fluid.

The oil should be temperature controlled to avoid thermal influences on spindle performance. Hydrostatic bearings are sensitivc to pollution, so the spindle must be protected against the intrusion of chips and coolant, especially in machining ceramics where very small, abrasive chips are generated. It is obvious that the design affords and the costs for producing a grinding machine with a hydrostatic spindle are high. On the other hand, hydrostatic spindles show good performance concerning the radial deviation, the maximum load characteristics, and the damping characteristics. The radial and axial run-out of such a bearing is principally low.

Aerostatic bearings are based on the same principle as hydrostatic ones but use air as a medium. This limits the maximum pressure and the damping characteristics of the bearing. The advantage of gas as a medium is the reduction of friction (compared to oil) which enables high rotational speeds. Aerostatic spindles have very low radial deviations of below 0.01 μm. This is the best value of all spindle types. Because air is compressible, the maximum load that can be applied to the bearing is lowered. This limits the acceptable forces on the spindle. The damping rate of air is lower than oil. This, along with the density characteristics of gas, is the reason for the sensitivity of aerostatic bearings to self-excited vibrations. This spindle type is used for precision grinding operations at low loads. Typical examples are inner diameter cut-off grinding machines for slicing single crystalline silicon wafers or internal grinding machines.

The most frequently applied spindle bearings are ball bearings. As standardized parts, they are available in many different modifications at reasonable prices compared to other bearings. So-called antifriction bearings for spindles in machine tools can either be roller or ball bearing types. Roller elements have linear contact areas which are larger than the contact zones of balls which have spherical contact. Ball bearings have decreased friction compared to roller bearings because of their reduced contact area. Therefore, for higher velocities, ball bearings are applied with a reduced maximum load capacity. Figure 4.62 shows the ranges of rotational speed for different grinding operations, and the maximum speed for the different bearing principles. For the majority of grinding machines, angular ball bearings are used which are also capable of carrying axial loads. The load in a ball bearing is transmitted by a thin lubricant layer between the rings and the roller elements. A breakdown of this layer causes fatal damages to the spindle bearing. The lubricant of roller bearings can be realized by grease, which under normal conditions lasts the whole life cycle, by oil injection, and by oil mist for very high speeds. High frequency spindles for small grinding wheel diameters generally use an oil mist because of its lower friction and decreased losses. Ball spindle bearings are regularly assembled with a preload to achieve close tolerances. Because of preloading and the thin lubricant layer, ball bearings have a good stiffness at low damping rates. The roundness deviation tolerances of these bearings depend on the quality of manufacturing and assembly of rings and rolling elements. For precision bearings, roundness deviations of below 3 μm are realized. For higher speeds, so-called hybrid ball bearings

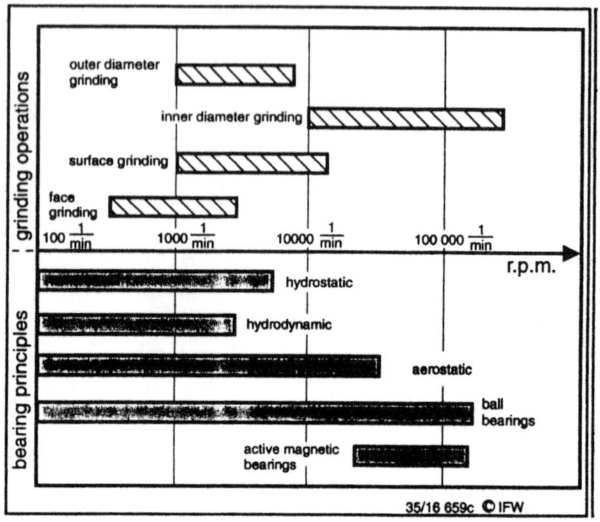

Figure 4.62: Speed Ranges and Spindle Bearings for Grinding Machines

can be applied. The rings of these elements are of conventional bearing steel - for instance 52,100 according to SAE - and the balls are made from ceramics with lower density and thus lower centrifugal forces and favorable tribological properties. For the grinding of ceramics, an air pressure protection can be used to avoid the intrusion of dust and chips.

Magnetic spindle bearings have been on the market for several years. The ferromagnetic spindle is carried by four magnetic fields. This kind of bearing is principally unstable, i.e., without further measures the spindle would run off the center and would contact the outer part. This effect must be suppressed by a closed loop control. Sensors measure the position of the spindle and, via the control circuit, influence the supporting magnetic fields. The position of the spindle is thereby stabilized. Additionally, the principle offers the possibility to measure forces or to tilt the spindle to compensate deflections. The main advantage of these bearings is very low friction. Figure 4.63 shows the design of such a magnetic bearing. Magnetic bearings are only used for high frequency spindles into which the motor is integrated. They can be applied for rotational speeds from 30,000 to 120,000. The pos-

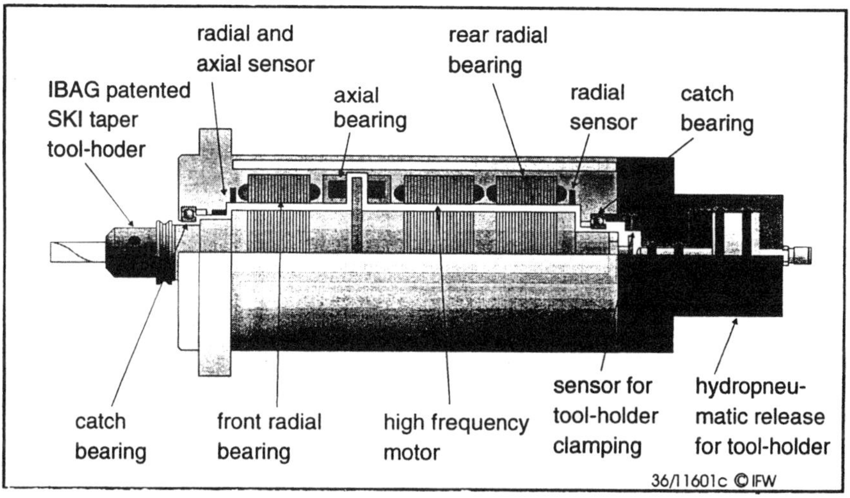

Figure 4.63: High Frequency Magnetic Spindle

sibility of adjusting the dynamic characteristics of the control loop provides the opportunity to fit the bearing stiffness to the process. Magnetic bearings have a constant maximum torque over the whole speed range. The main disadvantage of these types of bearings are minimum velocity (30,000 rpm) and the cost, which is several times higher than that of a ball bearing spindle.

Clamping Devices

The main function of a clamping device is to fix the workpiece in an exact position within the working area of the machine tool (Figure 4.64). Especially in machining of ceramics, the specific characteristics of the material require some special considerations.[91][92]

The demands presented in Figure 4.64 can be fulfilled in different ways, as shown in Figure 4.65, for external and internal machining operations.

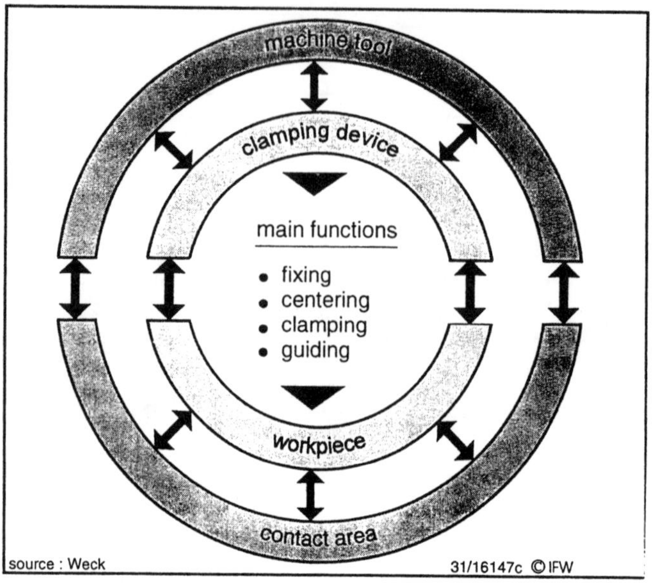

Figure 4.64: Main Functions of Clamping Devices

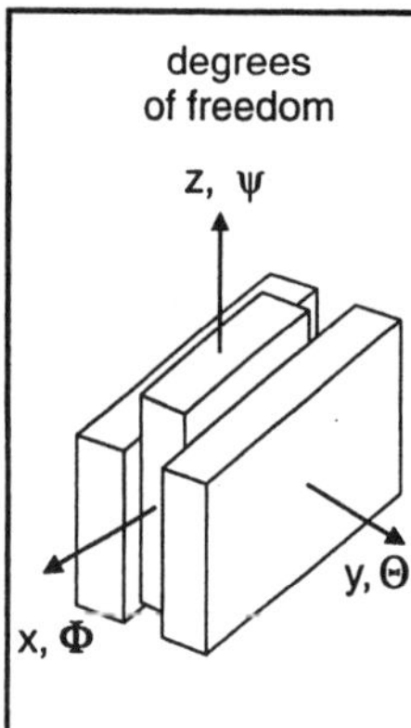

	clamping device	force locking	form locking
internal machining	o multiple-jaw chuck, adapter sleeve	z, ψ	x, y, Θ, Φ
	o extensive chuck	-	x, y, z, Θ, Φ, ψ
	o axial grip	x, Φ	x, z, Θ, ψ
	o magnetic chuck magnetic clamping plate	x, y, z, Θ, Φ, ψ	-
external machining	o multiple-jaw chuck, internal mandrel	z, ψ	x, y, Θ, Φ
	o expanding mandrel, clamping bolt, expanding cartridge	z	x, y, Θ, Φ, ψ
	o magnetic chuck, vacuum chuck	x, y, z, Θ, Φ, ψ	-

source : Sturghes 31/09594c © IFW

Figure 4.65: Reduction of Degrees of Freedom by Different Clamping Designs

The main focus of a clamping device is to ensure the workpiece position in translatory and rotary directions. This can be done either by a force-locking (magnetic chuck, vacuum chuck) or by a form-locking device. In most applications, the clamping device is a combination force and form locking clamping device (multiple-jaw chuck, mandrel, etc.).

Different from metals, ceramics can only be deformed elastically within small limits. Furthermore, most ceramic materials are not magnetic, thus excluding the application of magnetic clamping devices that are used for ferrous materials. With these features, several characteristic demands on clamping devices arise as shown in Figure 4.66.

A main criterion is to avoid local load peaks which may lead to workpiece damage by crack initiation. Due to the high process forces in the machining of ceramics, simultaneously high clamping forces must be produced. To fulfil the high quality demands on ceramic components, precise clamping devices with high repetition accuracy are required. Furthermore, easy operation is required as well as high flexibility. A solution to these requirements offer form locking clamping devices, which provide a flux of force over an extended area into the workpiece. Local load peaks can also be avoi-

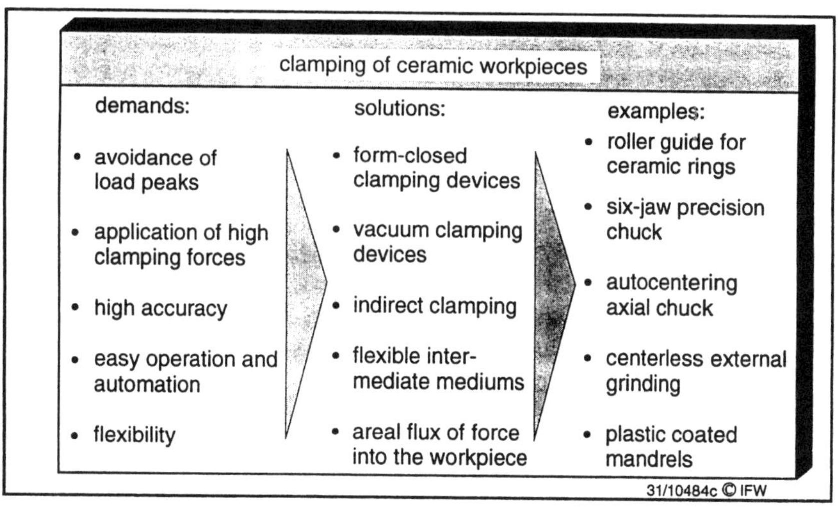

Figure 4.66: Clamping Device Development for Ceramic Components

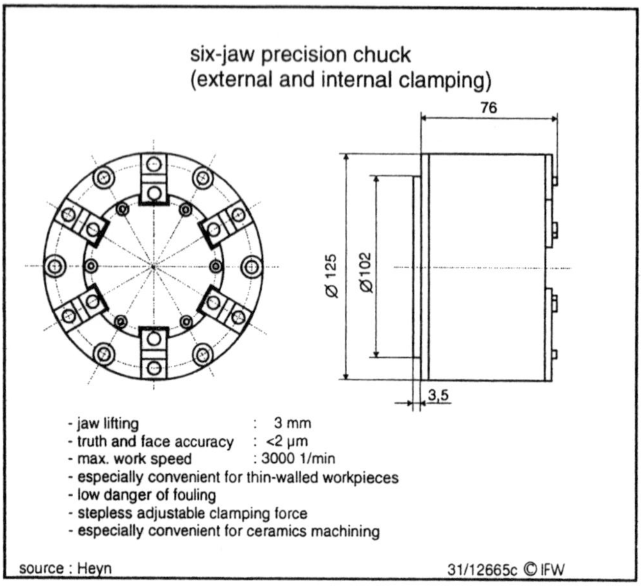

Figure 4.67: Precision Chuck

ded by flexible intermediate media. Other solutions are vacuum clamping devices and indirect clamping,[92] where the workpiece is glued to a metallic fixture or is clamped by well-known means.

In internal and external grinding of ceramic components, mostly multiple-jaw chucks are used for workpiece clamping. A six-jaw precision chuck is shown in Figure 4.67. It is characterized by a high radial and axial accuracy of less than 2 µm and is also appropriate for the clamping of thin-walled workpieces. Advantageous to the machining of ceramics is an adjustable pneumatic clamping force.[93]

Figure 4.68 presents different types of clamping devices for surface grinding operations of ceramics as well as their range of application. Well-known from the machining of metals are clamping shoes and bolt levers. Because of the danger of load peaks in ceramics, only a few are suitable and thus are predominantly used for single-part production.

Vises are convenient in clamping workpieces with plane parallel faces. For ceramic materials, soft jaw materials can be used such as aluminium or plastics. Complex workpiece shapes can be clamped by modular

devices	application
• clamping shoe • bolt lever	complex workpiece shapes (single-part production)
• vise	workpieces with plan parallel faces
• modulare clamping device	complex workpiece shapes (high flexibility)
• magnetic slide rail	flat workpiece shapes
• vacuum clamping device	flat workpiece shapes
• pouring in • pasting	complex workpiece shapes (quantity production)

31/16148c© IFW

Figure 4.68: Clamping Devices for Surface Grinding Operations

clamping devices providing high flexibility. Magnetic slide rails make it possible to clamp non-magnetic, flat workpieces on the magnetic clamping plate of a surface grinder by loading them with an additional retaining force. Vacuum-clamping devices offer similar possibilities like magnetic clamping plates and are convenient for clamping flat workpiece shapes. For complex shapes in large-scale production, pouring the workpiece in and/or pasting it on a magnetic carrier plate are two methods of fixing.

Figure 4.69 shows two different clamping devices for the external grinding of ceramic, slide-ring sealings. The flexible internal diameter device offers the advantage of center slide-ring sealings in a range of diameter of about 10 mm. By a hydraulic three-point centering of the inner diameter and by adjusting a nut in the axial direction, the slide-ring sealings are simultaneously clamped. The disadvantages of this system are high cost and difficult installation.

Internal diameter clamping devices fix the workpiece by the adjustment of a nut and an adapter ring. For slide-ring sealings with different internal diameters, only the adapter ring must be changed, while the same basic

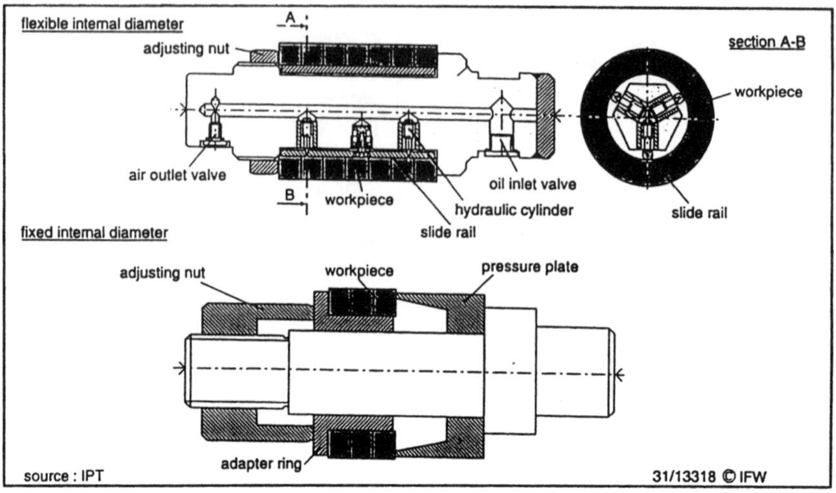

Figure 4.69: Clamping Devices for External Grinding of Slide-ring Sealings

system can be used. The advantages of this clamping device compared to the flexible internal diameter system are easy assembly and low production costs.

Grinding Machine Topology

Surface Grinding Machines

Surface grinders for ceramics can be divided into creep-feed grinding and reciprocating grinding types. Reciprocating grinding works with relatively small depth of cut and high feed speeds (up to 40 m/min) and requires particular specifications of the driving system such as high damping and acceleration. Unlike reciprocating grinding, in creep-feed grinding operations the depth of cut may be more than 1 mm. For this purpose, it is necessary to work with a low and smooth feed velocity of less than 60 mm/min.

Because of the large depth of cut, the contact length is high, the number of cutting edges is large, and thus the normal force acting between wheel and workpiece is relatively high. This means that surface grinders for creep-feed machining must have a high static stiffness against these normal forces.

Grinding spindle drives of surface-grinding machines require steplessly adjustable numbers of r.p.m. up to 10,000 1/min. In dependence on the grinding wheel diameter, cutting speeds of around 200 m/s are achieved so that high speed machining conditions are reached. The spindle power of surface grinders depends on the field of application and may amount up to 80 kW.

Coolant supply of a surface-grinding machine must provide high coolant flux (400 l/min) and high coolant pressure (30 bar) to guarantee effective cooling of the workpiece and cleaning of the wheel. In combination with the small particle size of ceramic chips, high requirements also have to be met by the coolant purification system. Since oil is the common cooling liquid in ceramic machining, a completely sealed working area combined with an efficient suction of oil mist is required, in order to avoid danger of fire or explosion.

Furthermore, surface-grinding machines can be subdivided into three main types of construction: the column, moving column, and transverse construction. In Figure 4.70, a column-type grinder is shown. The design depicts a few characteristics for ceramic machining and provides a high quality machined surface. By compact construction of the machine components as well as by using cast iron material for the machine bed, a high static, dynamic, and good thermal symmetrical behavior is attained. In addition, the machine is modified to a speed-stroke grinder and is provided with special accessories for grinding of ceramics.

A special device for trueing and dressing of diamond grinding wheels allows the application of diamond trueing wheels up to a diameter of 150 mm and a width of 40 mm. Coolant supply and complete pipework are prepared for high pressure and high throughput rates of 200 l/min. The spindle drive with a 25 kW motor is adjustable up to a r.p.m. of 10,000 l/min. To realize a workpiece speed of 60 m/min, which is required for the speed-stroke grinding concept, the following demands have been taken into consideration. By

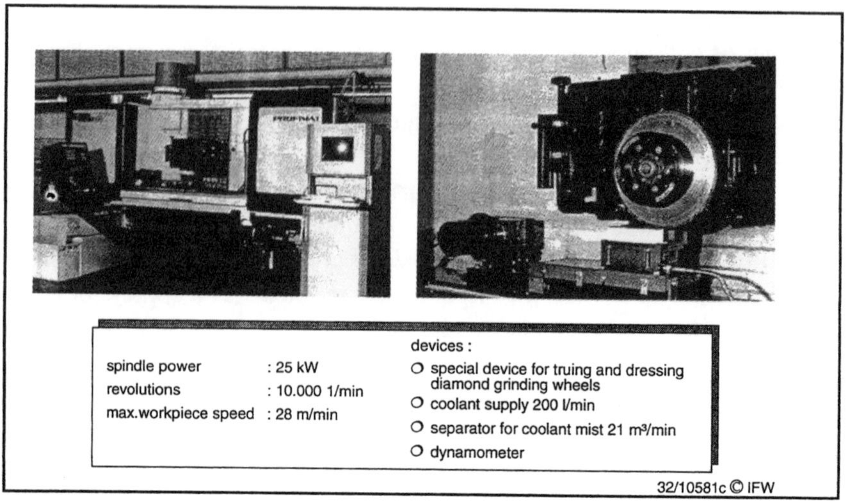

Figure 4.70: Speed-stroke Grinding Machine Blohm Profimat 307

increasing the table speed, the time needed for reversing the table rises; nevertheless, the time of engagement between the wheel and workpiece becomes shorter. As a result of this, the effectiveness of the grinding process may suffer. Only by simultaneously increasing speed and acceleration of the table drive can the ratio of engagement to cycle time be kept constant.

A surface grinder in moving column construction, which is also appropriate for ceramic machining, is shown in Figure 4.71. In addition, the machine is provided with an external cylindrical grinding device. The concept is based on the specifications required for utilizing CBN and diamond grinding wheels. The moving column travels on a central bed and carries the grinding spindle and the spindle motor. These machine components made of polymeric concrete assure high rigidity and good damping.

The external feed drive device controlled by a CNC axis is developed for creep-feed grinding operations. The rotation speed is adjustable up to 25 l/min. A specific trueing device mounted on the top of the grinding wheel allows continuous dressing (CD) during grinding. Intermediate dressing with a table trueing device is also possible. Furthermore, the machine is fitted with

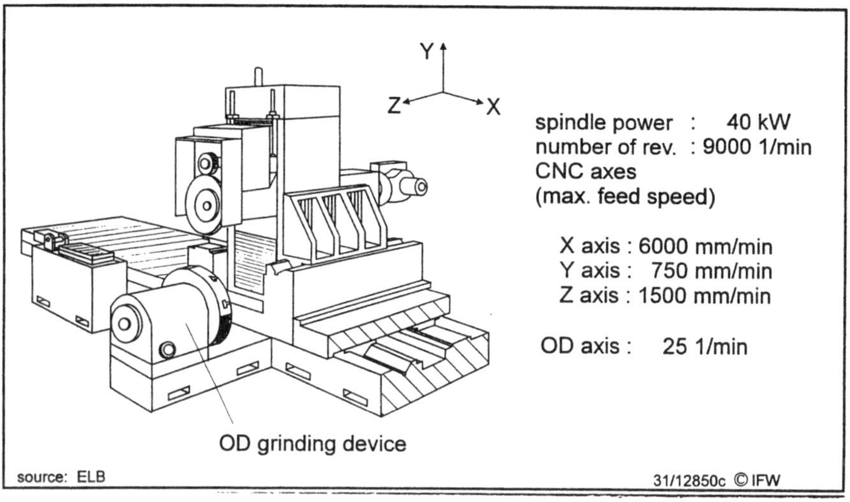

Figure 4.71: High-speed Grinding Machine with External-grinding Device

complete measuring devices to record information on the workpiece, tool, and machine system.[94]

Internal Grinding Machines

Internal grinding can be subdivided into longitudinal grinding and plunge grinding. In internal longitudinal grinding, the wheel and workpiece move in parallel relative to their superficies. The infeed in the radial direction is carried in increments at the reversal points. In internal plunge grinding, the radial infeed is continuous. Modern internal grinders are designed to do both, longitudinal and plunge grinding. The requirements on an internal grinder for machining of ceramics are equal to other ceramic grinding machines. Internal grinding machines, in general, are characterized by low static and dynamic stiffness of the spindle system and the mandrels, which in total account for up to 90% of the overall compliance. These attributes affect workpiece accuracy and the internal grinding process itself.

Since for typical machined surfaces of ceramic components, workpiece tolerances less than 2 μm are required, the positioning accuracy of the machine feed axes must be 0.5 μm. This can be reached by ball-screw feed drives combined with high accuracy measuring devices. The angle fault of machine swivel axes must be less than ± 0.001°.

The grinding spindles of internal grinders for the machining of ceramics should also allow sufficient wheel speeds up to 60-80 m/s. This requires adjustable grinding spindles with a maximum number of revolutions up to 60,000 l/min. (min. spindle power 7kW). These demands can be met by high-frequency spindles with conventional or magnetic bearings. Magnetic bearings also offer the possibility to balance the spindle and to compensate the spindle displacement.

The workpiece in internal grinding should be adjustable in the range of 50 to 2,500 1/min. The feed velocity of machine axes in the radial and axial directions should be in the range of 5 to 8 m/min. This is also adequate for the essential dressing feed velocity of grinding wheels used in the machining of ceramics. Internal grinders are usually built in flat-bed design. An example for a CNC controlled machine type is shown in Figure 4.72. This

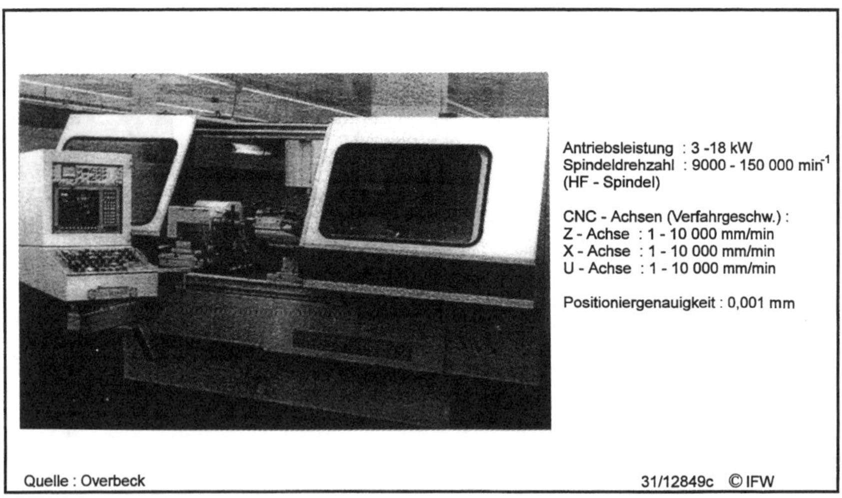

Figure 4.72: Internal Grinding Machine (Flat-bed Construction)

machine can also be used for external and face grinding operations. Its compact cast iron construction exhibits high static and dynamic stiffness, as well as thermal stability. Guideways of feed axes are free from backlash because of the thin, pretensioned, ball-screw design. Prestressed bearings for the guiding of feed axes and ball-screw feed drives with directly adapted servomotors guarantee high accuracy of feed motions.

Another construction principle is shown in Figure 4.73. The characteristic feature of this slant-bed construction is a 70° swivelled cross slide supporting the workpiece clamping device. The advantage of slant-bed constructions are high-guide length (guiding in the area of engagement), easy flux of force into the bench, and the avoidance of pitching moments.[95] These characteristics are advantageous especially in grinding ceramic components because they withstand the high machining forces in the normal direction and guarantee a high workpiece accuracy. The main disadvantages of a slant-bed construction are high construction costs and machine flexibility. These characteristics render this machine type more suitable for large-scale production.

A severe problem with internal grinding of ceramic workpieces is the machining of long bores with small diameters. For these special applications,

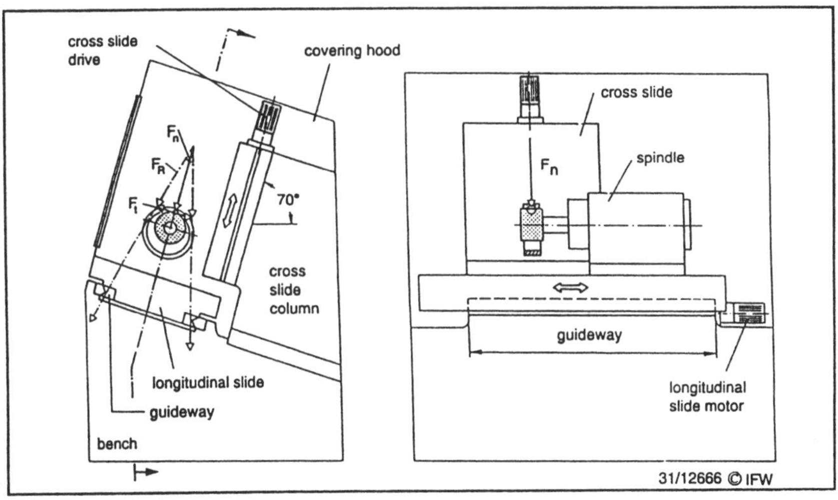

Figure 4.73: Schematic of a Slant-bed Construction

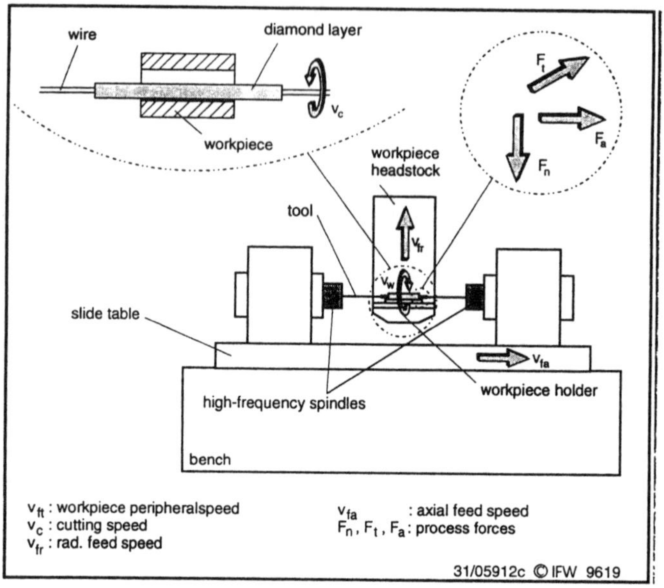

Figure 4.74: Scheme of the ILB-grinding Process

the so-called ILB (Internal Long Boring) grinding process was developed (Figure 4.74). In ILB-grinding, a diamond-coated wire is clamped between two synchron-driven high-frequency grinding spindles and put under axial tension up to 1200 N. To meet this load, a maximum number of revolutions up to 45,000 l/min hybrid bearings are required.

Compared to the conventional machining of small bores using grinding mandrels, the ILB-principle provides higher workpiece accuracy, especially in the longitudinal direction because of reduced tool deflection. Depending on the bore diameter, single-layer electroplated diamond tools (with wire diameters from 1 to 8 mm and a maximum length of about 150 mm) are used for ILB-grinding. The infeed is performed by the workpiece headstock, also carrying the workpiece clamping device. The moving slide table creates the oscillating axial feed.[96]

External Grinding Machines

The external grinding process can, as well as internal grinding, be subdivided into traverse and plunge grinding. The feed in the axial direction of the workpiece is carried out by the slide table or the grinding spindle headstock. The radial infeed is performed by the grinding wheel. Up to now, the industrial market does not offer any external grinding machine that is especially developed for the machining of ceramics. Therefore, the adaptation of single machine components will be the first step to fulfil the demands on the machining of ceramics.

Bench, guideways, and feed drives of an external grinder must provide the same features as shown above for the other machine types. Because of the high process forces in the normal direction, the critical component of an external grinding machine is the workpiece spindle characterized by its low stiffness. It can be raised by using pre-stressed, conical-roller bearings or multiple-area hydrodynamic bearings. Another critical component, because of the high compliance, is the tailstock for clamping the workpiece between center points. Therefore, especially in external grinding operations of ceramic workpieces, a stiff tailstock design is required. For that purpose, some machine manufacturers offer center sleeves with pre-stressed bearings or adjustable tailstocks. Another possible way to raise machine rigidity is to use workpiece spindle heads movable against a fixed tailstock.

Spindle drives for external grinding are equipped with hydrostatic, hydrodynamic, or aerostatic bearings. At present, hydrodynamic bearings are mainly used. To meet the demands on the machining of ceramics, the grinding spindle must provide a high spindle power and a load-independent number of r.p.m.

In Figure 4.75, an example of a high-speed external grinder is shown. High-performance grinding (HSG-technology) leads to specific demands on the system components. Due to high removal rates in the machining of ceramics, a high stiffness of the machine is required. This can be fulfilled by different bed materials such as cast iron, concrete, or even with welded structures. The choice of the appropriate construction material depends on economical factors and therefore differs among machine manufacturers.[97]

For HSG-technology, cutting speeds up to 180 m/s should be reached. This requires spindles with up to 9000 r.p.m., which are at the same time

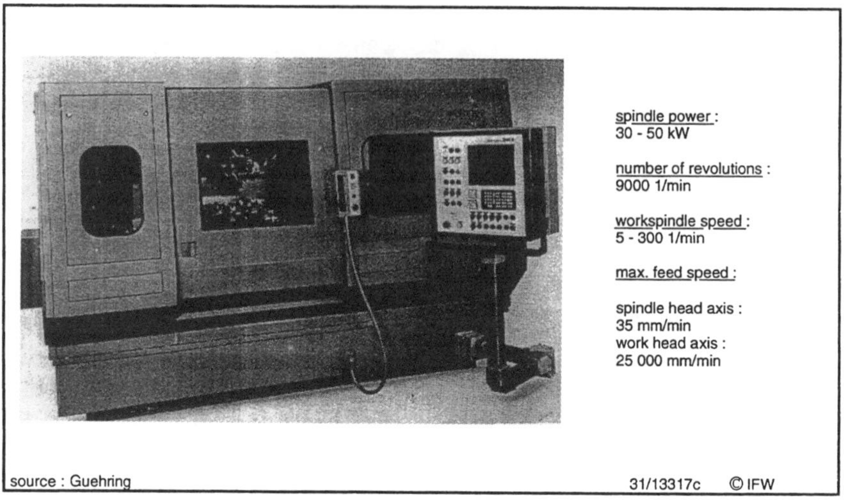

Figure 4.75: HSG External Grinding Machine

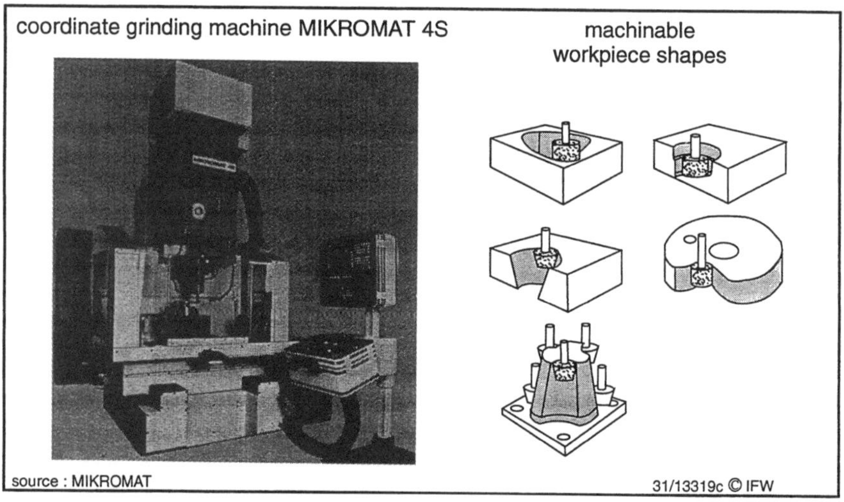

Figure 4.76: Coordinate Grinding Machine

able to carry high process forces. Therefore, ball-bearing spindles are used as well as hydrodynamic and hydrostatic spindles. In specific applications, high-frequency spindles are provided with hybrid bearings or magnetic bearings.

Coordinate Grinding Machines

Typical applications of this universal machine type (Figure. 4.76) are the grinding of bores and contours of different shapes in a single setting. Typical components are cutting tools, jet molds, cams, dies, and molds. By a modern control concept, an efficient production process in single and large-scale production is being assured. Using oil and emulsion as coolant lubricants gives the possibility to reach high grinding ratios. Because of the high accuracy of the machine components, good true-to-size, shape accuracy, and high surface quality can be achieved.

In coordinate grinding machines, steplessly adjustable high-frequency grinding spindles are used with between 5,000 and 12,000 r.p.m. The spindle power reaches from 0.4 up to 1.5 kW. The grinding spindle (D axis) can be pitched to the planetary spindle axis. Modern drives and guideways also assure positioning accuracy less than 2 μm.

Coordinate grinding machines come with up to 6 NC axes. The basic axes are slide table, cross slide, and the spindle head. Further axes are the fine adjustment of the spindle eccentricity, rotation, and oscillation of the planetary spindle. Additionally, the number of NC axes can be raised by a rotary table. The universal behavior of a coordinate grinder and the specific construction of the machine require extremely high demands on the stiffness of each single machine component.

Insert Grinding Machines

Another special purpose in the machining of ceramics is the grinding of ceramic cutting inserts. For these applications, special machine tools have been developed. Figure 4.77 shows a typical cutting insert grinder used for tool manufacturing.

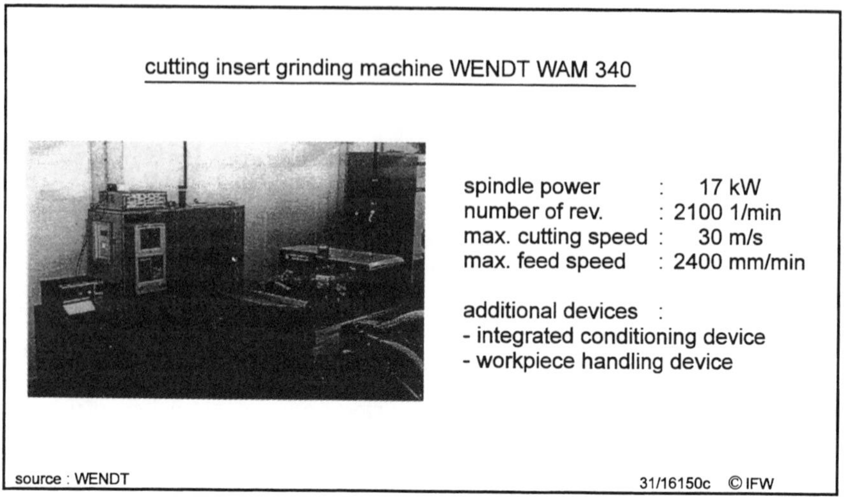

Figure 4.77: Insert Grinding Machine

The heavy grinding unit with a bearing spindle is mounted on a separated beam cantilever and is able to oscillate on the upper machine body in a wear-resistant pivot bearing. Automatic oscillation of the grinding unit is accomplished via an eccentric drive, resulting in smooth and shock-free operation. The workpiece unit is swivel-mounted on the machine bed by an axial bearing for setting the desired angles. The swivelling is infinitely variable by a DC-motor driven tooth belt; thus, the whole system is free of play. Exact positioning is ensured by an electronic positioning control system.

The machine is provided with 4 CNC axes, with a positioning accuracy of 1 µm. The grinding spindle drives with a 17 kW motor and up to 2,100 r.p.m. allows cutting speeds of about 30 m/s. The feed speed of the workpiece slide (max. 2,400 mm/min) is generated by a DC-motor via tooth belt and precision ball screw.[98]

In addition, the machine is equipped with a conditioning device for automatic dressing between grinding cycles, ensuring a high surface quality of the machined workpieces.

Cut-Off Grinding

In addition to grinding and finishing operations, the machining of ceramic components often requires cut-off processes. These processes can be subdivided into peripheral cut-off grinding, ID cut-off grinding, and wire cut-off grinding.[92]

The peripheral cut-off grinding process is subdivided into oscillation grinding with a stationary workpiece and rotational cut-off grinding where the workpiece is moving around its longitudinal axis. The main advantage of rotational cut-off in comparison to other techniques is the very small contact length. Machines for peripheral cut-off grinding operations are just the same as for cutting metals, hard metals, or glass. Only for special applications, an adaptation to the demands of ceramics is required. The machine settings in peripheral cut-off grinding depend on the work material. Cutting speeds from 20 up to 45 m/s are common, as well as cutting depths of about 40 mm with feed speeds up to 1,000 mm /min.[92][99]

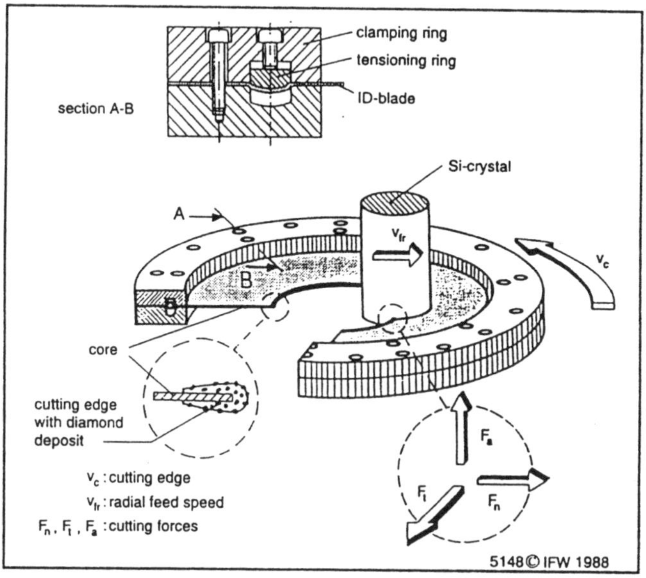

Figure 4.78: Scheme of ID Cut-off Grinding

ID cut-off grinding can also be subdivided into oscillating and rotational grinding. The characteristic feature of an ID cut-off grinder is the unconventional tool made of a high strength chromium-nickel steel ronde; the internal diameter is electroplated with an abrasive layer. In order to assure adequate axial rigidity of the tool, the blade is tensioned like a drum-skin on its outer rim by a special clamping head (Figure 4.78). During cut-off, the rotating tool plunges with its active grinding profile into the workpiece at a certain feed speed.[100]

Common applications for ID cut-off grinding are cutting-off of bar-shaped and tubular workpieces, especially in silicon wafer production. Depending on the workpiece material and its diameter, the feed speed can reach over 100 mm/min, while the cutting speed is in the order of 25 m/s. The main advantage of ID cut-off grinding is a kerf loss less than 0.5 mm.

In wire cut-off grinding, an endless wire tool electroplated with an abrasive layer is utilized. The wire is guided, driven, and pre-stressed in the axial direction by four guide rollers. This relatively new machining concept for ceramic components allows cutting speeds up to 70 m/s. The radial feed speed rises up to 15 mm/min. Wire cut-off grinding makes possible the cutting of complex contours. The disadvantage is the low stiffness of the wire vertical to the cutting direction; this causes wire deflection and reduced workpiece shape accuracy.[92][99]

REFERENCES

1. Tonshoff, H. K., Peters, J., Inasaki, l., Paul, T. : Modelling and Simulation of Grinding Processes, Annals of the CIRP, 41/2 :677-688.(1992)
2. Inasaki. l. :Grinding of Hard and Brittle Materials, Annals of the CIRP, 36/2 : 463-470.(1987)
3. Inasaki, l., Yonetsu, S, :Regenerative Chatter in Grinding, Proc.18th International MTDR Conference, 423-429.(1977)
4. Inasaki, l. :Monitoring and Optimization of internal Grinding Process, Annals of the CIRP, 40/1 :359-362.(1991)
5. Inasaki, l. :Speed-Stroke Grinding of Advanced Ceramics, Annals of the CIRP, 37/1 : 299-302.(1988)

6. Nakagawa, T.; Suzuki, K.; Uematsu, T; Highly Efficient Grinding of Ceramics and Hard Metals on Grinding Center. CIRP Annals 35 (1986) 1, pp.205-210.
7. Hagiuda, Y.; Karikomo, K.; Nakagawa, T.; :Cast Iron Bonded Diamond Cutting-off Wheel Made by Roll-forming. Bulletin Jap. Soc. of Precision Engineering 21 (1987) 1, pp.65-66.
8. Method of Manufacturing Diamond Abrasive Tools. United States Patent. Pat. No. US 3 894 673. July 15, 1975.
9. Chattopadhyay, A.K.; Chollet L, H.E.; Hintermann, H.E.; :On Performance of Brazed Bonded Monolayer Diamond Grinding Wheels. Annals of the CIRP, Vol. 40 (1991) 1, pp.347-350.
10. Grinding or Separating Tool and Method for Producing the Same. United States Patent. Pat. No. US 005 221 293 A. June 22, 1993. Holder: Ernst Winter & Son, Hamburg, Germany
11. Evert, L.; Suresch, S.; Crack Propagation in Ceramics under Cyclic Loads. Joumal of Materials Science, 22(1987), pp. 1173-1192.
12. Bearbeitungspotential durch werkstoffgerechte Prozeβgestaltung (Machining Potential by Tailoring the Process to the Material). Wettbewerbsfaktor Produktionstechnik pp.219-227, Ed.: AWK Aachener Werkzeugmaschinenkolloquium, Publ.: VDI-Verlag, Düsseldorf 1990.
13. Wemhtiner, J.:Angepaβte Endbearbeitungstechnologien für keramische Bauteile, Harte Werkstoffe richtig bearbeiten (Adapted Final Machining Technologies for Ceramic Components, the Right Machining of Hard Materials). Berichtsband zur Fachtagung der VDI-Gesellschaft Produktionstechnik (ADB), Stuttgart (1987) 2, pp.l-20
14. Uhlmann. E.:Tiefschleifen hochfester keramischer Werkstoffe (Creep-Feed Grinding of High-Strength Ceramic Materials). Reihe Produktionstechnik Berlin, Forschungsberichte für die Praxis, Vol. 129, Ed.: G. Spur, Hanser-Verlag Munich (1993).
15. Verlemann,E.:Proze ngesta ltung be im Hochgeschwind igke itsa u nenrundschleifen von lngenieurkeramik (Process Oesign in High-Speed Cylindrical Grinding of Engineering Ceramics). Berichte aus der Produktionstechnik, Vol. 17, Dr.-lng. Thesis, R1~H Aachen (1994).
16. Kiinig, W.; Popp, M.;Wemhiiner, J.:Keramik bearbeiten - wie? (Machining Ceramics How?). Proceedings of 1988 Material Research Symposium of BMFT, Hamm, 12-14.09.1988, pp.787806.
17. Tio,T.H.:Pendelplanschleien nichtoxydischer Keramiken, Prozenverhalten und Randzoneneigenschaften (Reciprocating Grinding of Non-Oxide Ceramics, Process Behavior and Surface Zone Properties). Reihe Produktionstechnik Berlin, Forschungsberichte fijr die Praxis, Vol. 80, Ed. G. Spur, Hanser-Verlag, Munich (1982).

18. Wobker,H.G.: Schleifen keramischer Schneidstoffe (Grinding of Ceramic Cutting Materials). VDI Fortschritt-Ber. Reihe 2, Fertigungstechnik No. 237, Dijsseldorf, VDI-Verlag (1991).
19. Spur,G Keramikbearbeitung- Schleifen, Honen, Läppen, Abtragen. (Ceramics Machining - Grinding, Honing, Lapping, Material Removal). Publ.: Hanser-Verlag, Munich (1989).
20. Böschke, K.: Der Kühlschmierstoff als Werkzeug. wt 83 (1993) 3, S. 62-64.
21. Brinksmeier. E.Aufgaben der Kühlschmierstoffe bei spanender Bearbeitung. In: Tagungsband zum Seminar "Kühlschmierstoffe in der spanenden Fertigung" des Dtsch. Industrieforums f.Techn. (DIF), Frankfurt, 21.-22. Oktober 1991.
22. Brinksmeier,E.; Schneider,C.: BausteinefürumweltverträglicheFeinbearbeitungsprozesse. In: Tagungsband zum 7. Internationalen Braunschweiger Feinbearbeitungskolloquium "Hohe Prozeβsicherheit, hohe Leistung, hohe Präzision", Braunschweig, 2.-4. November 1993, S. 19.01-19.32.
23. Brücher, Th.: Kühlschmierung beim Schleifen keramischer Werkstoffe. TU Berlin, Diss., Berlin, 1996
24. Keyser,W.; Kiihlschmierung beim Schleifen. IDR 4 (1970) 3, S. 158-161.
25. König, W; Fertigungsverfahren, Band 2. Düsseldorf: VDI-Verlag, 1980
26. Lingmann, H.: Gebrauch von Kühlschmierstoffen. wt-Z. ind. Fertig. 75 (1985) 9, S. 571-574.
27. VDI-Richtlinie 3396: Kühlschmierstoffe für spanende Fertigungsverfahren. Düsseldorf: VDI-Verlag, Oktober 1983.
28. Zwingmann, G.: Kühlschmierstoffe für die spanende Metallbearbeitung, Teil 1 und 2. Werkstatt und Betrieb 112 (1979) 6 und 7, S. 409414 und S. 483-487
29. DIN 51 385: Kühlschmierstoff- Begriffe. Berlin: Beuth, November 1981.
30. Die Reinigung der Kühlschmiermittel für den Schleifprozeβ, Teil 1 und 2. Fachberichte für Oberflächentechnik Juli/August und November/Dezember 1968.
31. Knobloch, H.; Kühlschmierstoffpflege in der Praxis. Reihe Kontakt und Studium, Bd. 29, Hrsg.: Dr.-Ing. W. J. Bartz. Grafenau: Lexika, 1979.
32. Briicher, Th.: Kühlschmierung beim Keramikschleifen. In: Vortragsband zum Symposium "Schleifen von Hochleistungskeramik", Kaiserslautern, Oktober 1994. .
33. Liebe, I.: Auswahl und Konditionierung von Werkzeugen für das Auβenrund-Profilschleifen technischer Keramiken. TU Berlin, Diss., Berlin, 1996.
34. Spur, G.; Liebe, I.: External Cylindric Grinding of Complex Ceramical Piece Parts. In: Proceedings of the ist International Machining & Grinding Conference, Detroit, USA, September 1995.

35. Uhlmann,E.: Tiefschleifen hochfester keramischer Werkstoffe. TU Berlin, Diss., Berlin, 1994.
36. Spur, G.; Uhlmann,E.; Brücher, Th.: Werkstoffspezifische Schleiftechnologie - Schlüssel für erhöhte Prozeβfähigkeit in der Keramikbearbeitung. In: Jahrbuch "Schleifen, Honen, Läppen und Polieren", 57. Ausgabe, Hrsg.: E. Saljé, E. Westkämper. Essen: Vulkan, 1993,S.387-419
37. Tio, T. H.; Brücher, Th.: Kühlschmierung bei der Schleifbearbeitung keramischer Werkstoffe. Vortrag anläßlich des Arbeitskreises "Keramikbearbeitung". Produktionstechnisches Zentrum Berlin, Oktober 1988.
38. Brücher, Th.: Kühlschmierung - Ein wesentlicher Faktor für wirtschaftliche Schleifbearbeitung. In: Tagungsband zum Seminar "Wirtschaftliche Schleifverfahren" des Dtsch. Industrieforums f. Techn. (DIF), Ratingen, 23.-24. März 1993.
39. Roth, P.; Wobker, H.-G.: Schleifbearbeitung keramischer Werkstoffe. Sprechsaal 124 (1991) 4, S. 254-262.
40. Wobker, H.-G. Schleifen keramischer Schneidstoffe. Fortschr.-Ber. VDI, Reihe 2, Nr. 237. Diisseldorf: VDI-Verlag, 1992.
41. Spur, G.; Brücher, Th.: Optimization in the Cutting Fluis System for Grinding of Advanced Ceramics. In: Proceedings of the ist International Machining & Grinding Conference, Detroit, USA, September 1995.
42. Heuer, W.; AuBenrundschleifen mit kleinen keramisch gebundenen CBN-Schleifscheiben, Fortschr.-Ber. VDI, Reihe 2, Nr. 270. Diisseldorf: VDI-Verlag, 1992.
43. Spur, G.; Brücher, Th.: Cutting Fluids in Grinding of Advanced Ceramics. In: Technical Papers of the 23rd North American Manufactoring Research Institution of SME, Houghton, Michigan, May 1995.
44. Spur, G.; Uhlmann, E Tiefschleifen - Fine Verfahrensvariante zur wirtschaftlichen Endbearbeitung keramischer Werkstoffe. In: Jahrbuch "Schleifen, Honen, Läppen und Polieren", 56. Ausgabe, Hrsg.: E. Saljé. Essen: Vulkan, 1990, S. 104-124.
45. Byme, G.; Scholta, E.: Environmental Clean Machining Processes - A Strategic Approach. Annals of the CIRP, Vol. 42/1/1993, S. 471- 474.
46. Byme, G.: Environmentally Clean Machining Processes. In: Proceedings of the 6th International Machine Tool Engineers Conference, Osaka, Japan, 31. Oktober-1. November 1994, S. 181-193.
47. König, W. et al; Kühlschmierstoff-Eine ökologische Herausforderung an die Fertigungstechnik. Wettbewerbsfaktor Produktionstechnik, Sonderausgabe fiir AWK. Düsseldorf: VDI-Verlag, 1993.
48. Tönshoff, H. K. Brunner, G.Minimalmengen-Kühlschmierung bei Schleifprozessen. Maschinenmarkt 101(1995)45, S. 30-37.

49. Lingmann, H.: Hygiene und Arbeitsschutz helm Einsatz von Kühlschmierstoffen. In: Tagungsband zum Seminar "Kühlschmierstoffe in der spanenden Fertigung" des Dtsch. Industrieforums f. Techn. (DIF), Frankfurt, 21.-22. Oktober 1991.
50. Geretzki, P.: Erkrankungen durch Kühlschmierstoffe in der metallverarbeitenden Industrie. Dermatosen in Beruf und Umwelt 31 (1983) 1, S. 10-14.
51. Studt, P; Kerner, W.; Win, T.: Untersuchung der mikrobiologischen Schädigung wassergemischter Kühlschmierstoffe mit dem Ziel der Verbesserung der Arbeitshygiene, der Minderung der Geruchsbelastung und der Menge zu entsor-gender Emulsionen. Forschungsbericht 170, Bundesanstalt für Material-forschung und -priifung, Berlin. Bremerhaven: Wirtschaftsverlag. 1990.
52. Howes, T.D.; Tönshoff, H.K.; Heuer, W.: Environmental Aspects of Grinding Fluids. Annals of the CIRP, Vol. 40/2/1991, S. 623-630.
53. König, W; Rummenhöller S; Ecological Manufacturing. Production Engineering, Vol. I/2, München, Wien: Hanser, 1994, S. 1-4.
54. Tönshoff, H. K.; Kroos, F.; Spinting, W.; Brandt, D.: Reducing Use of Coolants in Cutting Processes. In: Production Engineering, Vol. I/2. München, Wien: Hanser, 1994,S.5-8.
55. Dopatka, J.; Obst, M.; Siegfried, F. Vermeidung von Abfällen durch abfallarme Produktionsverfahren -Kühlschmierstoffe -Zerspanende Fertigung in Groβb-etrieben. Studie im Auftrag des Ministeriums für Umwelt Baden-Württemberg, Stuttgart, September 1992.
56. Dopatka, J.; Obst, M.; Siegfried, F.: Vermeidung von Abfällen durch abfallarme Produktionsverfahren -Kiihlschmierstoffe -Zerspanende Fertigung in mittleren und Kleinbetrieben. Studie im Auftrag des Ministeriums für Umwelt Baden- Württemberg, Stuttgart, Mai 1993.
57. Grohmann, J.: Umweltschutz in der Zerspanung - Probleme und Verbesserung der ökologischen Situation im Betrieb. wt 83 (1993)11/12, S. 60-63.
58. Heisel, U.; Lutz, M.: Investigation of Cooling and Lubricating Liquids. In: Production Engineering, Vol. I/1. München, Wien: Hanser, 1993, S.23-26.
59. Heisel, U.; Lutz, M.: Probleme der umwelt- und humanverträglichen Fertigung am Beispiel der Kühlschmierstoffe, Teil 1 und 2. dima 47 (1993) 8/9 und 10, S. 81-83 und S. 35-40.
60. Heisel, U.; Lutz, M.: Spath, D.; Wassmer, R; Walter, U. Application of Minimum Quantity Cooling Lubrication Lutz, M.; Technology in Cutting Processes. In: Production Engineering, Vol. II/1. Miinchen, Wien: Hanser, 1994, S. 49-54
61. Tönshoff, H. K. Wobker, H.-G.; Brunner, G.; Kroos, F.: Limitations for Lubrication-Free Grinding of Hardened Steel. In: Advancement in Intelligent Production, The Japan Society of Precision Engineering. Amsterdam: Elsevier Science, 1994, S. 503-509.

62. Kienzle, 0. 1956. Die Grundpfeiler der Fertigungstechnik. VDI-Z 98 (1956) 23, pp. 1389-1395.
63. Tojnshoff, H.K. 1994. Survey of Manufacturing Processes. in Dubbel: Handbook of Mechanical Engineering, Springer 1994, pp. kl-k2.
64. Tonshoff, H.K., F. Klocke, R. Egger. 1996. Environmental Aspects of Electro-physical and Electochemical Processes. Annals of the CIRP 45 (1996)
65. Howes, T.D., H.K. Ttinshoff, W. Heuer. 1991. Environmental Aspects of Grinding Fluids. Annals of the CIRP 40 (1991) 2, pp. 623-630.
66. Bennet, E.O. 1983. Dermatitis in the metalworking industry: Factors that influence the incidence of cutting fluid dermatitis within the plant. ASLE Publication SP-11.
67. Gadian, T. 1983 Skin problems in chemical industry. Chemistry and Industry, 17(January): 52-54.
68. Emmet, E. 1984. The skin and occupational diseases. Archives of Environmental Health 39 (3): 144-149.
69. Zugerman, C. 1986. Cutting fluids: Their use and effects on the skin. Occupational Medicine: State of the Art Reviews 1 (2): 245-258.
70. Warne, T.M. and C.A. Halder. 1986. Toxity of lubrication oils. Lubrication Engineering 42 (2): 97-203.
71. Jarvholm, B., B. Lavenius and G. Sällsten. 1986. Cancer morbidity in workers exposed to cutting fluids containing nitrites and amines. British Journal of industrial Medicine 43:563-565.
72. Remmerswaal, J. L. 1990. Round Table: Report on the Round Table held in Berlin at the 40th General Assembly 1990. Annals of the CIRP 39 (2): 589-593.
73. Alting, L. 1990. Lifecycle design of products. Institute of Manufacturing Engineering, University of Denmark.
74. Eckhardt, F. 1983. Kijhlschmierstoffe fijr die spanende Metallbearbeitung Tell 1-11, tz fiir metallbearbeitung 77 (no. 1-11).
75. Vahle, H. 1984. Recycling fluids. Production Engineering 10. 64-68.
76. VDI-Richtlinie 3397, Blatt 1, Pflege von Kiihlschmierstoffen.
77. Knobloch, H. 1979. Optimale Kiih(schmierstoffpflege in der Praxis. Kontakt und Studium Band 29, Expert Verlag: Grafenau Wijrtenberg
78. Evans, C. 1990. Treatment of used cutting fluids and swarf. Tribology international 80 (no.3): 248-250.
79. Triesch, J. 1990. Umwelffreundliche Filtiersysteme. wt Werks-tatttechnik 80 (no. 3): 248-250.
80. Joseph, J. 1987. Coolant Filtration. E. Syracuse N.Y.: Joseph Marketing

81. Bennet, J. and J. Creps. 1990. An overview of filtration technology. In waste minimization and wastewater treatment of metalworking fluids, edited by R. Kelly, 63-79. Alexandria, VA: Independent Lubricant Manufacturers Association
82. McCoy, J.S. 1978. A practical approach to central system control. Lubrication Engineering 34 (no. 4): 113-118.
83. Deutsche Norm DIN 52292 Teil 1: Prüfung von Glas und Glaskeramik, Bestimmung der Biegefestigkeit, Doppelringversuch an plattenftirmigenProbenmit kleinen Prüfflächen, Entwurf, Beuth, Berlin 1983
84. Deutsche Norm DIN 51110 Tell 1. Prüfung von Hochleistungswerkstoffen, 4-Punkt-Biegeversuch bei Raumtemperatur (1990), S. 1-19
85. Weibull, W.: A statistical distribution function of wide applicability, J. Appl. Mech. 18(3) 1953, p 293-297
86. Uhlmann, E.G.: Tiefschleifen hochfester keramischer Werkstoffe Carl Hanser Verlag, München, Wien, 1994
87. Wobker, H. G.: Schleifen keramischer SchneidstoffeVDI-Verlag Dijsseldorf 1992
88. Tönshoff, H.K.: Grundlagen der Werkzeugmaschinen Springer Verlag, Berlin, Heidelberg, 1995
89. Jürgenharke, B.: Untersuchungen zur Technologie und Dynamik des Innenrun-dschleifens Dr.-lng. Dissertation, Universität Hannover, 1978.
90. Tellbüscher, E.: Konstruktion von Diimpfern und deren Einsatz an Rundschleifmaschinen Dr.-lng. Dissertation, Universität Hannover, 1985.
91. Week, M.: Werkzeugmaschinen, Band 1 VDI-Verlag, Düsseldorf, 1991.
92. Spur, G.: Keramikbearbeitung Carl Hanser Verlag, München, Wien, 1989.
93. N.N.: PMH-Präzisionsspannfutter Firmenschrift der Präzisions Mechanik Heyn, Pforzheim, 1992.
94. Warnecke, G., U. Rosenberger: Hochleistungs- Schleifrentrum für praxisnahe Forschung VDI-Z 136 (1994) 1/2, S. 39 ff.
95. Althaus, P.G.: Leistungssteigerung beim innenschleifen durch kubisches Borni-trid (CBN) und neue Maschinenkonzeptionen Dr.-lng. Dissertation, Universität Hannover, 1982.
96. Tönshoff, H.K., H.-G. Wobker, W. Buchholz:Innenrundschleifen langer Bohrun-gen mit galvanisch belegten Diamantwerkzeugen Industrie Diamenten Rund-schau 27 (1993) Nr. 3, S. 163-168.
97. N.N.: HSG-Tecltnologie Firmenschrift der Guehring Automation GmbH, Stetten a.k.M., 1991.
98. N.N.: WENDT-ReportFirmenschrift der Fa. WENDT, Meerbusch-Strümp, 1991.
99. König, W., E. Verlemann, A. Wagemann: Trennschleifen von Hochleistungskeramik Industrie Diamanten Rundschau 26 (1992) 3, S. 165-171.

100. V. Schmieden, W..Werkzeugbeanspruchung beim Innendurchmesser-Trennschleifen Dr.-lng. Dissertatian, Universität Hannover, 1990.

5

Honing and Superfinishing

Gunther Spur

1.0 TYPOLOGY OF THE HONING PROCESS

Honing is a technology for mechanical finish-machining of brittle-hard ceramic and metallic parts. According to N.N. Honen,[1] honing is defined as cutting with geometrically undefined cutting edges, during which multi-edge tools conduct a cutting motion consisting of two components. One component of the cutting motion is the stroke motion, so that the machined surface shows defined crossing traces. Shape elements machinable with honing are determined by the technological variants: plane, round, screw, hob, profile, and form honing. The shape elements are machinable both inside and outside. An additional subdivision of the technology into long-stroke honing and superfinish is done in relation to the cutting motion. While during superfinish, also described as short-stroke honing, a rotary motion is overlapped by a short-stroke oscillating motion, the cutting motion during long-stroke honing consists of a rotary motion and a long-stroke, there-and-back motion.

Honed surfaces shows good tribological properties. High profile bearing length ratios and low roughness in connection to the defined crossing canals, which can absorb lubricants, improve the running-in behavior and increase the life of tribologically stressed parts. Apart from the high surface quality, the technology of long-stroke honing makes it possible to produce internal cylinder surfaces with high dimensional and form accuracy as well as surface quality. Fields of application include the finish machining of slide bearings, cylinder liners, and piston cylinders. For short-stroke honing, fields of application open up where highly stressable parts with high surface quality and special form accuracy are required. Here, friction and sliding surfaces, parts of antifriction bearings and slide bearing pivots, as well as sliders and seats of packing rings, and collars are machined in detail.

During production of the mentioned parts from ceramic materials, the suitable parameters that were determined for honing of metallic materials are not transferable due to their principally different, ceramic-specific material properties, and cutting mechanisms. The part quality required for tribological stress, but also the gaining of high tool lives as well as of short machining times for the minimization of costs, calls for an adjustment of input quantities on the machined ceramic materials, which are comprised for long-stroke honing in Figure 5.1.

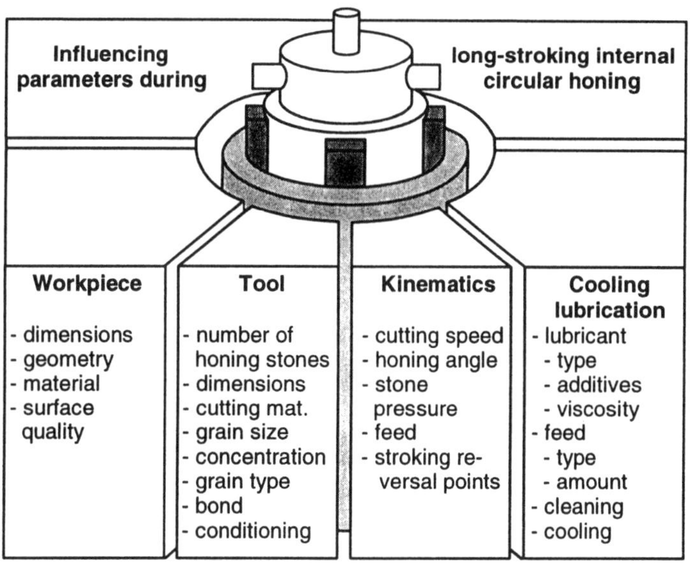

Figure 5.1: Process Parameters of Long-stroke Honing[2]

Kinematics of Long-Stroke Internal Circular Honing

The cutting motion during long-stroke internal circular honing is created by a stroke and rotary motion of the tool (Figure 5.2).

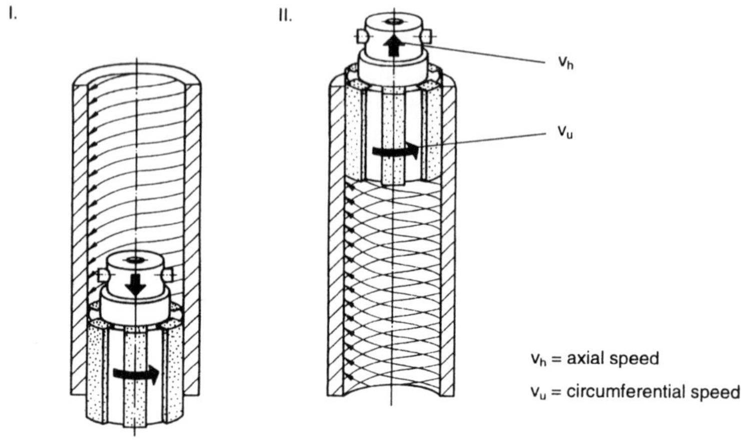

Figure 5.2: Kinematics of Long-stroke Honing[2]

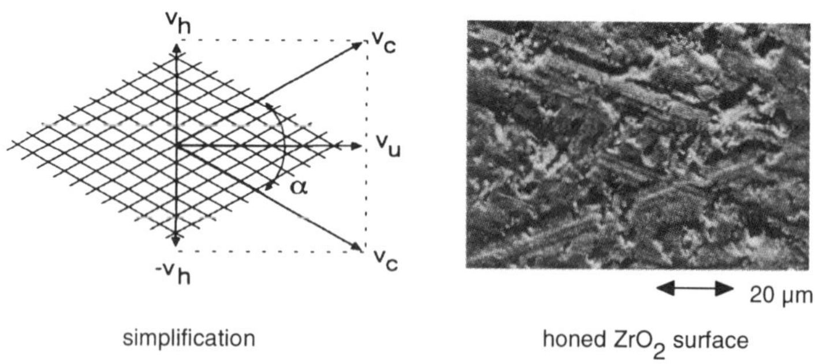

Figure 5.3: Cross-traces during Long-stroke Honing

The superposition of stroke and rotary motion leads to the forma tion of cross-traces, which are characteristic for honing and intersect under the honing angle α (Figure 5.3).

The cutting speed v_c and the honing or intersecting angle α are calculated based on the circumferential speed v_u and the axial speed v_h with:

$$v_c = \sqrt{v_h^2 + v_u^2} \quad \text{and} \tag{5.1}$$

$$\alpha = 2 \cdot \arctan(\frac{v_h}{v_u}). \tag{5.2}$$

The circumferential speed v_u is calculated based on the workpiece diameter d_{wst} and the rotational speed of the workpiece n with:

$$v_u = \pi \cdot d_{wst} \cdot n. \tag{5.3}$$

The chosen honing angle influences the ability of the surface to absorb lubricants and hence, at tribological stress, direct the part conduct. Extensive empirical tests in engine construction have led to a definition of suitable honing angles for metallic functional parts. For ceramic materials, different cutting mechanisms are active due to their different material properties and thus, the created surface topography cannot be compared to the one of metallic materials. For this reason, suitable honing angles, which were determined for metallic surfaces, are not necessarily applicable to ceramic surfaces. Publications on the determination of a suitable honing angle in relation to the tribological conduct of ceramic surfaces are not yet known.

Stroking Reversal Points

The position of stroking reversal points, also described as honing stone overflow, has a significant influence on the cylindricity of the honed surface. In order to achieve cylindrical shapes of bore, honing stones should come out of the workpiece to one third on the upper and lower stroking reversal point during machining of both metallic and ceramic materials.[3]

Kinematics of Superfinish

During short-stroke honing, a close-grained honing stone with an encircling angle of 60-70° is pressed on the circulating workpiece, and thus made to oscillate parallel to the machining surface (Figure 5.4A). The stroke length of the honing stone hereby is 1-6 mm at 1000 to 3000 oscillations/min, which are produced by compressed air or electro-mechanical means.[4] The

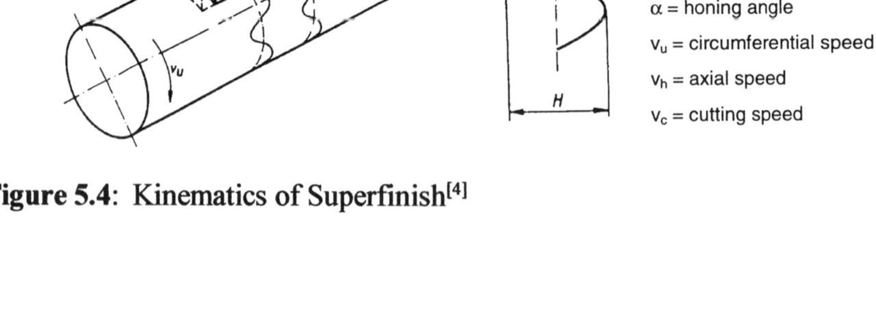

Figure 5.4: Kinematics of Superfinish[4]

stone pressure is chosen relatively low between 10 and 120 N/cm^2 and is usually created by compressed air. This way, the material removal is carried out gently without an alteration of the surface structure by contact forces and temperature.

Sinusoidal machining traces result from the rotation of the workpiece and the oscillating motion of the honing stone, which cross at the change of direction during the stroke. Figure 5.4B shows the development of a path curve of a cutting grain relative to the workpiece. Hereby, the resulting cutting speed v_c is created by the oscillation or axial speed v_h and the circumferential speed v_u:

$$v_c = \sqrt{v_u^2 + v_h^2}\ . \tag{5.4}$$

Investigations proved that a higher economy as well as better results with regard to roundness and surface quality can be reached by application of higher circumferential speeds of the workpiece, which are today at 10 to 600 m/min. The honing angle α is calculated as follows:

$$\tan\alpha = \frac{v_h}{v_u}\ . \tag{5.5}$$

It becomes smaller with increasing circumferential speed and decreasing oscillation speed and thus, the sinusoidal machining trace becomes more flat.

Mandrel Honing

The machining of ceramic parts is, in addition to conventional honing, also possible with mandrel honing. During mandrel honing, a honing tool adjusted to its final dimensions is used. Therefore, the feed system of honing stones can be dismissed. The cutting process is usually performed with one to a maximum of three double strokes. While the stroke speed is adjusted lower compared to conventional honing by the factor 5 to 10 to v_h = 2-6 m/min, the values for circumferential speed are with v_u = 30-50 m/min in the range of conventional honing.[5][6] The higher form stability of the firmly adjusted tool is the advantage of mandrel honing. Hence, compared to conventional long-stroke honing, it is now possible to achieve a higher dimensional and form accuracy. Small drillings up to a diameter of 1.5 mm, as well as interrupted drillings and those which lie on top to each other, can be machined with high accuracy. An adjustment of honing stones for the purpose of wear compensation is also possible; it is carried out manually or automatically by a feed-back regulation. In order to prevent an overstress of the honing stones, only a few micrometers of material can be removed with one tool during mandrel honing. Consequently, it is necessary for the achievement of the required material removal to design the mandrel honing process as a multistage process, or to have a conventional honing process precede the one of mandrel honing.

2.0 HONING AND SUPERFINISHING TOOLS

Interior honing tools fill the bore and lean equally over the honing stones at the circumference of the bore (Figure 5.5). The feed of honing stones is done by a double cone, which converts the axial feed motion of the pressure bar into a radial motion of the honing stone support. These honing stone supports, on to which the honing stones are soldered, adhered, or clamped, are situated in the radial slots of the tool base. Multiple-stone honing tools are available in series for the diameter range of 5-1000 mm. The potential machining diameter of a honing tool is determined by the cone angle as well as the height of honing stone support and the honing stone itself. It is only a few millimeters. A conditioned honing tool has a very limited operating range, because the outer radius of honing stones is coordinated with the radius of the honed bore.

Figure 5.6 shows the schematic description of the feed system and the taking-up of the honing stone onto the oscillating device during short-stroke honing.

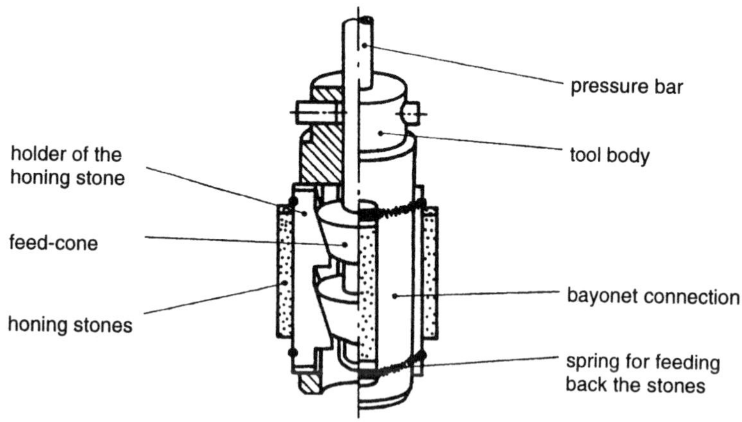

Figure 5.5: Structure of a Long-stroke Internal Circular Honing Tool[2]

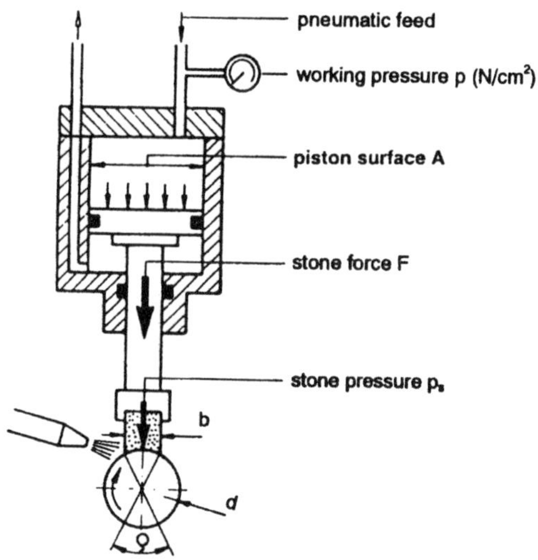

Figure 5.6: Tool and Feed System for Short-stroke Honing[4]

Structure of Honing Stones

Diamond and CBN-honing stones are used for the honing of ceramic materials. The characterization of honing stones with regard to grain and binding specification is done in the same way as is the characterization of diamond and CBN grinding wheels. The grain sizes of diamond grains are not designated in Mesh, in the number of meshes per inch screen length, but in thin mesh size of test screens in micrometers. For the designation of the quantity of diamond grains in the honing stone, the diamond quantity is given in percentage related to 4.4 Karat/cm^3. Hence, a honing stone of 100% concentration contains 4.4 Karat/cm^3, respectively 0.8 g/cm^3 diamond. In addition to bronze-bonded honing stones, ceramic and epoxy resin-bonded honing stones are also applied.

Dimensions of Honing Stones

An improvement in form accuracy of the workpiece surface is usually another objective of honing. Hereby, the length and width of the used honing stones play a significant role. Whereas long honing stones reduce the cylindricity deviation, wide honing stones can diminish the roundness deviation.[7] Since the chip transportation is aggravated with increasing width of honing stones, additional scavenging grooves are also created in the lining during application of wide honing stones. At a tool overflow of 1/3 of the length of honing stones, the maximum length of honing stones is theoretically restricted to 4/3 of the height of the honed bore. Under this condition, the total length of honing stones engages during one stroke. Practically however, the length of honing stones is often in a range of ½ to 1/5 of the workpiece height.

Conditioning of Honing Stones

The profiling of diamond honing tools is normally done by grinding with bakelite-bonded silicon carbide grinding wheels. The following grinding parameters have proved to be favorable:[3]

- cutting speed: 15 m/s,
- grinding wheel: SC 80, hardness K,
- wheel dimensions: 500 x 50 mm,
- rotational speed of workpiece: ca. 20 min-1 (synchronism), and
- feed speed: ca. 720 mm/min.

The grinding of honing stones leads to a strong grain outburst and thus, to a minor grain overflow. For this reason, a sharpening of honing stones is generally necessary prior to application. The honing stone bond is hereby moved back by manual lapping or taken off with SiC-sharpening stones.

Conditions for Cooling Lubrication

Conditions for cooling lubrication influence to a significant extent the work result reached with honing. The cooling lubricant must be matched with the honed material as much as the honing stone specification. The task of the cooling lubricant is, on the one hand, to clean the workpiece and tool by rinsing off chips and burst out diamond grains, as well as to eliminate the rising heat. On the other hand, the lubrication prevents the heat development between the two drift surfaces. Honing oils, water-mixed cooling lubricants, or emulsions can be used as cooling lubricants for ceramic machining. However, the effects of additives used during honing of metals cannot be applied to ceramic machining. This is especially true for additives such as chlorine, phosphorus, and sulfur,[8] whose activation is only reached by a chemical reaction with the metallic material.

The cleaning of the cooling lubricant is very significant for the achievement of constant work results. By means of a filter device, it must be controlled such that no abrasive ceramic chips and burst out diamond grains are flushed back into the machining surface to cause scratches. At the same time, the units of the cooling lubricant circulation themselves are to be protected against abrasive chips by filters.[9] A temperature regulation of the cooling lubricant is inevitable because larger quantities of heat are set free during honing of ceramics, particularly during realization of high material removal rates. This way, the viscosity of the cooling lubricant can be kept constant. In addition, the work result of the honing process, which depends on this, can be reached independent of the machining time.

3.0 HONING AND SUPERFINISHING MACHINES

The great variety of variants of honed units and parts, which is caused by the market, led to various construction forms of honing machines. Measuring and controlling tasks in various honing machines, too, became much more complex under the aspects of automation and process security during production of highly accurate bores. Thus, feed of honing stones, measuring and verifying systems, stroking and rotating drive, positioning axes, as well as the tool changer and the tool magazines can be integrated for instance in numerical controls (NC/CNC).

The diverse construction forms can be divided according to the machined workpiece dimensions such as length and diameter of bore, or to the number and position of the main spindles. Machines with a vertical hone spindle are favored in most cases because here, the gravity of tool and driving rod is without influence, and flexible honing tools and cardanic workpiece uptaking devices can be used.[3] Moreover, the feed of honing oil and the chip removal are facilitated. Machines with a horizontal spindle are also used, particularly for very small and very big workpieces.

Horizontal manual and lifting beam honing machines (Figure 5.7 a and b) are considered to be simple honing machines, which are especially applied to the production of small or medium series[3] in workshops. They are characterized by their simple structure and wide opportunities for application of problem-free operation. The manual feed of tools is done by a pedal or a feed nut.

Vertical production honing machines (Figure 5.7 c and d) are offered in C-structure or portal-construction, and are generally considered to be typical construction forms of long-stroke honing machines. The degrees of freedom between workpiece and tool, together with the forces active during honing, determine the selection of the more simple and better available C-structure, or of the more stiff portal construction. In the case of the C-structure, the spindle is taken up in a protruding upright, whereas in portal-constructions, it is taken up in a case which is supported on both its sides. The machine table is designed for both construction forms as either fixed, round, or NC-table. Particularly for multiple spindle honing machines, manual or automatic loading devices and transfer systems serve the purpose of rational feed, while industrial robots are also used more frequently.

Vertical pipe honing machines (Figure 5.7 e) are used for lifting heights up to 4500 mm.[3] In machines designed in flask mold with spindle support, the workpiece looms into an engine pit underneath the honing machine. This underfloor construction keeps the overall height within normal limits. Slide times can be shortened by application of slide tables with two uptaking devices.

Heavy honing machines are also installed over an engine pit. The machines, which are often designed in portal construction, serve to machine large, bulky workpieces such as cylinder liners of Diesel engines and press cylinders. The ability for automation becomes particularly important for special honing machines: even more so for multiple spindle machines, where the safe feed of workpieces is especially significant in addition to the automatic functional process with measuring devices.

Horizontal production honing machines serve, for instance, to machine crankshaft bearing bores and cooling compressors. Smaller workpieces such as planet toothed wheels are many times honed together in a package.[3] The horizontal long-pipe honing machine (Figure 5.7 f) represents another construction form, which is built today for honing lengths of up to 12000 mm and machining diameters of up to 1000 mm. Long steel pipes, gun pipes, and chills are produced with these machines. In order to develop the machining independent of gravity, the workpiece moves opposite to the tool. Moreover, a purposeful feed of the cooling lubricant is necessary. High driving powers of ca. 40 kW each for stroking and rotary motion are characteristic for these machines.

With regard to short-stroke honing machines, one distinguishes between raising apparatus, which are used on carrier machines, mostly center barring machines, and partially or fully automatic short-stroke honing machines.[3] The machined dimensions of workpieces depend firstly on the carrier machine for raising apparatus; the opportunities for application range from

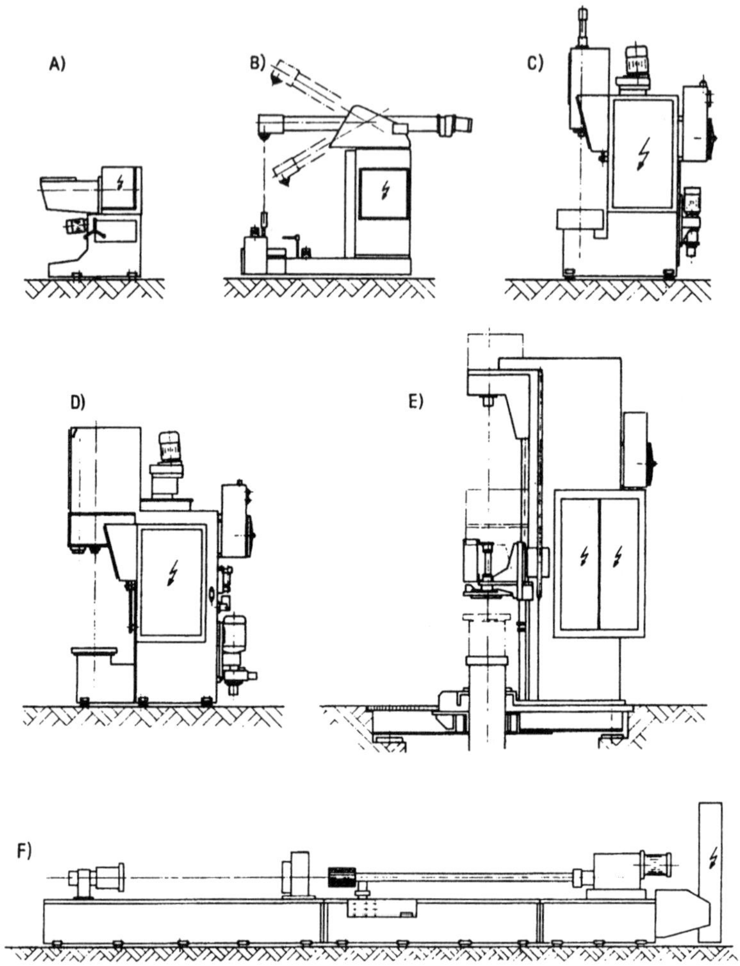

Figure 5.7: Designs of Long-stroke Honing Machines:[3] a) horizontal manual honing machine; b) lifting beam honing machine; c) production honing machine (light construction); d) production honing machine (heavy construction); e) vertical pipe honing machine; f) horizontal long-pipe honing machine

the machining of small bearings with diameters of 5-10 mm to high-finished and absolutely channel-free rollers of more than 1000 mm in diameter for the production of sheets, films, and qualitative bands.

Automatic short-stroke honing machines in series production are mostly equipped with automatic feed devices. Hereby, one distinguishes between the particularly economic continuous machining and plunge-cut machining. During continuous machining, workpieces are brought into a rotary and feed motion between tilted rollers rotating in the same direction, and transported under the oscillating honing stones.[10] Due to the gradation of grain size of the cutting material bound in the honing stones, significant material removal as well as a high surface quality can be reached. The use of powerful oscillating heads allows continuous speeds of up to 6 m/min; the application includes needle bearings, axes, axles, rollers, and piston pins of all kinds. Workpieces with different shape elements on the circumference, as for example, piston rods of telescopic legs, lead to machines with automatic removal of honing stones. Hereby, workpieces are scanned by sensors and the honing stones are only lowered on places which are to be machined.

Tapered rollers and gear shafts are machined, for example, with plunge cut machining, either centerless or between centers. Since the automation of the process is costly, plungecut technology is only possible for smaller piece numbers.

Band honing machines (Figure 5.8), too, belong to short-stroke honing machines and are characterized by a number of side-by-side, simultaneously honable machining places.[3] This is achieved by various flexible honing arms, each of which press a honing band which embraces the machining places and can be tightened. Furthermore, a short stroking motion in the axial direction is overlapped during rotation of the workpieces. Gear, crank, and camshafts can be machined favorably with band honing machines, which are often designed in portal construction.

Degrees of Freedom Between Tool and Workpiece

The degrees of freedom between tool and workpiece are decisive for the opportunities of correction of form and position during the honing process; they are determined by the positioning of tool and workpiece. Potential

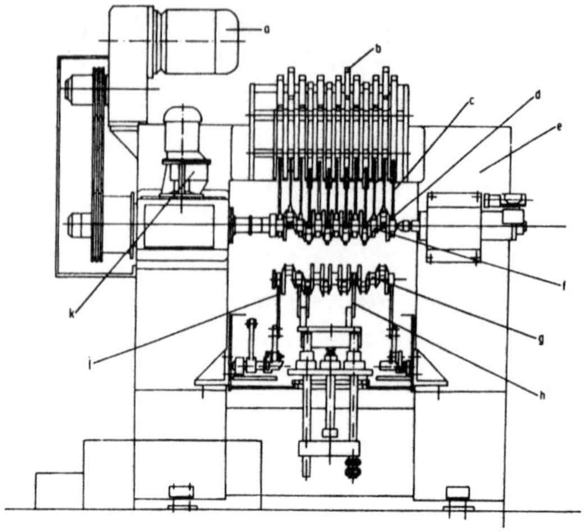

Figure 5.8: Band Honing Machine:[3] a) driving motor; b) reeled honing bands; c) honing arms; d) capstan idlers; e) upright portal; f) workpiece in machining position; g) workpiece in transport or loading position; h) induction stroke device; i) transport device; k) aiming gear for workpiece loading and unloading position

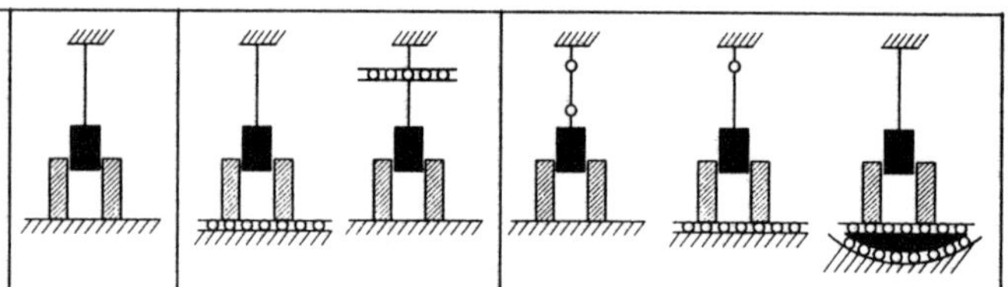

Correction of position	●	○	○
Improvement in the right angularity	●	●	○
Improvement in roundness and cylindricity	●	●	●

● possible ○ not possible

Figure 5.9: Degrees of Freedom between Tool and Workpiece[11]

forms of position are the double cardanic, cardanic, swimming, or solid position. A restriction in the degrees of freedom leads, starting with an improvement in roundness and cylindricity by the honing process, to an improvement in the right-angularity of a bore. Moreover, it results in an increase in the accuracy of position of a bore at further restriction (Figure 5.9), which, however, presupposes a high stiffness of the system.

Feed Systems of Honing Stones

The radial feed of honing stones against the workpiece serves on the one hand to adjust the removed add-measure of the material and on the other, to compensate the wear of honing stones. One distinguishes between force-activated hydraulic and way-activated mechanical feed systems for long-stroke honing, (Figure 5.10).

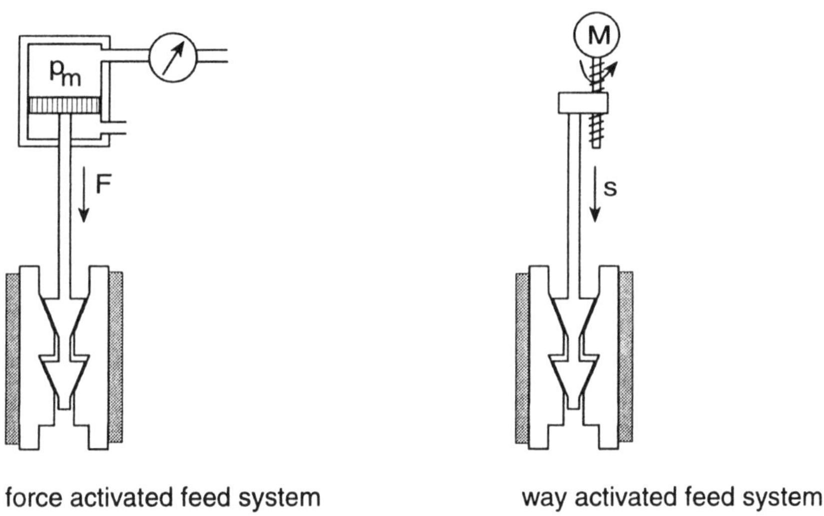

Figure 5.10: Feed Systems of Honing Stones for Long-stroke Honing[2]

Hydraulic feed systems transform oil pressure into feed force by means of a piston. This feed force is transferred to the honing stones via pressure bar, double cone, and honing stone support. The stone pressure p_s is a parameter for the description of the amount of feed force. It describes the surface pressure between honing stone and workpiece. Recommended values for diamond honing stones for metal machining are at $p_s = 1\text{-}8$ N/mm^2.[3]

At a force-activated hydraulic feed, the removed add-measure cannot be adjusted directly. The selected honing parameters lead to a constant material removal referred to as the honing time. The removed add-measure is regulated by the honing time. In order to produce narrow diameter tolerances, it is necessary to determine the workpiece diameter during the process. The honing process ends when the nominal diameter is reached. Special honing tools can be used in order to detect the actual measure of the bore. For these tools, the quantity of air which escapes from pneumatic jets on the tool circumference serves to define the actual bore diameter.

During application of way activated feed systems, honing stones are fed with defined feed steps in a range of ten to a hundred micrometers.[5] The feed usually progresses in a number of partial feed steps.

4.0 HONING TECHNOLOGY

Honing-In Performance

During the cutting of ceramics, constant cutting conditions are often impossible to achieve. The honing process is characterized by a honing-in behavior of the honing stones. The initially sharp cutting grains blunt generally with increasing machining time, which consequently leads to a decreasing material removal rate. Depending on the machined material and the used honing stone specification, constant cutting conditions appear after a certain time; the honing stones operate in the self-sharpening range, or they lose their cutting ability until no further removal can be reached and the stones must be sharpened. Figure 5.11 shows the honing-in behavior during machining of Si-SiC. While close-grained honing stones D3 and D7 show a constant

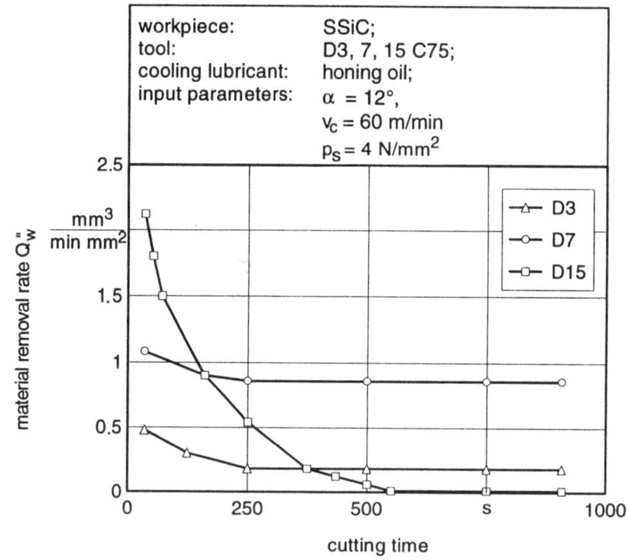

Figure 5.11: Honing-in Behavior of Various Honing Stone Grain Sizes[12]

performance after a short honing-in time, the material removal rate at grain size D15 goes back to zero in the first 500 s. Yet, only a relatively low material removal rate can be reached with close-grained honing stones, which is not sufficient for economic machining.

Honability of Ceramics

As one can see from Figure 5.12, the material removal rate related to the surface depends on the size of the diamond grain and of the stone pressure during machining of $A1_2 0_3$ and $Zr0_2$. While diamond grain size D3 does not allow an efficient material removal during the machining of the chosen Al_2O_3-material, all diamond grain sizes above D3 can be used. However, the application of diamond grain sizes exceeding D10 leads to a strong heating up of the tool and of the workpiece. In order to avoid damages to the tool and workpiece, the stone pressure is not increased above 2 N/mm². Here, it turns out

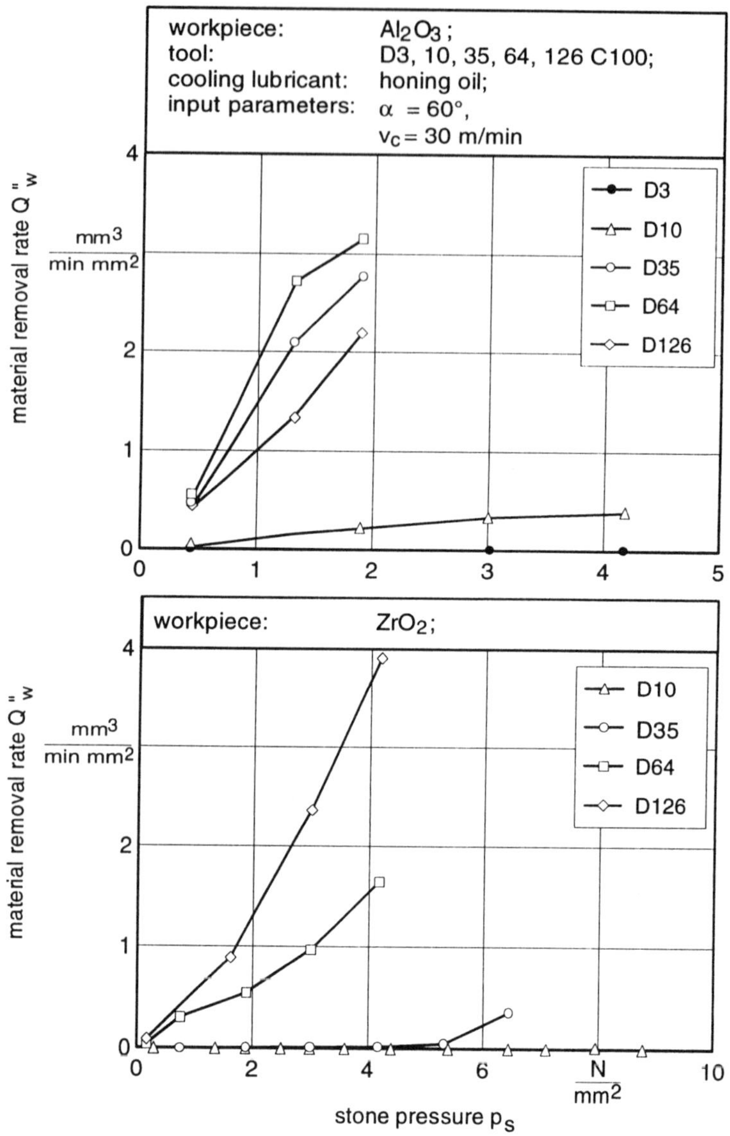

Figure 5.12: Honing Al_2O_3 and ZrO_2:Material Removal Rate depending on Grain Size and Stone Pressure[13]

that the material removal rate increases with increasing stone pressure because the normal force that presses the single diamond grain onto the material increases. The influence of the diamond grain size on the material removal rate, however, does not support such a general statement. At diamond grain sizes below D64, the material removal rate increases according to the size of the grain. Especially during the transition from the diamond grain size D10 to D35, there is a considerable increase in the material volume removed per time unit. The transition from diamond grain size D64 to grain size D126 leads to a reversion of this tendency and thus to a decrease in the material removal rate. This decrease can be explained by a decrease in the number of active diamond grains, while the size of the diamond grains increases. The reduction of the number of grains is connected to an increase in the normal force at each diamond grain. If the increase in the normal force at each grain leads to an under-proportional increase in the material volume removed from each grain, there can be a decrease in the material removal rate.

In contrast, material removal from zirconia is only possible with a diamond grain size above D36. At this grain size, it becomes obvious that an initial pressure must be surmounted in order to remove material. This pressure decreases with the growing size of the diamond grain. This is due to the fact that an increase in the grain diameter at constant diamond concentration is connected to a decrease in the number of diamond grains. This leads to high normal forces between each cutting grain and the ceramic material already at a low stone pressure. Accordingly, a stationary material removal during machining with small sizes of diamond grains should be obtainable by reducing the diamond concentration, thus also reducing the number of diamond grains.

Figure 5.13 shows the influence of stone pressure and the size of the diamond grains on the removed material volume during honing of SSN. As one can see, the removed material volume is approximately two decimal powers smaller than during the machining process of oxides. At all honing stone specifications, the material removal only starts at stone pressures higher than 1 N/mm^2. The removed material volume increases according to the size of the diamond grain. However, the increase in the material removal rate is very low at the transition from diamond grain size D64 to D125.

During honing of Si-SiC, no constant cutting conditions could be determined at stone pressures below 2 N/mm^2. With diamond grain size D3, no significant increase in the material removal rate could be reached with incr-

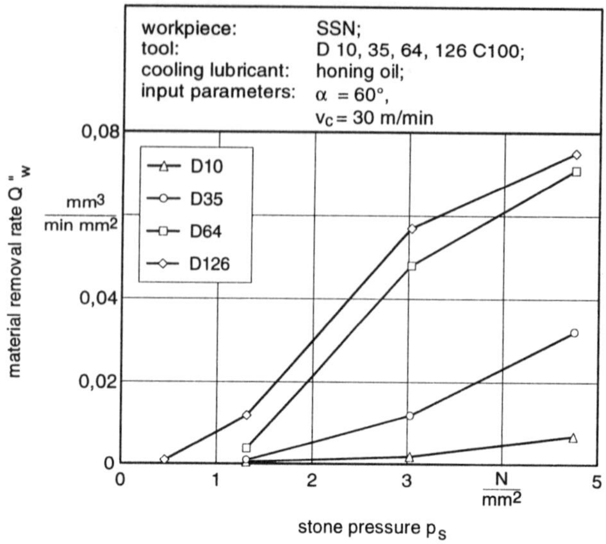

Figure 5.13: Honing SSn; Material Removal Rate depending upon Grain Size and Stone Pressure[13]

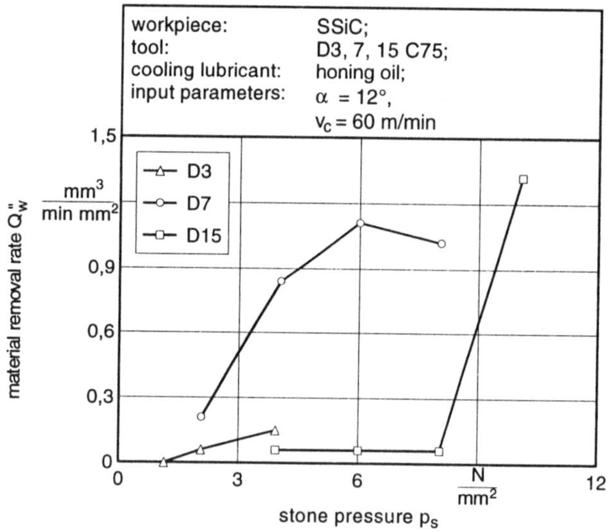

Figure 5.14: Honing Si-SiC; Material Removal Rate depending upon Grain Size and Stone Pressure[12]

easing stone pressure. In contrast, an increase in stone pressure by 2 N/mm^2 for grain size D7 resulted in a quadruple material removal rate. For grain size D15, stone pressures of 10 N/mm^2 were necessary in order to achieve constant cutting conditions (Figure 5.14).

Influence of Diamond Concentration

During short-stroke honing of the materials SiC, Si_3N_4, and TiN (Figure 5.15), an increase in the wear of honing stones becomes evident with increasing diamond concentration. This can be explained by a reduced strength of the honing layer bond, whose mass is diminished at increasing diamond concentrations, and by a bigger number of free, chipped-off grains which intensify the abrasive wear of the tool. Using low diamond concentrations, the roughness of workpiece surfaces can be reduced. For long-stroke honing, no examinations are known.

The results of the honing stone wear and the roughness of workpiece surfaces, however, do not lead to a uniform dependence for all materials. Too diverse are the peculiarities of structure and mechanical properties of the tested ceramics. Especially for titanium nitride, deviating results are reached due to its increased plasticity.

Influence of Cutting Speed and Honing Angle

Figure 5.16 emphasizes the increase in material removal with increasing cutting speed. This is caused by the longer cutting path covered by the single grain in the same time with increasing cutting speed. It proves that the decrease in material removal known for honing of metallic materials takes place above $v_c = 60$ m/min for the investigated honing stone material combination, due to hydrodynamic effects during exceeding speed limits.[15]

Investigations of $A1_20_3$ show that, in contrast to metallic materials, no honing angle can be determined at which the material removal rate reaches a maximum. The change in material removal rate at an alteration of the honing angle is very low when compared to a change in cutting speed, stone pres-

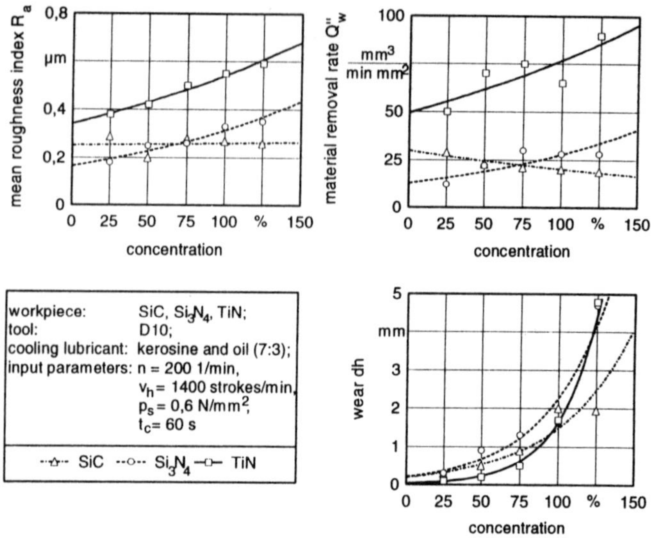

Figure 5.15: Influence of Diamond Concentration on Material Removal Rate, Roughness, and Honing Stone Wear[14]

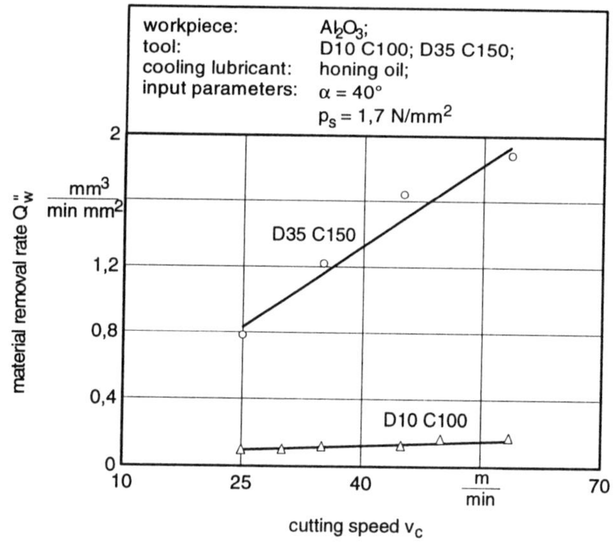

Figure 5.16: Material Removal Rate in relation to Cutting Speed[11]

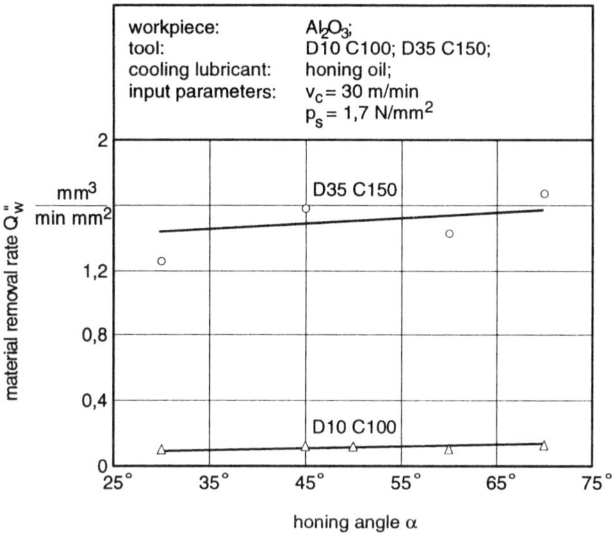

Figure 5.17: Material Removal Rate in relation to Honing Angle[11]

sure, or diamond grain size (Figure 5.17).

Influence of the Material Specification on Honing of Alumina

Indentation hardness tests and scratch tests with different diamond geometries, as well as the actual machining, revealed a clear relation of the working result to machining conditions and properties of the ceramic material. Particularly, the grain size of the ceramic material had a significant influence during tests with $A1_20_3$ and different diamond grain sizes.

Figure 5.18 shows that during honing of $A1_20_3$ the diamond grain size D10 can only be used for material removal of smaller $A1_20_3$-grain sizes. Using diamond grain size D35 and D64 during the machining process, the removed material volume increases at all examined material specifications compared with diamond size D10. In contrast, at all examined alumina specifications, the material removal rate decreases considerably under application of diamond grain size D126 compared to D64. A relation of the removed mat-

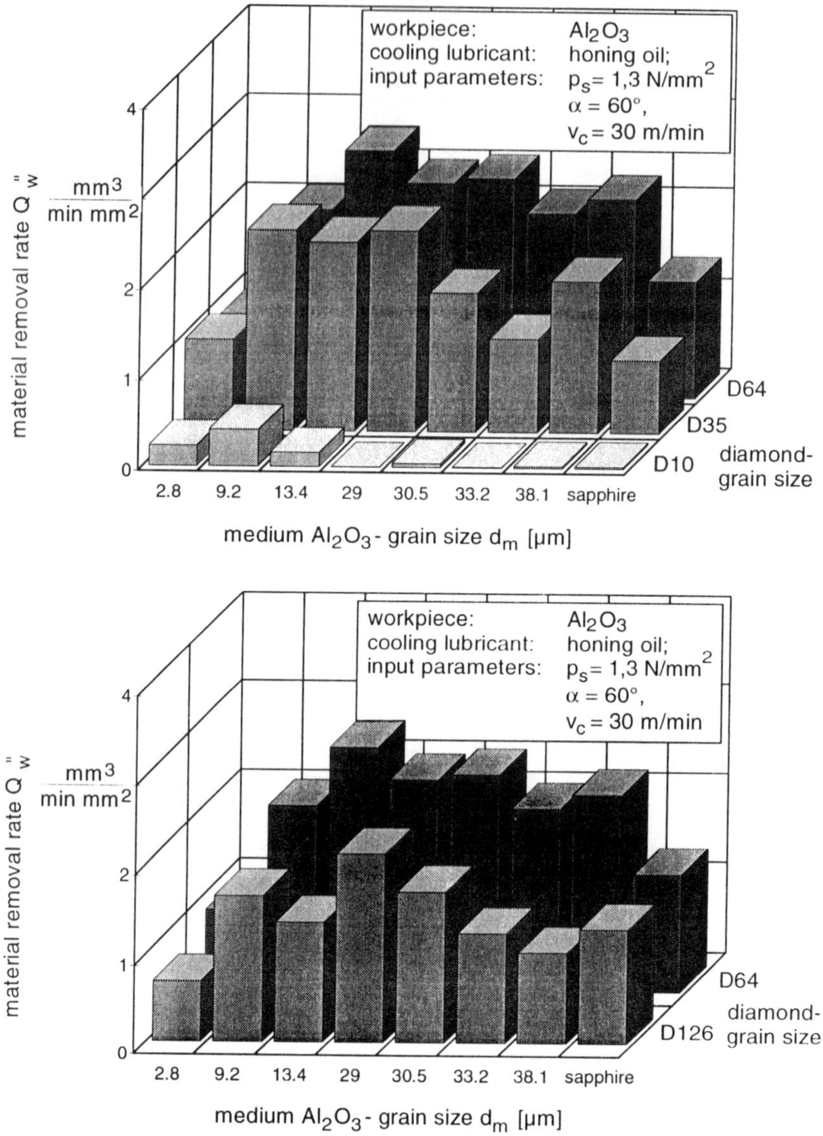

Figure 5.18: Influence of the Grain Size of Alumina on the Material Removal Rate [13]

erial volume to the Al_2O_3 grain size and to other determined material parameters could not be detected for diamond grain sizes D35, D64, and D126.

The next section shows that it is possible to explain the relation of the material removal rate to the size of the alumina grain under application of small diamond grain sizes based on removal mechanisms that occur during honing.

Influence of Machining Conditions on the Surface Formation of Alumina

Due to the higher normal forces on the individual cutting grain, Figure 5.19 demonstrates that intercrystalline grain chippings are predominant for honing Al_2O_3 with the bigger grain sizes D35 and D64. The low normal force active for diamond grain size D10 leads to intercrystalline grain chippings only at low alumina grain sizes. Bigger alumina grain sizes cannot be removed during the active lower stress; instead, as REM-shots show, a smoothening of the surface is caused here by microplastic deformations. Therefore, roughness values decrease under usage of diamond grain sizes D3 and D10 with increasing alumina grain size, while at the same time, the material removal rate related to the surface diminishes. For diamond grain sizes D35, D64, and D126, microplastic deformations emerge as a result of the increasing normal force of the single grain, which leave traces along the direction of motion of single grains. Simultaneously, intercrystalline grain chippings also occur at bigger Al_2O_3-grain sizes. These chippings cause an increase in the roughness with the Al_2O_3-grain size because bigger, chipped-off grains leave bigger flaws.

5.0 SUMMARY

It is evident that, as during honing of metals and ceramics, the demands for a high material removal rate as well as high surface quality are not met with a single honing stone specification. The realization of the honing process as a multistage process is necessary, because for ceramics, the decre-

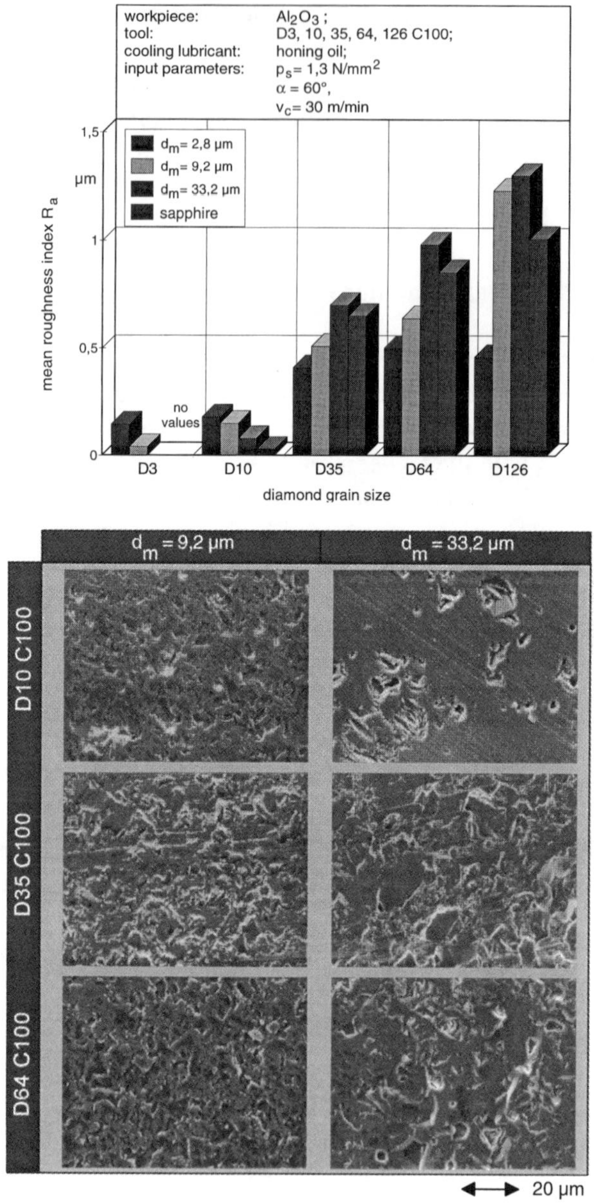

Figure 5.19: Roughness and Surface Generation depending on Alumina Grain Size[16]

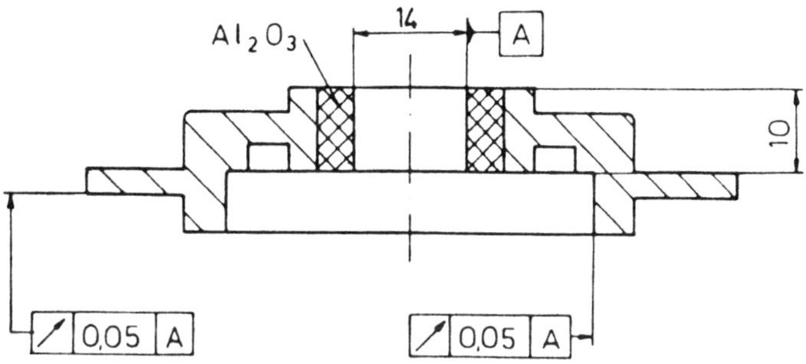

Rough-machining:	ground
Number of honing operations:	2
Cutting material:	diamond honing stones 1. operation: D15 2. operation: D7
Tool life per set of diamond honing stones (mean value):	1. operation: 60,000 bores 2. operation: 80,000 bores
Cooling lubricant:	emulsion (concentrate 5 to 8 %)
Cutting speed:	30 to 50 m/min
Stroking time:	10 s

	required	prior to honing	after honing
diameter of bore	14 + 0.004 0	13.940-13.960	14.000-14.004
cylinder shape	< 2 µm	< 10 µm	< 2 µm
roundness	< 1 µm	< 10 µm	< 1 µm
surface	< 20 R_a	0.4-1.0 R_a	< 0.20 R_a

Figure 5.20: Honing of Ceramic Bearings made of $A1_20_3$[17]

ase in stone pressure does not lead to a significant improvement in the surface quality. While the necessary material removal is achieved in a first stage, the required surface quality can be produced in the following stages. The finish-machining of circulation pump bearings made of aluminum oxide is one example of such a multistage process (Figure 5.20).

This multistage process can either be realized on a multi-spindle honing machine or by application of a honing tool which is equipped with two sets of honing stones. By means of two separately operated pressure bars, honing stones are led in series to engagement during rough-and finish-honing.

REFERENCES

1. N. N.; Honen. DIN 8589 T14(12/85).
2. Weigmann, U.P.; Honen - ein Verfahren zur Endbearbeitung keram. Bauteile. Tagungsberichtband turn Fortbildungsseminar "Bearbeiten keram. Werkstoffe" der DKG, 18.-19. März 1993, Berlin.
3. Haasis, G.; Honen., Handbuch der Fertigungstechnik; Bd. 3/2: Spanen. Hrsg. G. Spur u. Th. Stöferle. C. Hanser Verlag, München, Wien, 1980.
4. Baur, E.; Grundlagen und Anwendung des Kurzhubhonens. VDI Bildungswerk, BW 9807.
5. Klink, U.; Fachgebiete in Jahresübersichten Honen. VDI-Z 131(1989)9, S. 94-102.
6. Spur, G.; Honen und Läppen keramischer Werkstoffe.
7. König, W.; Fertigungsverfahren. Bd. 2: Schleifen, Honen, Läppen. VDI-Verlag, Düsseldorf, 1980.
8. Boor, U.; Kühlschmierstoffe zum Honen. VDI-Z 131(1989) 5, S. 70-75, Teil 1; 131(1989)6, S. 74-79, Teil 2.
9. Gehring; Superharte Schleifmittel. Ostfildern Nellingen (2/72), Firmenschrift.
10. König, W.; Stand und Entwicklungen beim Kurzhubhonen. Honen-Anwendung [Hrsg.], Jahrbuch Schleifen, Honen, Läppen und Polieren 46, S. 299-314, 1974.
11. Weigmann, U.P.; Honbearbeitung von $A1_20_3$: Einfluβ der Honleistenspezifikation auf die Korrelationen zwischen Werkstoffeigen schaften und Arbeitsergebnis. Vortragsband zurn Industriearbeitskreis "Keramikbearbeitung," 15. März 1993, Berlin.

12. Höhne, L.; Der Werkstoffabtrag beim Honen von Keramik. Honen in Forschung und industrieller Anwendung, Tagungsband S. 258-271, 4.-5. December 1989, Braunschweig.
13. Spur, G.; Weigmann, U.-P.; Influence of the Material Properties on the Honing of Ceramics. XXIII-NAMRC-Conference, May 24-26, 1995, Houghton, Michigan, USA.
14. Galkov, A. V.; u.a.; Untersuchung des Kurzhubhonens von Werkstücken aus Keramik auf Karbid- und Nitridbasis mit Diamanthonleisten. IPO. Nauk. dum. 3, S. 54-57, 1992.
15. Juchem, H. O.; Entwicklungsstand beim Honen von Bohrungen in metallischen Werkstücken mit Diamant und CBN., IDR 18(1984) 3, S. 174-184.
16. Spur, G.; Weigmann, U.-P.; Beeinflussung des Honprozesses durch die Zusammensetzung und die Eigenschaften keramischer Werkstoffe. Beitrag zum DFG-Abschlußkolloquium im SPP "Keramische Hochleistungswerkstoff," 17.-18. Februar 1994, Stuttgart.
17. Klink, U.; Vollautomatische Honanlagen für Kurbelgehäuse aus AISi 17, Honen von Lagerungen aus $Al_2 0_3$. Honen in Forschung und industrieller Anwendung, Tagungsband S. 129-157, 4.-5. Dezember 1989, Braunschweig.

6

Lapping and Polishing

Toshiro K. Doi, Toshio Kasai,
and Hans K. Tonshoff

1.0 INTRODUCTION

Lapping and polishing occur by the sliding frictions between particles and a surface. The lap or polisher travels across a work surface against which particles of sand or mud-type slurry are forced to the point of contact.

The origin of these processing methods dates from tool making in the primeval age. They were inherited and developed from generation to generation to a level of advanced technology.

Classification of lapping and polishing techniques by generation is shown in Table 6.1[1] Presently, the processing technology which belongs to the 4th generation, an improved version of the 3rd generation technology, has been applied to the fabrication of optical, electronic, and mechanical devices. A study on an automatic or computer aided machining system for both processing and ultra-precision polishing has been undertaken for practical use.

In this chapter, principles of lapping and polishing for hard and brittle materials, like optical glass and crystals belonging to the group of ceramics, are discussed including the ultra-precision processing technique.

Table 6.1: Classification of Lapping and Polishing Techniques by Generation[1]

Generation		Property	Application
1st		Matte finish	Stone Implements (matte)
2nd		Mirror finish	Beads, Bronze mirrors
3rd		Accuracies	Optical lens, prism mirror Block gauge, Plate
4th	-01	Ultraprecision polishing Damage-free	Quartz filter plate Si wafer, GaAs wafer Laser rod , mirror Magnetic head, Standard
	-02	Automatic machine tool	Mass–production for lens Si wafer, Photomask glass plate Magnetic memory disk plate
	-03	Computer control optical manufacturing	Large scale lens for laser nuclear fusion X-ray optics, Optical dies

2.0 TYPOLOGY OF PROCESSES WITH LOOSE ABRASIVES

There are fixed and loose abrasives for processing using abrasives. Processing with the loose abrasives includes buffing and abrasive shot machining in addition to lapping and polishing. Lapping and polishing are, however, considered two of the most outstanding processing methods because of their capability to secure high accuracy. The processing method is achieved by causing sliding frictional forces between the work surface and the tool. Slurry is constantly fed into the tool during this process.

The main difference between the lapping and polishing processes is obvious in glass lens manufacturing. Lapping produces a rough surface processed with coarse abrasives and a hard plate tool while polishing produces a mirror-like surface processed with fine particle abrasives and soft pads.

Figure 6.1 shows the relationship between lapping and polishing. Each is based on a different combination of abrasives and tools.[2] However,

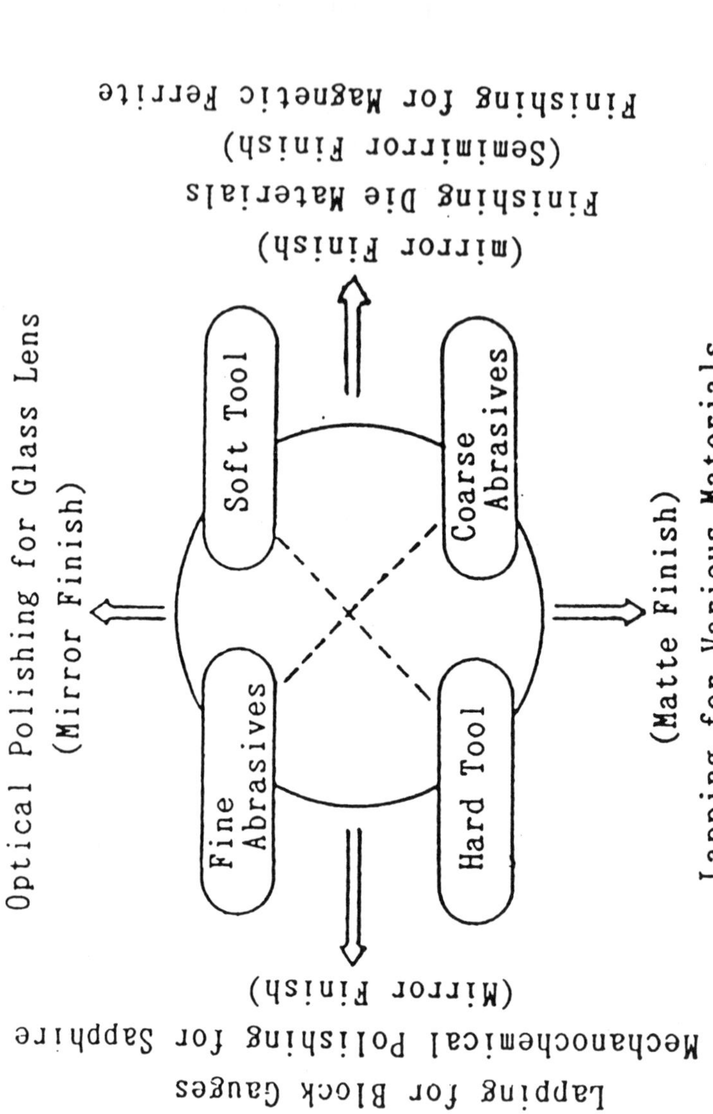

Figure 6. 1: Classification of Lapping and Polishing on the Basis of Combinations of Tools and Abrasives[2]

depending on the type of workpiece materials, the choice of abrasives and tools is not severely fixed. As a result, the finishing appearance of surfaces may vary. As a matter of fact, the definition of lapping and polishing is not used with severe distinguishment.

The following are the characteristics of lapping and polishing techiques:

1) Processing is easily performed, through the relative motions of the sliding workpiece against the tool, using slurries.
2) Almost all solid materials, such as metal and non-metal materials, can be processed.
3) Since processing advances with the micro-action of abrasives, processing efficiency is generally low.
4) Since the processing efficiency is so low, it is possible to approximate gradually to the desired level of precision; this is a great advantage for high precision processing.
5) A very sophisticated processing machine is not required.
6) Control of slurries is complicated due to some problems such as slurry-splashing and accumulation of chips in slurries.

Figure 6.2 indicates the relative positioning of lapping, polishing, and ultra-precision polishing for the device fabrication of hard and brittle materials.[3] The horizontal axis is a scale to indicate the units of processing, i.e. depth by abrasives, size of chips, surface roughness, etc.[4] The minimum unit of the ultra-precision polishing is shown in the order of an atom.

3.0 LAPPING

Precision devices with mirror-like surfaces are not manufactured by the polishing method alone. Prior to the finishing process, lapping is applied to make the surface of the desired geometry and to condition the surfaces.

Therefore, understanding lapping mechanisms is of foremost importance to establishing polishing conditions, not to mention the margin for stock removal.

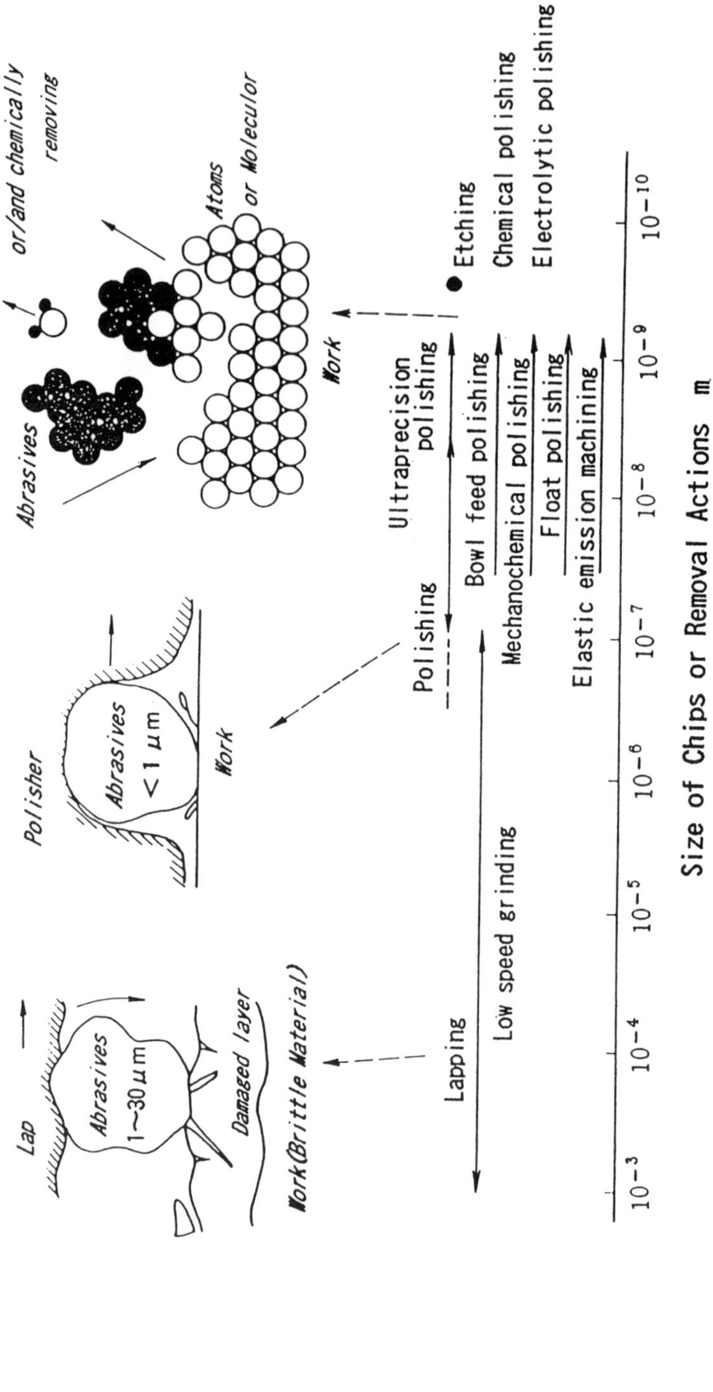

Figure 6.2: Models of Lapping, Polishing, and Ultraprecision Polishing for Hard and Brittle Materials.[3]

Abrasives contained in slurries, during the lapping of hard and brittle materials, induce new fissures in the already-existing cracks and irregular features as indicated in the lapping model of Figure 6.2. This subsequently generates chips through the cracks by the pushing, rolling, and scratching behaviors of abrasives. A description of lapping is given below with concrete examples.

Lapping Process

In device fabrications, the processing flow is established with device performance and properties, shapes, accuracy, quality and quantity of work materials. An example, of the manufacturing process-flow of glass lens,[5] is shown in the following:

1) Lens-shape-pressing of raw glass materials,
2) Grinding by curve generator, and
3) Lapping and polishing.

The above process is supported by peripheral technologies such as waxing, washing, and measuring of workpieces.

In the mass-production of lens, lapping, which corresponds to the smoothing of workpieces in a pre-polishing process, has replaced low speed grinding with a pellet-like diamond stick stone. Recently, this was applied to the processing of a small quantity production of specific items. Each single face of the lens is usually finished to a desired spherical shape by a single-sided lens lapping or polishing machine. Copying a plane or curved surface of the lap or polisher is shown in Figure 6.3.

Lapping and polishing for glass or quartz substrates and achieving high flatness and parallelism on such substrates is critically important. As a typical manufacturing method, a large number of small workpieces are processed with the above single-sided machine. Workpieces are to be adhered to a plate jig with a thin wax layer and are then finished to be flat and parallel. Plate jig made of glass or stainless steel was previously finished to an optical parallel.

Lapping and polishing make very practical use of their characteristics since stock of removal is fundamentally proportional to the relative speed between the workpiece and tool, pressure, and time. This means that a suitable

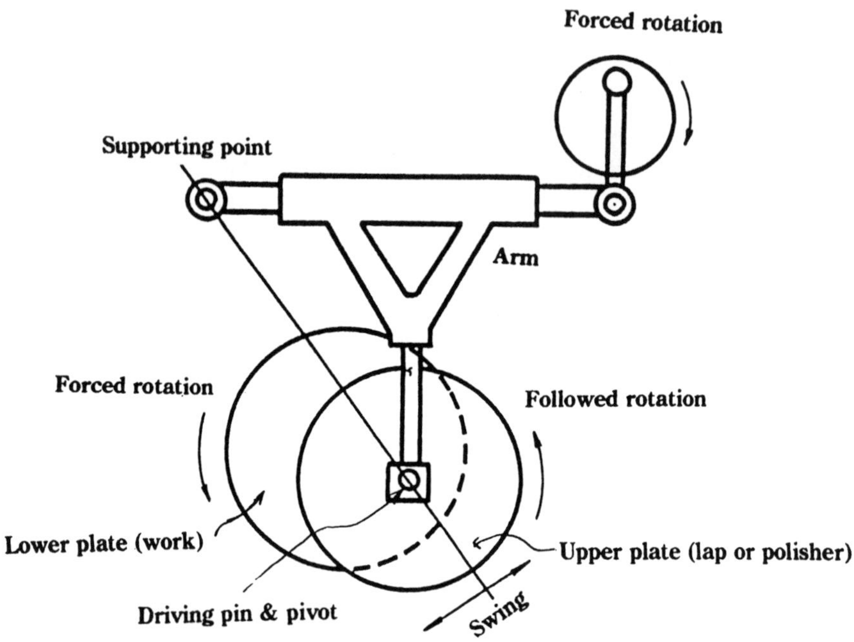

Figure 6.3: Basic Motion of a Lens Lapping and Polishing Machine

processing time can be set under uniform speed and pressure or eccentrical pressure in order to finish to a desired thickness and parallelism.

Figure 6.4 and 6.5 illustrates two examples that provide eccentrical pressure: (a) eccentrically concentrated loading method[6][7] and (b) eccentrically distributed loading method.[8] An eccentric weight has a simple structure like a cylinder that is cut obliquely. This is a significantly reasonable method to correct parallel error, since theoretical values of pressure distribution and experimental values in lapping and polishing are in good agreement. Figure 6.6 shows results of the varying thickness measured from 8 points on 5 mm inside the edge of a ⌀ 100 x t 20 mm glass substrate. This is lapped during a given time by rotating the position of maximum height of three kinds of eccentric weight by 180 degrees every time.[8]

During the improvement process of parallelism in mirror polishing, the optical interference fringe patterns from the reflecting light of the surfaces, both sides of which are mirror finished, are visibly measured. This proves the effectiveness of eccentrically distributed loading.[8]

Both lapping and polishing methods are applicable to the correction of the angle and crystal orientations. This can be achieved with additional equipment like an autocolimeter for measuring angles, X-ray, Raue, or diffraction.[9]

For mass production, a double-sided lapping or polishing method is applied due to the absence of induced waxing errors. Both faces of the workpieces are kept and processed between two laps or polishers which are each rotating in the opposite direction as shown in Figure 6.7. Lapping and polishing methods are important to obtain parallelism and desired thickness for quartz and glass substrates as well as for silicon wafers and ceramic plates after the slicing process. In this case, well-distributed placement of workpieces in order of thickness on the lap surface is very important.

Lapping Factors

Main factors relating to the surface quality, accuracy, and working efficiency of lapping and polishing are shown in Table 6.2.

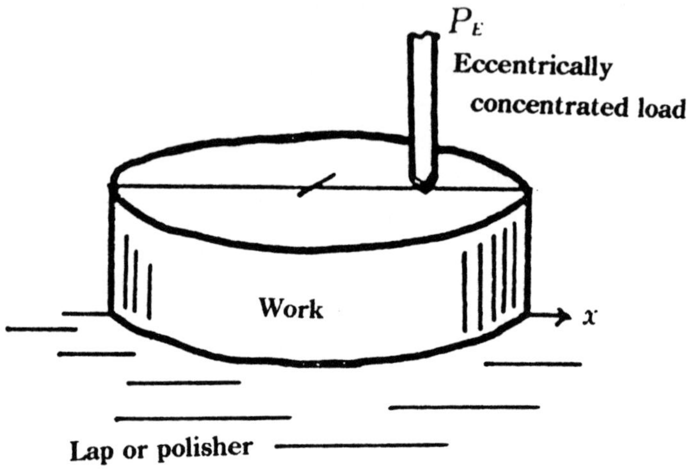

Figure 6.4: Eccentrically Concentrated Loading Methods for Correcting Parallelism.[6]

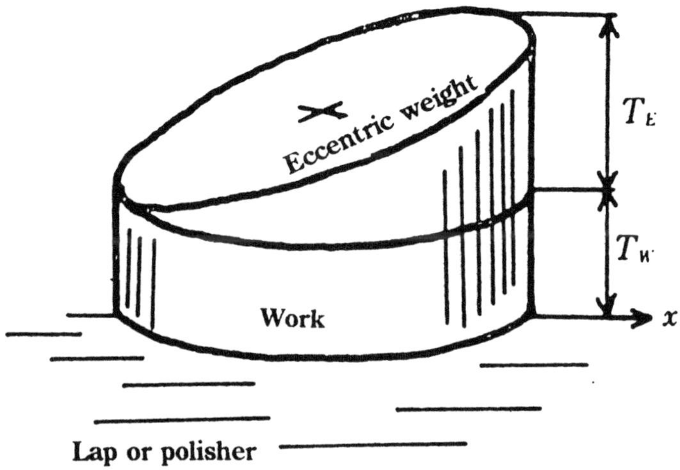

Figure 6.5: Eccentrically Concentrated Loading Methods for Correcting Parallelism[8]

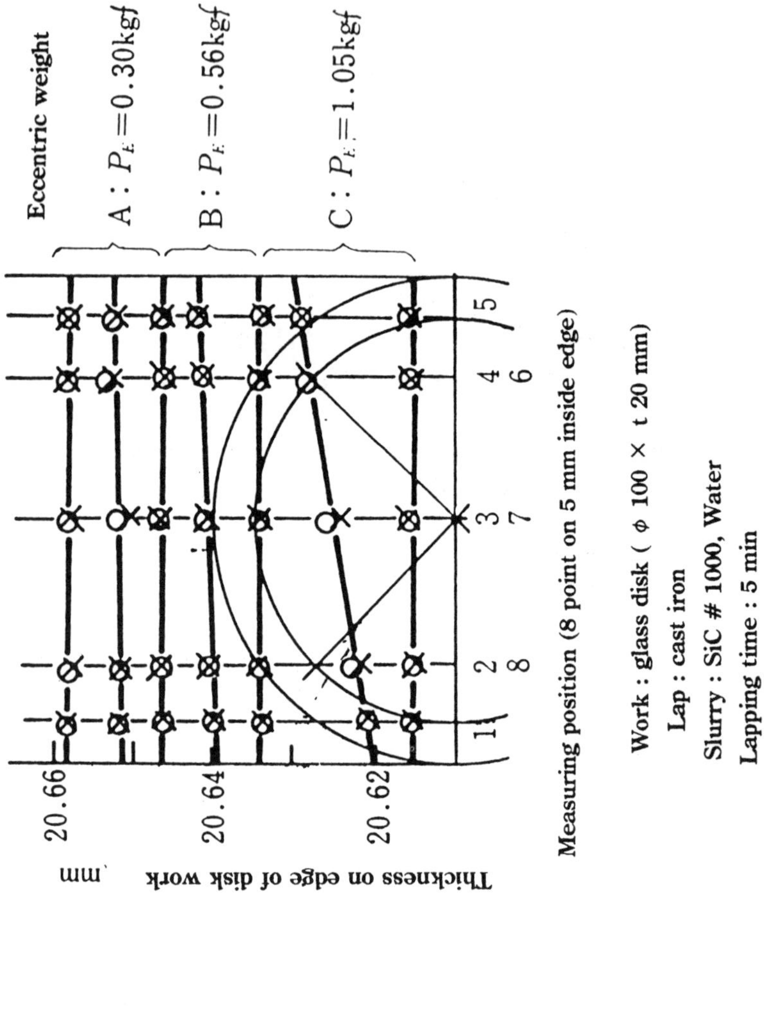

Figure 6.6: Correcting Effect of Parallelism in Glass Lapping with Eccentric Weights[8]

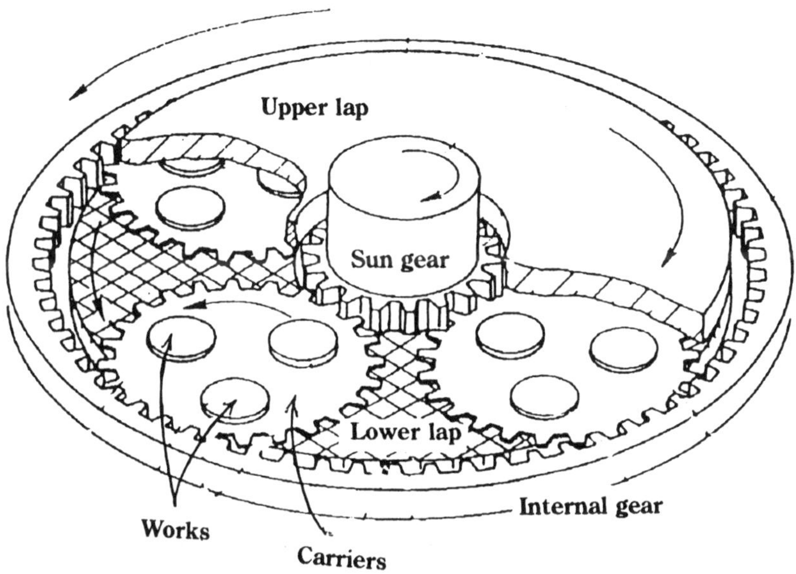

Figure 6.7: Double-sided Lapping or Polishing Machine

Table 6.2: Main Factors of Lapping and Polishing

		Lapping	Polishing
Abrasives	Material	Metal oxide, Carbide, Diamond	
	Property	Hardness, toughness, Activity	
	Size	1- 30μ m	< 1μ m
Tools	Material	Cast iron	Plastic, Pitch
	Property	Hardness, Elastoplasticity, Wear	
	Surface	Groove pattern, Texture, Shape	
Fluids		Water- type, Oil- Type, Additives	
Machine	Type	Single- sided, Double-sided	
	Motion	Rotation, Reciprocation	
Speed (m/mim)		5 -30	
Pressure (kPa)		1 - 100	

Abrasives

In lapping, Al_2O_3 and SiC abrasives are usually used because of their low cost and stable supply. These abrasives are classified by size and are commercially available from mesh number #300 (= 50 mm) to #8000 (= approx. 2 mm) for general lapping. For the final lapping of glass lens and silicon wafers, abrasives #1000 or #1200 are used.

For the effective use of abrasive sizes during the processing of glass lens or prisms, the mesh number of abrasives is doubled in every step to finally obtain the desired surface roughness across the work. Moreover, there are no clear explanations for the above. It appears that such lapping conditions have been determined through experience. Surface irregularities and damaged layers formed during the lapping step can be completely removed to produce the desired accuracies of shape and size. Under such conditions, doubling the mesh number in every step can also be applied to the lapping of ceramic materials.

Magnetic abrasive finishing was recently proposed as a new lapping method. Magnetic abrasives, made of ferromagnetic material and sintered/

crushed abrasive powders, interact with the magnetic field of the work or workpieces, therefore necessitating a specially designed machine for processing.[10]

Lap

Cast iron is commonly adopted as a lap material because of its low cost and consistent accuracy on the lap surface. According to metallography, cast iron is a type of polycrystal metal, consisting of hard pearlite iron, soft ferrite iron, and free carbon. Such pearlite iron has a layered micro-structure with the hard cementite and soft ferrite layers. It is said that fine irregularities are easily formed on a lap surface which allows it to hold abrasives in the recessed regions and makes lapping advantageous .

On the other hand, there are some cases where the tempered steel is used to meet the requirements for a hard lap. Glass or ceramics are also used to avoid the contamination of metal ions.

Lapping Speed and Pressure

In usual lapping, the relative speed and pressure between work and polisher are not great. Abrasives in the slurry should be fed in between the work and lap, and broken abrasives as well as chips of work should be discharged from the gap after lapping. A suitable lapping speed and pressure should be adopted for such processes to be carried on smoothly. Under high lapping speed, the slurry tends to splash. At high pressure, less abrasives seep into the gap. For gentle lapping conditions, a lapping speed of about 10 m/min and a lapping pressure of 10 kPa are adopted in practice. For precision working, it is important to adopt such lapping conditions so that stock removal is proportional to the lapping speed, pressure, and time.

Lapping Characteristics

Various lapping characteristics have been obtained by changing combinations of work and lapping materials.

When the stock removal is proportional to the lapping speed (m/min), pressure (Pa) and time (min), high precision lapping can be achieved. The proportional constants ($\mu m \cdot m^{-1}/Pa$, the specific stock removal in lapping) are recognized as a processing efficiency.

Figure 6.8 shows the relationship among average size or mesh number of silicon carbide abrasives, processing efficiency, surface roughness, and maximum depth of crack across the lapped surface of $LiTaO_3$ single crystal.[12] These values increase in proportion to the increase of the average size of abrasives. These mimic the mechanical action of individual abrasive grains and thereby show the same tendency.

In Figure 6.9, these values are indicated when lapped. With three different kinds of abrasives, each value is shown in order of (SiC) > (Al_2O_3) > emery, which implies the dependence upon the hardness of abrasives.[12] On the other hand, a different tendency arises with fragile lapping material. Processing efficiency and surface roughness of TeO_2 single crystal which is more fragile and softer than $LiTaO_3$ crystal are shown in Figure 6.10. There is little difference of lapping efficiency when four kinds of abrasives are used, namely: SiC, Al_20_3, emery, and garnet. Slight differences in the surface roughness seem to be caused by the distribution of abrasive size.[13]

Since abrasives are crushed, it is necessary to use harder abrasives for the lapping of harder and lower efficiency materials. Such abrasives are unnecessary for soft materials with high lapping efficiency.

Surface roughness on lapped work is usually interpreted as the result of the generation of chips. It is desirable that the extra energy, supplied to make chips, is converted into heat energy and is removed with the chips. However, such energy is actually kept in the lapped surface in the form of strain, stress, and damaged layers. There is no divergence in the relation between chip generation and surface roughness. In lapping hard and brittle materials like glass, the presence of plastic deformation on a top surface layer, micro-crack towards the inside, and plastic and elastic deformations at the extension of such cracks have been detected. These have been clarified by lapping single crystals such as quartz and silicon.[12][14]

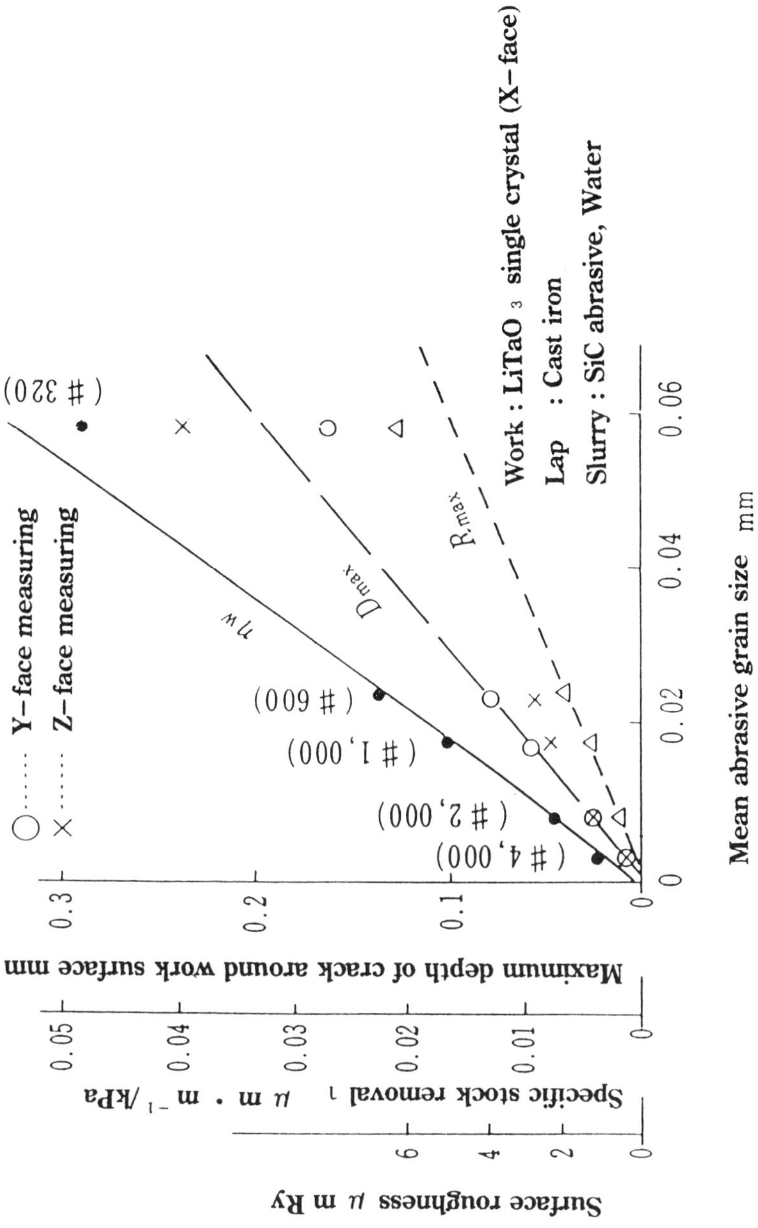

Figure 6.8: Relationship among Average Size or Mesh Number of Abrasives, Specific Stock Removal, Surface Roughness, and Maximum Depth-of-crack around Work[12]

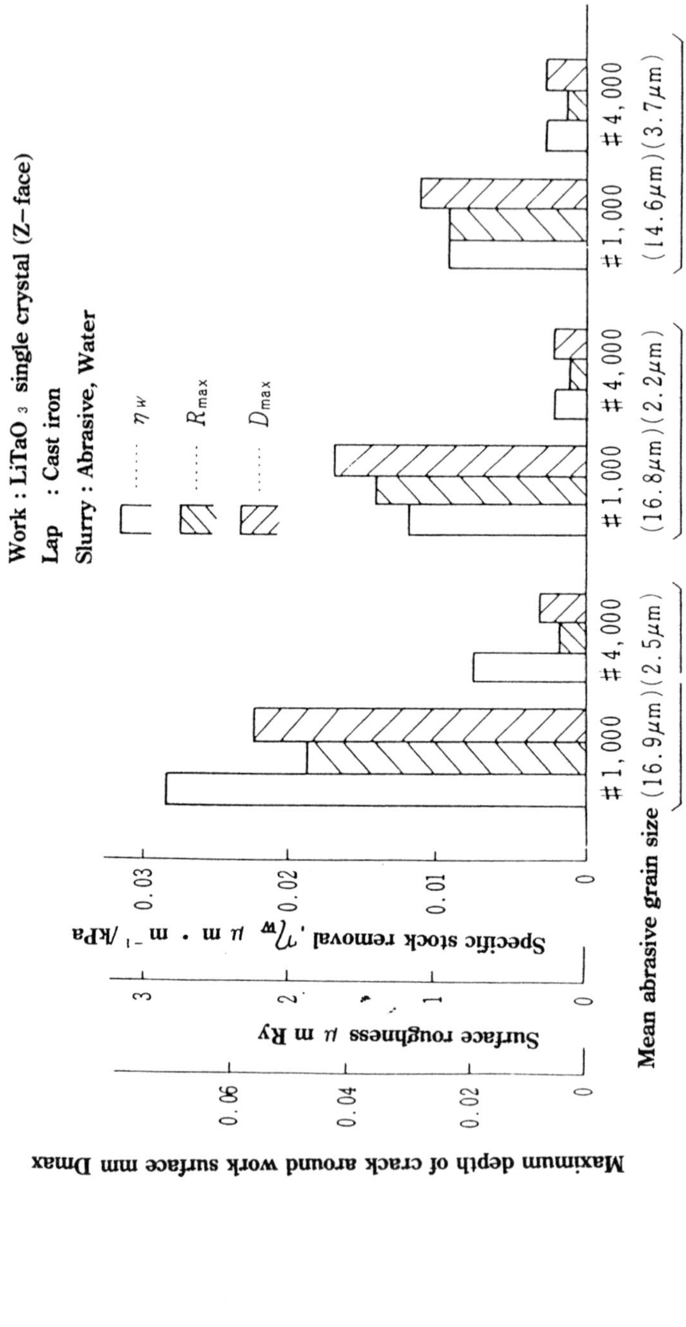

Figure 6.9: Relationship between Kinds of Abrasives and Specific Stock Removal of $LiTaO_3$ Single Crystal[12]

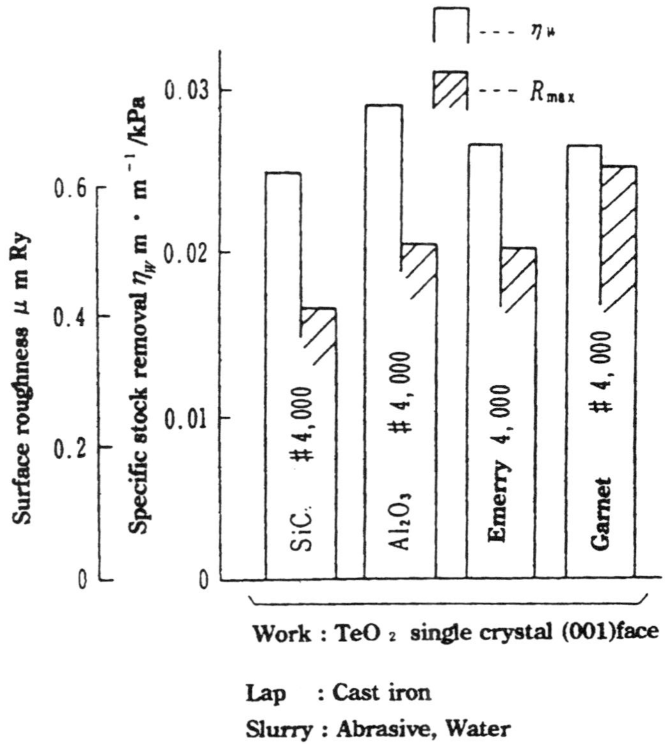

Figure 6.10: Relationship between Abrasives Stock Removal and Surface Roughness of TeO_2 Single Crystal Work[13]

The depth of a damaged layer on a lapped surface is estimated to be 6 to 15 times as large as surface roughness and in the same order of dimension of the abrasives used. On the basis of such estimation, a reasonable stock allowance for lapping and polishing can be determined for device fabrications.

Lapping Machines

Since a mirror-like, high-accuracy surface has been easily obtained with a simple lapping machine, no need has arisen for an advanced precision machine tool.

One of the simplest lapping equipments uses a lap of ϕ 200 - 300 mm. The workpieces are manually abraded against the lap at 300 - 1000 r.p.m. while simultaneously supplying slurry and pressing the workpieces against the lap surface in a to-and-fro motion in the direction of the lap diameter. Such a lapping method is similar to the metallurgical polishing using Cr_2O_3 abrasive powder and a felt sheet polisher; this polishing method is used to observe the microscopic structure of metal specimen.

Some improvements were made on lens lapping and polishing machines. For example, manual lapping operations were shifted to the machine, with increasing relative speeds between the work and laps while maintaining high working pressures. In a working operation of the lens and polishing machine, workpieces, with a driving pin at the center of the back, are forced against the rotating lap surface. The pin allows the workpiece to make reciprocating motions and to rotate in conformity with the revolution of the lap as shown in Figure 6.3.

In making a theoretical analysis for precision lapping, it is necessary to consider the motion between the work and lap, the deviated distribution of abrasive size, and the surface roughness of the work caused by the destruction of abrasives. Therefore, this is a complicated area that has not yet been attempted.

When lapping flat surfaces and spherical surfaces of a long focal point, a conditioning ring-type lapping machine, shown in Figure 6.11, has been employed. The conditioning ring and conditioning plate are forced down against the rotating ring lap surface thus maintaining the lap surface in good condition.

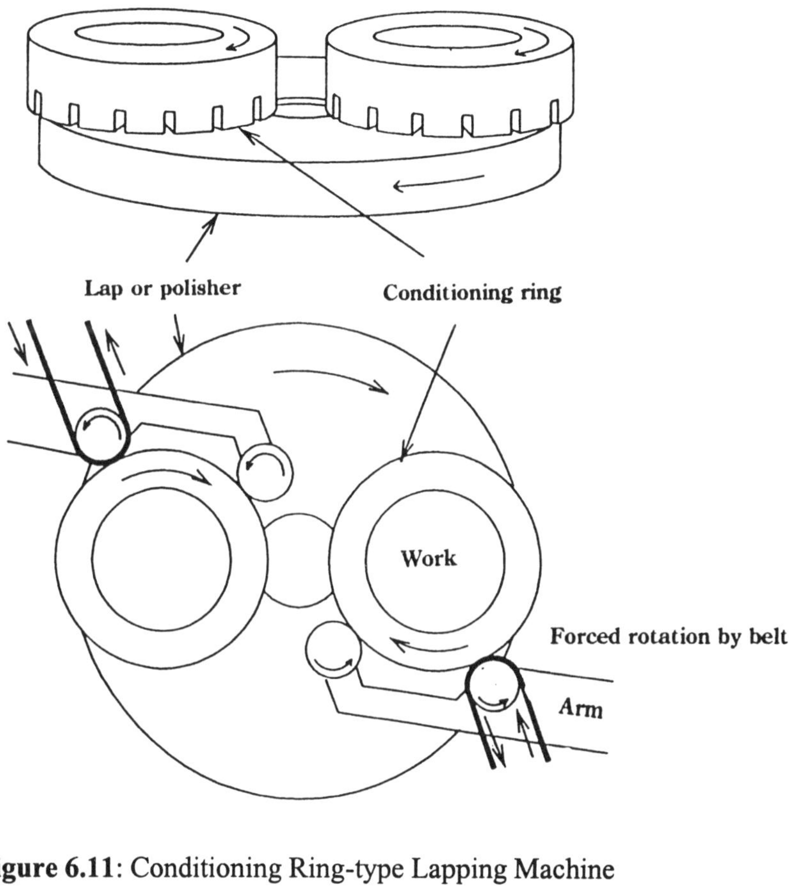

Figure 6.11: Conditioning Ring-type Lapping Machine

4.0 POLISHING

The stock removal unit in the polishing machine is small as is shown in Figure 6.2. Abrasives only act on the top layer of a work surface since much finer abrasives and softer tools are adopted than in the case of lapping.

In glass polishing, fine metal oxide abrasives, such as cerium oxide powder (CeO_2) and red rouge powder (Fe_2O_3) of less than 1 μm, are used along with soft polishers like pitch, wax, plastic, and synthetic leather. The abra-sives elastoplastically sustained in the polisher are presumed to produce extremely small scratches on the polisher. Unsubstantial chips in glass polishing accompanied by the hydration phenomenon are shown. However, it is impossible to observe such chips under a wet polishing condition.

Some of the following explanations were given for the mechanism of making a mirror finish for glass: (a) mechanical stock removal, (b) plastic flows, and (c) chemical etching. Mirror finishing methods have various characteristics for different materials as summarized in Figure 6.12. As mechanical and chemical (electric) actions, there are (a) mechanical removal by abrasion, (b) friction to disorder the atomic arrangement, (c) etching and dissolving, and (d) formation of a thin film. The combination of these actions at different ratios constitute various polishing methods which are listed outside the circle.[15] In glass polishing and mechanochemical polishing of silicon wafers, work surfaces are covered with a soft film by chemical reactions; this is mechanically removed by abrasives. These processes are repeated in order to remove surface irregularities and to gradually yield mirror-like surfaces.[16][17][18]

Polishing Process and Factors

Polishing is a very important processing method in the fabrication of the following devices:

- passive optical devices: lens and prisms of various sizes, polarizer, and wave-length constant,
- active optical devices: solid-state laser rod, optical deflector, and electro-optic modulator,

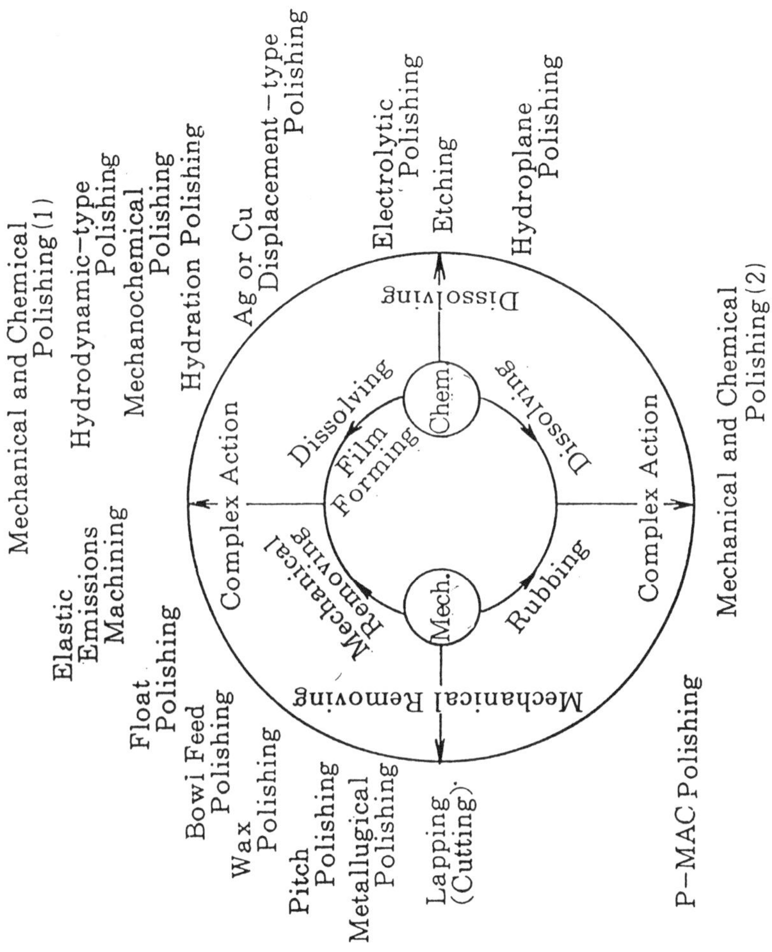

Figure 6.12: Classification of Polishing Methods based on Stock Removing Mechanism[15]

- piezoelectric devices: quartz oscillator and filter,
- semiconductor devices: LSI, the magnetic devices as magnetic memory head and magnetic bubble memory, and
- mechanical devices: ceramic plate and ball, etc.

There are a wide range of work materials: from fragile glass and crystal to the hardest material like diamond. High quality and high precision polishing is performed to meet requirements for a higher grade of device function and performance.

The basis of the device manufacturing process, on lapping using the process of a glass lens, was briefly described in Section 3.0. Polishing factors are shown next to the lapping factors in Table 6.2. A high quality mirror-like surface is required for polishing, for which there are some points to be noted. It is important to keep the polishing area free of sand particles, dust, and to completely separate it from the lapping area. Clean surroundings, a pure water/ultra pure water supply system, and washing machines for the polishing equip-ment, jigs, polisher, etc., are necessary.

With regard to the polishing slurry, careful attention should be paid to the adoption of fine abrasives so as to avoid contamination by large particles and dust.

Abrasives for Polishing

Abrasives contained in the polishing slurry are metal oxide fine powders such as CeO_2, Fe_2O_3, Al_2O_3, and SiO_2; these are generally used by being suspended in water. Whether it is suitable is judged by the stock of removal and polishing quality accumulated through polishing experiments. Abrasive hardness is one of the criteria for abrasive selection; however, since the values are obtained from single crystals or sintered materials, they do not represent the value of the abrasives themselves. It seems that quite complicated chemical reactions occur with the slurry.

In glass polishing, for instance, CeO_2 powder is superior to others in terms of stock removal. On the other hand, Fe_2O_3 is considerably better on surface quality, and pure SiO_2 powder is excellent in avoiding contamination from the transition elements.

Diamond powders of single substances are the hardest abrasive materials ranging from sub-μm to 4 μm. These powders are employed for polishing as a water- or oil-type slurry and paste. Since these powders are quite expensive, the amount fed is controlled in order to consume less. Usually, spraying or coating it with a finger is most economical.

Polisher

Polishers made of a hard-type polyurethane foam sheet are currently employed for glass lens manufacturing because of low wear and simple usage. As innumerable pores of the foam supply and discharge paths of slurry, it is crucial to make the work surface moderately rough in an earlier stage of polishing, in order to enable it to hold abrasives.

For manufacturing specific optical devices, the pitch polisher, which has been in use for three hundreds years, is utilized with mesh grooves on its surface. The effect of mesh grooves is similar to the pores in the polyurethane polisher. Concerning the pitch polisher, its surface roughened by the work surface holds abrasives that make both polisher and work surfaces smooth as polishing advances.

The wear and softness of the polisher can cause deterioration of geometric accuracy, such as flatness, and generation of turned-down edges. The correct selection of a polishing machine and its operation is important to satisfy the requirements for high precision polishing.

Polishing Characteristics

Figure 6.13 shows the relationship of glass BK6 with polishing time, stock removal, and polisher wearing amount, lapped to a mat surface, and polished with CeO_2 powders and an acrylic resin polisher. In the early stage of polishing, the work surface still remains rough producing small stock removals and large degrees of polisher wear. However, once a mirror-like surface is obtained, stock removal increases significantly in proportion to the polishing time. On the other hand, polisher wear turns out to be small; this also

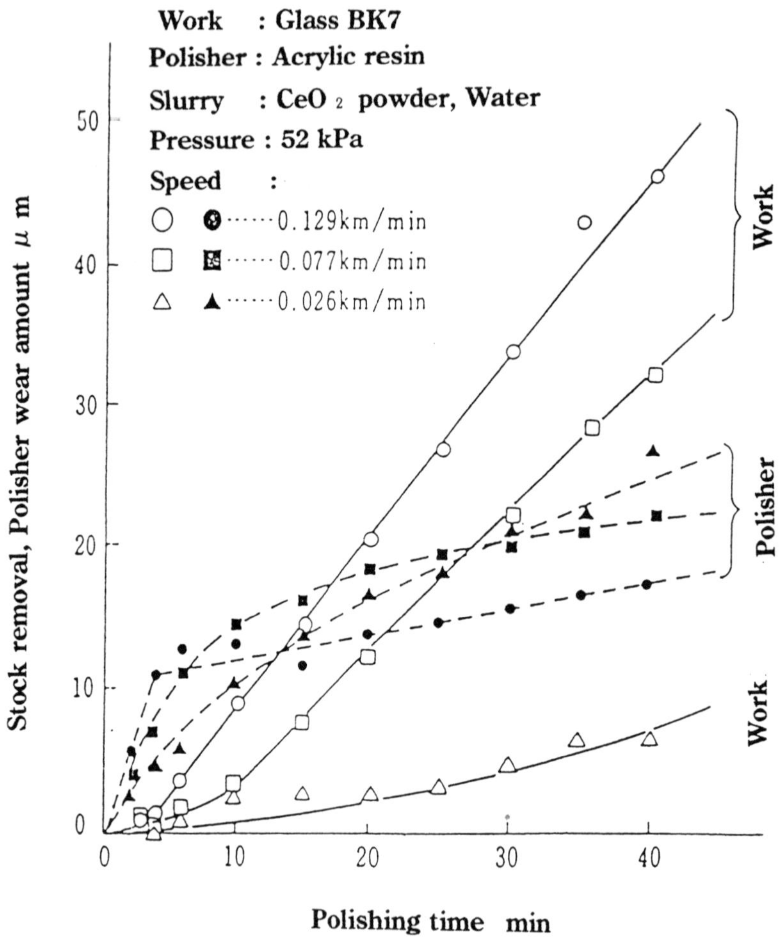

Figure 6.13: Relationship between Stock Removal and Polisher Wear in Glass Polishing.[19]

increases in proportion to the polishing time.[19]

Irregularities of the work surface, generated during the prior processing, rub against the polisher like an abrasive stone or file and causes the polisher to wear. Stable polishing conditions continue after such irregularities are removed and a smooth surface is obtained. Stock removal is proportional not only to the polishing time but also to the speed and pressure, thus satising Preston's principle.[20][21] Proportional constants of stock removal and polisher wear, which indicate the specific stock removal as processing efficiency and the specific wear amount of polisher respectively, are important factors to provide polishing conditions for theoretical analysis.

The profile of surface roughness of zirconia ceramics polished with a polisher of acetate fiber is shown in Figure 6.14. Diamond and Cr_2O_3 abrasive powders of similar dimension were used, and a mirror-like surface was obtained in both cases. However, a big difference was observed on the profiles of surface roughness. Sintered ceramics are heterogeneous in view of its structure and are made up of fine crystals, boundaries, and pores. Smooth surfaces are obtained with diamond powders that function superiorly on stock removal and uniform removal of irregularities across a work surface. Cr_2O_3 powders, which are inferior to the diamond in stock removal, only attack weak parts of the work surface and make emphatic pits or dents as appeared in the profile of the surface roughness.[22] Some improvements can be expected by changing polishing materials to relatively hard metals like tin and copper.

The concept of the highest grade mirrors with optical polishing methods, namely, super smooth mirror-like surface, has been defined. Such surfaces may have a wavelike profile formed with regularly arranged atoms like STM (scanning tunneling microscope) images of a silicon cleavage face. If high density atoms are arranged, the peak and valley of the surface roughness will be one tenth or one hundredth of the atomic diameter. Although the surface roughness is different depending on the difference of work materials, the peak and valley of the surface roughness on a super smooth mirror surface is expected to be small.

Actual mirror polishing progresses through material removal ranging from the minimum scale of an atom or molecule to the large scale of clusters of atoms/molecules. If material removal progresses along the motion locus of abrasives and projecting areas of the polisher holding abrasives, it is possible to produce surface textures similar to the directional ridges in a cultivated field as in a diffraction grating. Taking the crossing of polishing motion into acco-

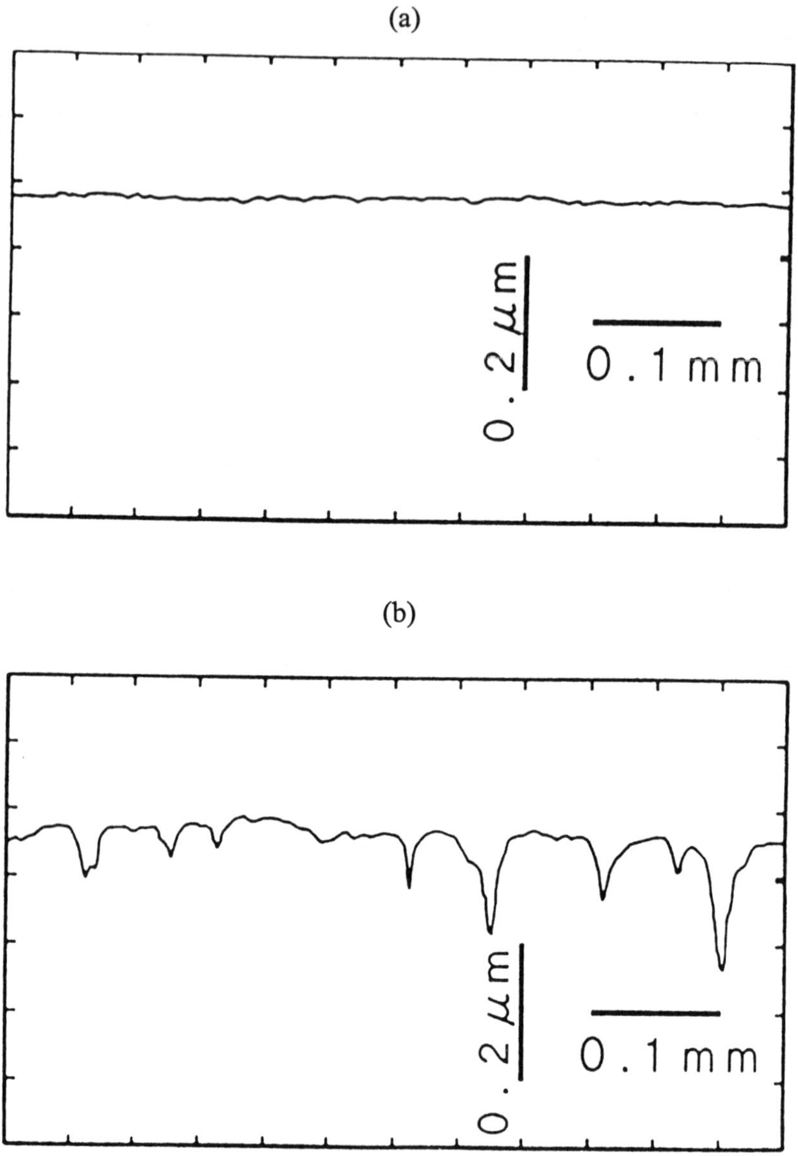

Figure 6.14: Surface Roughness Profiles of Polished Zirconia Ceramics: (a) polished surface with diamond powder and (b) polished surface with chromium oxide powder slurry.[22]

unt, it is also possible to change it into pyramid-like or cone-like arrangements. There may be traces of such material removal on polished surfaces, beneath which, there are damaged layers of disordered atomic arrangement; this may affect atomic layers from several layers to a few hundred layers down from the surface. Surface quality could be improved by making the surface texture finer. It is very important to use fine abrasives and fine textured polishers. As mentioned above, the existence of large particles in the slurry and irregularities on the polisher incurs the deterioration of surface quality. Figure 6.15 shows a conceptualized relation between large particles and surface roughness. When a polishing machine is installed in a dusty environment, the quality of the worked surface deteriorates as polishing advances because of the dust-drenched conditions.[23]

There are materials, for the polisher, where the surface is irregular in the beginning but becomes smooth in the finishing stage of polishing. Typical materials for such poishers are pitch and wax.

With regard to the conventional pitch polishing for glass, this surface roughness has remained 50 nm Ry for a long time and is attributable to dust and large particles as mentioned above. As a counter measure, a coated polisher with pitch was prepared on a metal plate that was previously made to the desired shape with a 1 mm x 1 mm (width x depth) groove on 4 mm patterned grids and flatness. During the cooling process of the metal plate that was heated to the melting point of the pitch, the thickness of the polisher becomes uniform across the surface with the aid of the pitch surface tension; this leads to the realization of an ideal polisher if used in a clean atmosphere. With 60 nm SiO_2 abrasives, the surface roughness of 1 - 0.3 nm Ry is achievable.[24]

Polishing Machine

Polishing machines are basically much the same as lapping machines with respect to the progress of the work enhanced by polisher movement. Theory-based analysis and controls of the wear and deformation of polishers become important in order to secure accuracy and a mirror surface.

Lapping and polishing machines are sorted out in Table 6.3. With a motor as their power source, the majority of workpieces and tools move by revolution.

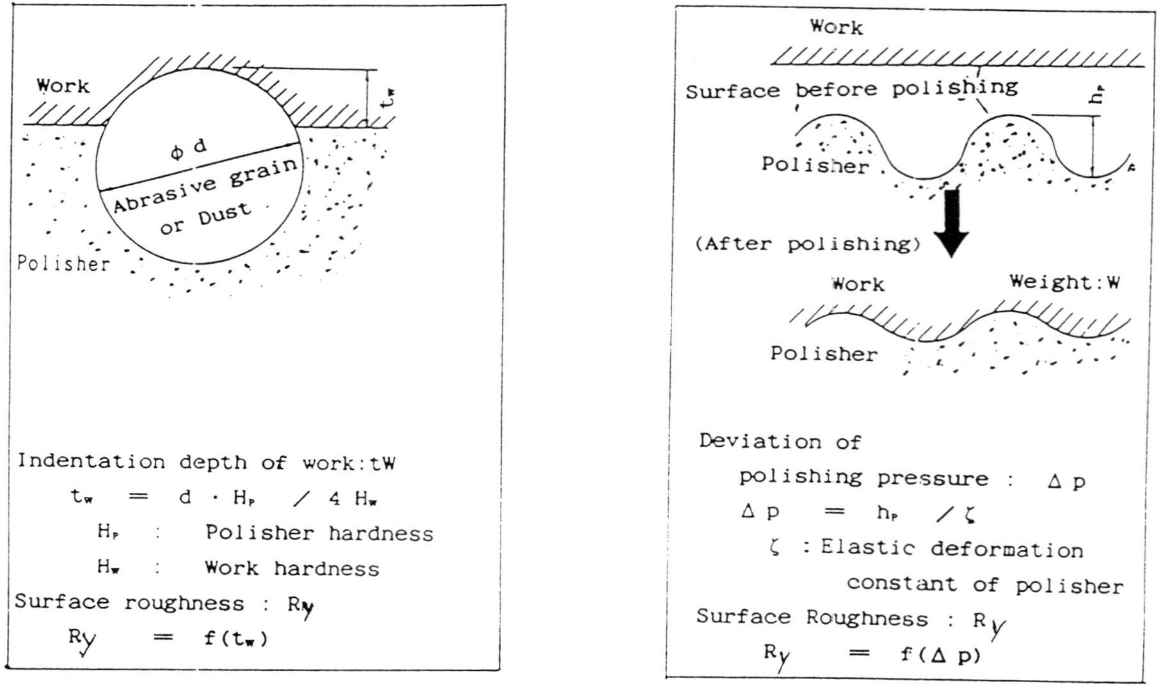

(a) (b)

Figure 6.15: Basic Formation Models of Surface Roughness: (a) relation between work, polisher, and abrasive or dust; (b) relation between surface texture of polisher and irregularity of polishing pressure.[23]

Table 6.3: Classification of Lapping and Polishing Machines

		Work	Lap or Polisher	
Single-sided machine	Flat	• Forced rotation • Upward	• Disk plate • Downward	Lens lapping & polishing machine
		• Forced rotation • Conditioning ring • Downward	• Swing & rotation • Ring (disk) plate • Upward	Conditioning ring & type lapping polishing machine
		• Followed rotation • Conditioning ring • Downward • Forced rotation	• Forced rotation • Ring plate • Upward • Forced rotation	
		• (Conditioning ring) • Downward • Rotation & swing	• Ring plate • Upward • Forced rotation	
		Motion- free	Oscillation	
	Spherical	Lens lapping & polishing machine		
		Conditioning type lapping & polishing machine		
		• • Upward • Forced rotation	• Ring (plate) • Downward • Swing & rotation	
		• • Sideward • Forced rotation	• Ring (plate) • Sideward • Swing & rotation	Advanced curve generator
	Aspherical	Lens lapping & polishing machine		
		• • Upward • Forced rotation	• Aspheric tool • Downward • Swing & synchronous rotation	
		• • Upward • Computer control rotation	• Small tool • Downward • Circular swing	Computer controlled lapping & polishing machine
		• • Sideward • Forced rotation	• Small tool • Sideward • Forced rotation & cam drive	
Double-sided machine	Flat & Parallel	• Supported by carrier • Rotation on carrier • and tool center	• Lower & upper ring plates • Fixed	Two motion type double-sided lapping machine
		• Supported by carrier • Rotation on carrier and tool center	• Lower & upper ring plates • Either rotation	Three motion type double-sided lapping machine
		• Supported by carrier • Rotation on carrier and tool center	• Lower & upper ring plates • Opposite rotation	Four motion type double-sided lapping machine
		• Supported by carrier • Circular swing of carrier	• Lower & upper disk plates • Fixed	
		• • Sideward • Followed rotation by roll supporting	• Disk plate • Sideward • Both forced rotation	Magnetic memory substrate double- sided lapping machine

In the theoretical analysis of polishing, the proportional constants of Preston's law are used for the calculation of stock removal as well as the wear amount. These constants represent specific stock removal and wear amounts of the polisher; there dimensions are the same as that of the specific wear amount in the friction and wear. These values are influenced by the characteristics of workpieces, abrasives, slurries, and polishers and become one of the polishing conditions in the theoretical analysis.[21]

Polishing speed and time are given by kinematic analysis between the work and polisher. In this case, diameters, revolutions, and true contact ratio are critically important factors. Moreover, if polishing starts under such initial polishing conditions that the shape of both surfaces are expressed in a mathematical formula, the relationship between the behaviors of both surfaces and pressure distributions can be obtained with the following assumptions:

1) The abrasives and machined traces are considered to be in close contact when they are distributed uniformly across both surfaces.
2) The pressure distributed on the work surface is given by the elastic deformation amount of the polisher.
3) The inelastic behavior of the polisher should be included in the wear.
4) There is no negative pressure across the entire surface.

Stock removal hw (μm) and wear amount of polisher hp (μm) at a random point on the work or polisher surface are expressed as follows:

$$\text{hw or hp} = \eta\cdot(\text{v})\cdot(\text{p})\cdot\alpha\cdot\Delta t \tag{6.1}$$

where, Δt (= t/m, min) is a very short interval, η ($\mu m \cdot km^{-1}$ /Pa) stands for the specific stock removal or specific wear amount of polisher, (v) (m/min) or (p) (Pa) are the average velocity or pressure on machined traces of every random point, and α is a ratio of actual processing time at random points against apparent polishing time. It is important for the development of theoretical analysis that the sum of pressures, at each point, corresponds to the loading weight values on work. Figure 6.16 shows a relation between the work and polisher. Figure 6.17 shows calculation results of stock removal, wear amount of the tool, and variation of flatness when obtained. Pitch polishing was performed for glass work from their very initial conditions as being

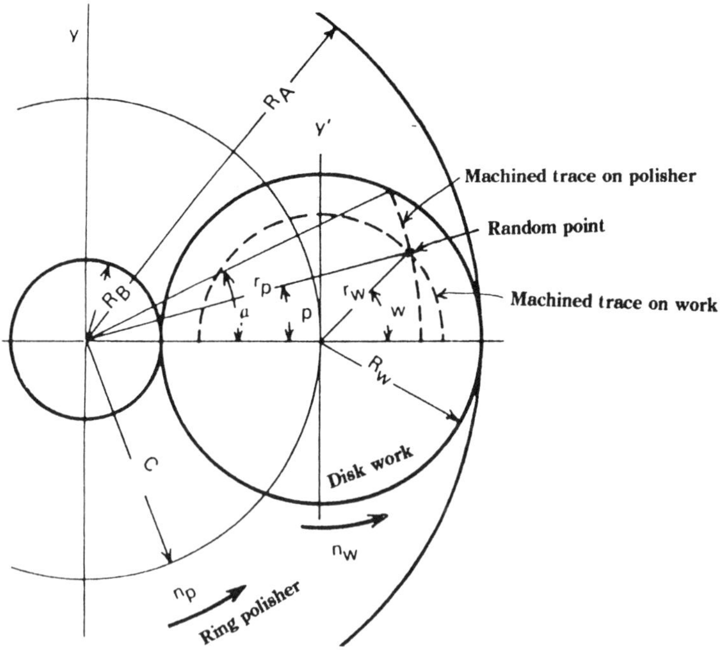

(a)

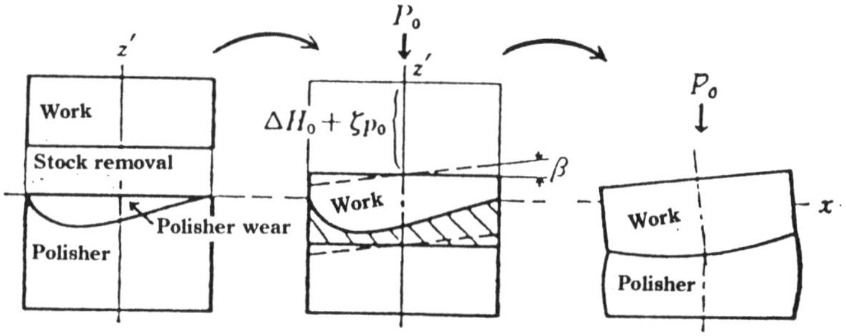

(b)

Figure 6.16: Relationship between Work and Polisher: (a) relation between disk work type polisher and (b) distributed pressure model taking the figure of a random point on work and polisher into account.[21]

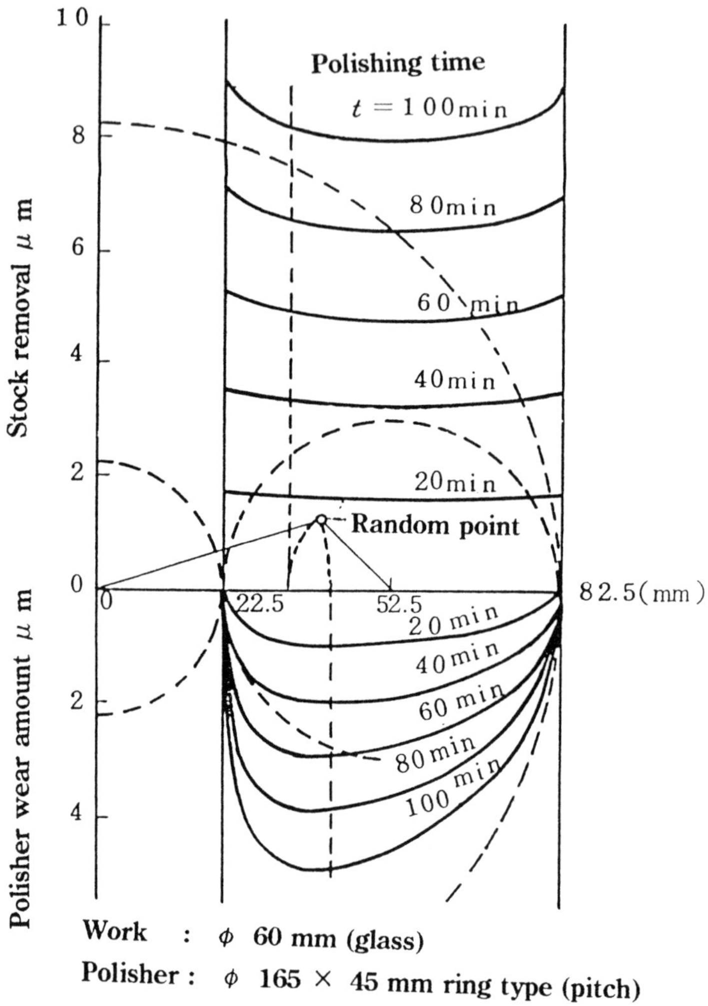

Figure 6.17: Calculation Results of Stock Removal, Wear Amount of Tool, and Variation of Flatness[21]

completely by plane. According to numerical calculations, as polishing advances, the flatness markedly deteriorates with a large specific wear amount of polisher. The degree of flatness deterioration seems rather small when elastic deformation constants are large.[21]

Advanced Polishing Methods

Recently, various polishing methods originating from the conventional lapping and polishing methods were newly proposed for device fabrications that were sorted out and indicated in Table 6.4.

Improvement of Optical Polishing

In the manufacturing of advanced optics, it is necessary to avoid dust, and to adopt fine abrasive powders, soft polishers, (ultra) pure water, and a clean environment.

In bowl feed polishing, the works and polishers are soaked in slurry while polishing. High quality and high accuracy were obtained under such special polishing conditions which are excellent in classifying fine abrasives, cooling them, and in shock absorotion by fluid. A superior surface quality was obtained with a pitch or fluorocarbon coated polisher.[25][26]

As often pointed out, a great deal of skill is required to accomplish high grade polishing, and therefore, a rapid shift to a computer controlled polishing method has been anticipated. For instance, a correcting polishing method for flat, spherical, or aspherical surfaces has been established, where a small-size polisher which performs small circular motions travels across the work surface with the aid of a computer as shown in Figure 6.18. This polishing method is supported by the measuring technology.[26][28]

Table 6.4: Advanced Polishing Methods

		Property of Processing	Application
	[Improvement of Lapping]		
	Ultraprecision low speed grinding	Mechanically stock removing with abrasive stone and oil- or water- type solution	Block gauge End of scale Plate
	Ultraprecision lapping	By using fine abrasives and metal or ceramics lap, and filtered oil- or water- type solution	
	Conventional. lapping Conventional Polishing	By using Al_20_3 or SiC abrasives and cast iron lap, or CeO_2 powders and pitch or plastic polisher, and lens lapping & polishing machine and conditioning ring type lapping machine	Glass lens or prism
	[Improvement of Optical Polishing]		
Mech.	Ultraprecision polishing	Suppressing mechanical action and dust influence by adopting very fine abrasive powders, soft polisher and (Ultra) pure water	Quarts plate Laser rod Optoelectronic devices X- ray mirror Large tele-scope mirror
	Bowl feed polishing	Work and polisher being in slurry for suppressing mechanical action or impact and temperature rise	
	Computer aided polishing	Making a small polisher to travel on work surface by using computer	
	[Using Flow Effect of Slurry]		
Complex.	EEM	Mild stock removing action with abrasives in hydrodynamic flow of slurry around ball tool	X-ray mirror Optical grade dies Glass Plate Ferrite
	Float polishing	Stock removing action with abrasives inhydrodynamic flow of slurry on flat polisher	
	[Removing Chemical Reactive Products]		
	Mechanochemical polishing (wet)	Growing and removing reactive products as hydration film under wet condition	Si wafer GaAs wafer
	(dry)	Peeling off reactive product film with 0.01 μm SiO_2 powder under dry friction condition	Sapphire Si_3N_4 ceramic
Chem.	Hydration polishing	Growing hydration film under 50 - 200°C water vapor condition and wiping off with wood or carbon tool	Sapphire ZnSe lens
	[Using chemical removal]		
	Disk- type chemical polishing	Chemically removing with friction by using non-abrasive chemical solution and non- woven fiber sheet	GaAs wafer InP wafer CdTe wafer ZnSe wafer
	Hydroplane polishing	Chemically removing with non- abrasive chemical solution layer between work and polisher	
	P- MAC polishing	Chemically removing by using non- abrasive chemical solution and fluorocarbon foam polisher. Working conditions change as polishing advances	

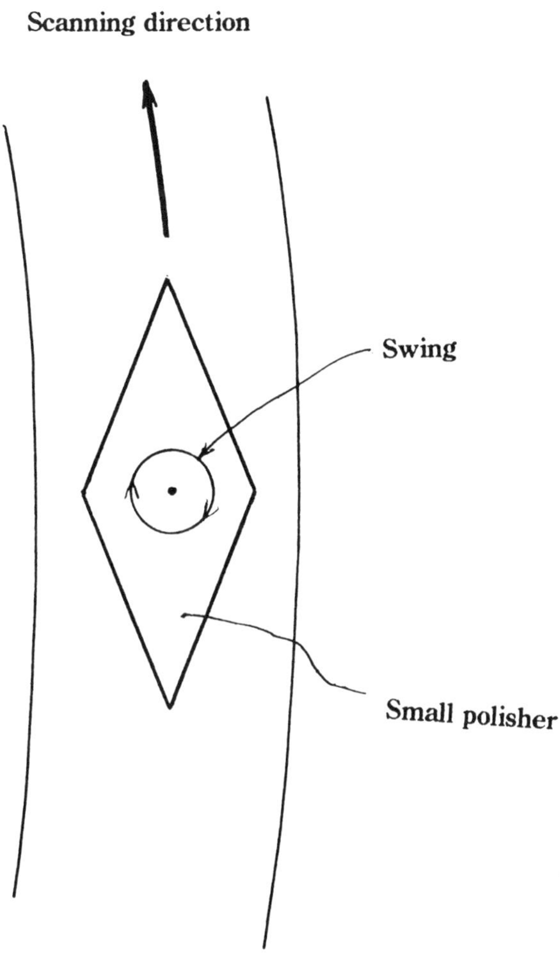

Figure 6.18: A Small Size Tool used in Computer Aided Polishing[27][28]

Non-contact Polishing Method

The polishing method described here is a method whereby the work and polisher do not come in contact during polishing. In conventional polishing, damaged layers on finished works are induced not only by the actions of polishing abrasives but also by mechanical actions produced directly by the polisher surface. The newly proposed polishing method makes good use of solutions or abrasives suspended in a slurry in order to control the influence of such mechanical actions.

Elastic emission machining (EEM)[29] and float polishing[30] are typical of this method. In EEM, a urethane rubber ball tool is forced against the work surface. When the spring-controlled ball rotates, a film of slurry is formed between the ball and the work surface. Polishing advances as the abrasives between the ball and the work surface attack the work surface. Deviation from the streamline is caused by centrifugal force since the specific gravity of abrasives is larger than that of the solution, as shown in Figure 6.19. This method provides a smooth surface, thus making it possible to accomplish high accuracy polishing by computer scanning the work surface with the ball.

In float polishing, a special machine is used that has ultra-precision rotary bearings of a hydrostatic type. For a flat work surface finishing, a tin polisher machined to a perfect flat mirror is used, against which work surfaces are forced and polished while a sufficient amount of slurry is fed. When the polisher revolution is suitable, the flow of slurry between the polisher and workpieces becomes turbulent. As a result, abrasives in the slurry attack the work surfaces. Such abrasive actions were observed during the research of hydrodynamic bearings.[31] Although the surface quality obtained with these polishing methods is outstanding, the stock removal rate is rather low; therefore, these are used for specific finishings.

The above method is also used for chemical polishing: in hydroplane polishing[32] and progressive mechanical and chemical polishing (P-MAC polishing)[33][34] of compound semiconductor wafers using bromine methanol solution without abrasives. Both methods make efficient use of the non-contact condition between work and polisher through improved machine structure or selection of dummy materials.

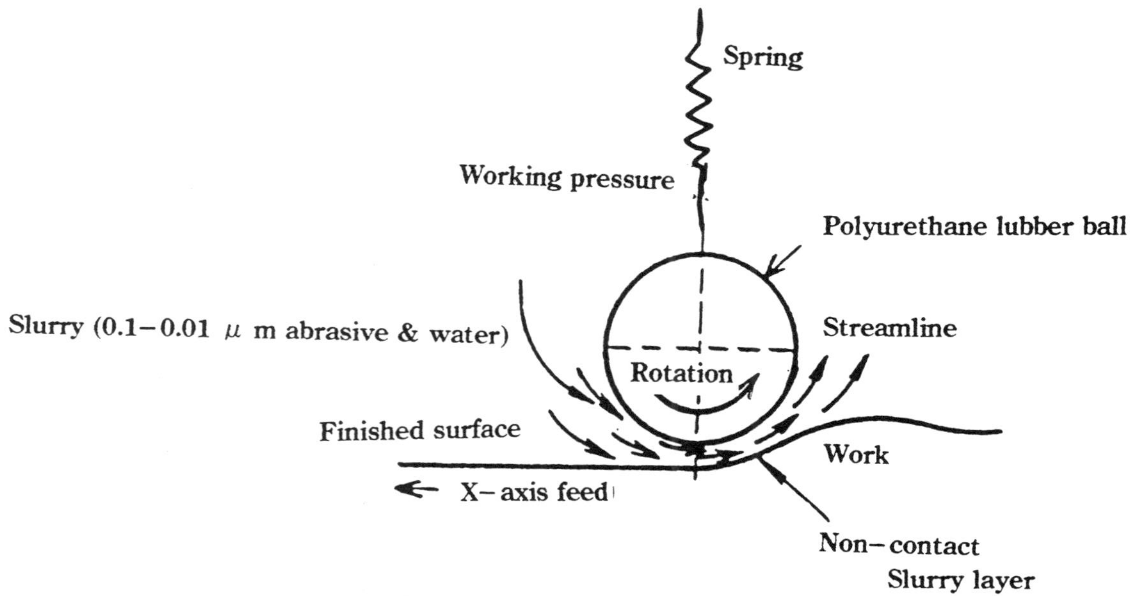

Figure 6.19: Ball Tool Action in EEM (elastic emission machining)[29]

5.0 CHEMICAL COMPOUND POLISHING

With an increasing demand for high function and high performance optomechatronics components, ultra-precision polishing, a final finishing process of the fabrication, has been recently recognized as the most critical step throughout the processes. For electronic components, for instance, with a growing tendency towards high density, extreme importance has been placed on the introduction of technologies to produce accurate topography and shapes in the order of 10^{-4} (micrometer) to 10^{-9} (nanometer). It is also indispensable from the viewpoint of material properties to produce a completely smooth mirror-like surface. By creating such high precision and high quality surfaces, the characteristics inherent to each functional material will be made available for full use, leading to the realization of high performance components. Meeting the above requirements, therefore, is of utmost importance for ultra precision polishing.

Recently, a variety of ultra-precision polishing methods were introduced. Regarding high precision polishing for glass, accurate and strain-free polishing has become available due to the improvements of processing conditions. Along with this, various processing methods by compounding chemical or other actions with mechanical action have been developed; these are used in agreement with the requirements of each material to be processed. Various areas including semiconductors, wet-type mechano-chemical polishing and colloidal silica polishing of crystals, are typical examples of chemical compound polishing.

In this article, various mechanical and chemical compound polishing methods will be reviewed such as wet-type mechano-chemical polishing and colloidal silica polishing. These permit the formation of a completely smooth crystal surface, i.e., a damage and strain-free mirror-like surface, that is indispensable for the opto-mechatronics components. Processing characteristics of crystal materials will be described here along with its processing mechanism.

Ultra-precision Chemical Compound Processing Methods

Processing Principles of Ultra-precision Polishing and its Positioning

In ultra-precision polishing, it is fundamental to produce a smooth mirror-like surface to the atomic order without disturbing the configuration of the surface crystal lattice (this is called "strain-free mirror-like finishing"). In other words, it is essential to make a processing unit in the order of an atom or molecule in order to remove only a few atomic layers from the surface of a workpiece. In the case of physical removal, the processing of a strain-free mirror-like surface is difficult as the region of strain propagation should be minimized to the atomic or molecular order.

Removal methods of workpiece surfaces include chemical and physical actions such as dynamic, mechanical, thermal, and electric energies. However, rather than using the above actions individually, better processing efficiency and surface quality are obtained by compounding physical and chemical actions together. Figure 6.20 illustrates the currently available physical (mechanical) and chemical compound processing methods.

Material removal in the polishing process is affected primarily by mechanical and chemical actions. Mechanical actions are considered to cover removal and frictional actions whereas chemical actions encompass dissoluble action and reaction products/film forming actions. By compounding these actions, novel ultra-precision polishing methods can be discovered, followed by the establishment of various characteristic processing conditions.

Table 6.5 summarizes the main processing factors, taking into account typical ultraprecision polishing methods and processing mechanisms. The following is a brief description of the typical ultra-precision polishing methods included in Table 6.5.

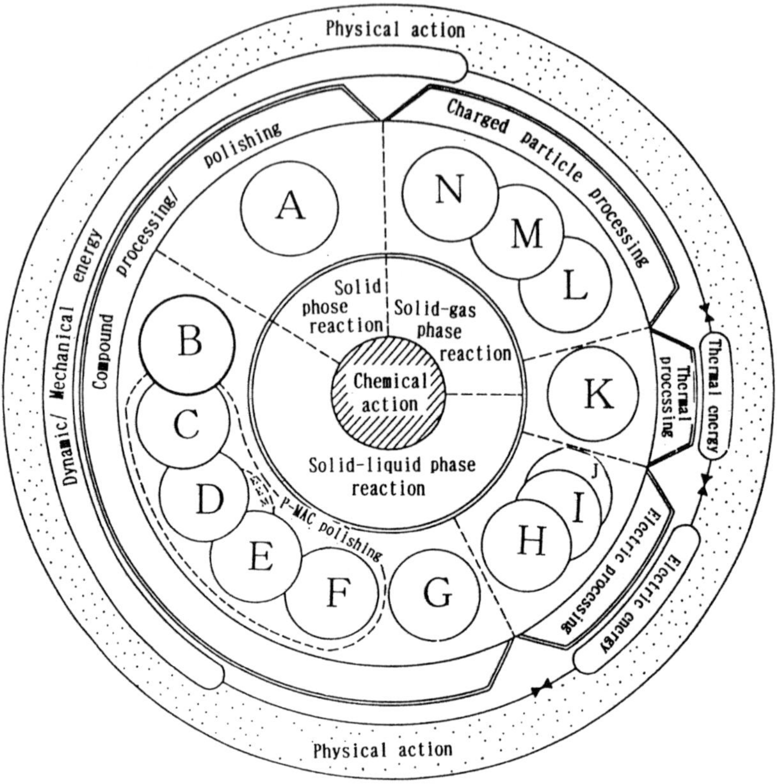

A: Dry type mechanochemical polishing
B: Wet type mechanochemical polishing
C: Replacement reactive type mechanochemical polishing
D: Non-contact type float polishing
E: Hydro-plane polishing
F: Disc type chemical polishing
G: Colloidal silica polishing
H: Electrolytic polishing
I: Compound processing under DC voltage
J: Compound processing of electrical discharge & electrolytic machining
K: Laser-asist etching
L: Reaction ion etching
M: Reaction ion beam etching
N: Ion beam asist etching

Figure 6.20: Physical/mechanical and Chemical Compound Processing Methods

Table 6.5: Main Factors for Polishing

Processing mechanism	Efficiency	Accuracy	Quality (low damaged layers)
[A] Mechanical Polishing	particle dia., hardness: large (diamond, alumina, cellium oxide, etc) pressure, relative speed: high	plate shape accuracy, system motion accuracy, particle dia. & distribution, polisher materials	opposite to what listed in "Efficiency"
[B] Wet-type Mechano chemical Polishing	reagent (chemicals) concentration, particle density: high atmosphere temp.: high	plate shape accuracy, system motion accuracy, plate/chemicals temp. control	reagent concentration: high atmosphere temp.: high particle density: low soft abrasives/soft polisher
[C] Dry-type Mechano chemical Polishing	particle material: reactivity against workpiece (solid-phase reaction), soft abrasives	plate shape accuracy, system motion accuracy	particle dia.: small particle material: reactivity against workpiece, soft abrasives
[D] Colloidal Silica Polishing	particle density: high, gelling speed, high pressure particle/workpiece: correlation, colloid chemical action	plate shape accuracy, system motion accuracy	particle density: high particle dia./distribution: small, particle/workpiece: colleration, soft particle, colloid chemical action
[E] Non-contact Polishing	particle dia., kinetic energy: large, number of particles: large, particle/workpiece: correlation	particle kinetic energy control, system motion accuracy, particle dia. & distribution	particle dia, kinetic energy: small, number of particles: large, particle/workpiece: correlation

[A] Mechanical Polishing. This corresponds to pitch polishing, wax polishing, and metallographic polishing. This polishing method progresses with a purely mechanical action by removing micro-regions of the surfaces by micro-cutting edges of individual abrasives that are basically slightly harder or similar to that of workpieces. This is the processing mechanism applied for the polishing of glasses, ceramics, ferrites, etc., with hard abrasives like diamond or alumina. Polishers, that is polishing pads, are usually made of soft metal such as tin, lead, copper, a pitch, or resin. A large number of abrasives get buried in a polisher in such a way that they are sustained by the polisher to enable its micro-cutting edges to act on the surfaces of workpieces thus producing fine scratches across the surfaces. Figure 6.21 illustrates the processing mechanism of this polishing.

[B] Wet-Type Mechano-chemical Polishing. Wet-type mechano-chemical polishing is a composite polishing activated by the mechanical action of abrasives and the chemical action of a polishing reagent; this is sometimes distinguished from the mechanical and chemical polishing that gains polishing rates similar to the total of mechanical and chemical removal rates. Ultra-precision polishing uses a slurry with $Si0_2$, fine abrasives dispersed in a mild alkali solution, and worked on Si wafers for LSI to produce a smooth, strain-free, mirror-like surface; this is a typical example of wet-type mechano-chemical polishing.[35]

By using chemicals as a polishing reagent to chemically induce etching on a workpiece material, fine abrasives promote chemical reactions (solid-liquid phase reaction) between the reagent and material, and simultaneously make surface irregularities more planar. Figure 6.22 shows a chart of the polishing promotion factors for wet-type mechano-chemical polishing. This issue will be discussed later.

[C] Dry-Type Mechano-chemical Polishing. Mechano-chemical polishing under dry conditions is activated by chemical reactions (solid-liquid phase reaction) between the workpieces and abrasives; this causes removal of the work surface with particles softer than the workpeice material. The typical example of this is the polishing of sapphire with fine abrasives of SiO_2.[36] When $Si0_2$ particles act upon sapphire ($A1_20_3$) surfaces, the contacting area

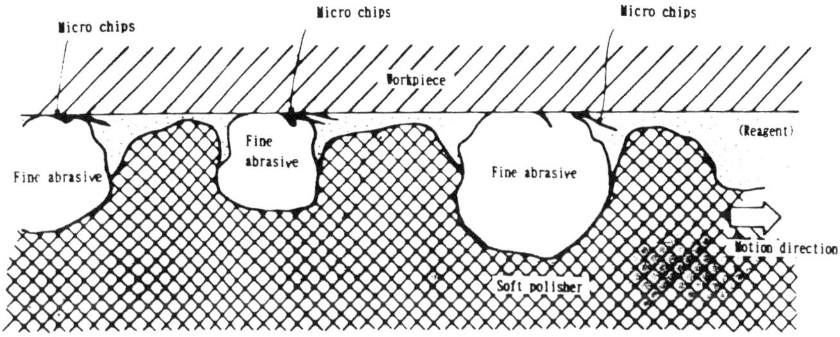

Figure 6.21: Processing Mechanism of Mechanical Polishing

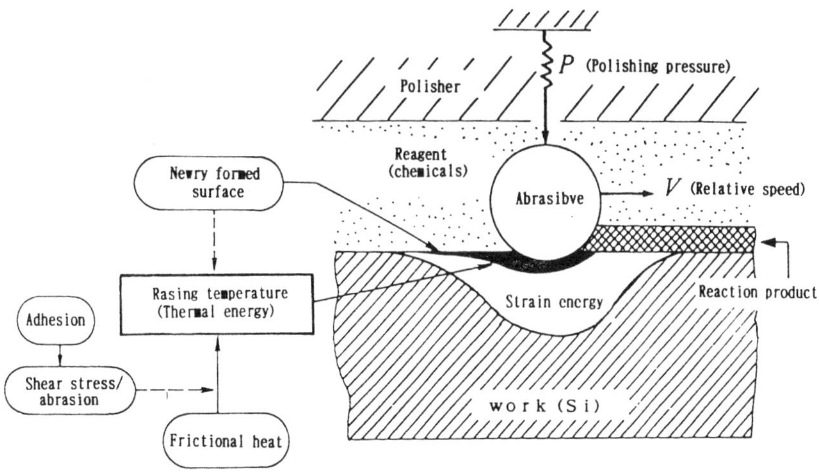

Figure 6.22: A Chart of the Polishing Promotion Factors for the Wet-Type Mechano-Chemical Polishing

becomes high pressured under high temperature, and $Si0_2$ and $A1_20_3$ initiate a solid-phase reaction; this forms what are softer mullites than $A1_20_3$ in the contacting area which will be later removed by $Si0_2$ particles. By repeating the cycle of generation, removal, and generation of reactive products, dry-type mechano-chemical polishing produces a more efficient polishing without inducing damaged layers as in the diamond particle polishing mentioned in [A].

[D] Colloidal Silica Polishing. Colloidal silica polishing is a processing method based on the characteristic colloidal phenomenon through the use of a slurry with super fine particles of several 10 to several 100 Å suspended in a pH controlled processing solution (e.g. alkali solution of pH.~12). Very thin colloidal silica exists between an artificial leather-like soft polisher and the workpiece. Once a shearing force is set to work on a horizontal direction under high pressure, the gelation phenomenon inherent in colloids is induced temporarily (this is called "dilatancy phenomenon"). In this mechanism, polishing is accelerated by the mechanical micro-removal action of an atomic/molecular order mainly through cohesive and abrasive actions of $Si0_2$ fine particles that are held in the apparently gelated solutions (Figure 6.23).[37]

Since the before-mentioned [B] and [C] are dependent on the chemical properties of workpieces, it can only be applied to a limited range of materials. However, this polishing method can process almost all kinds of materials to a strain-free, mirror-like surface. This depends, however, on the hardness of workpieces that sometimes cause degradation of processing efficiency. There are other polishing methods such as Si wafer processing that combines the polishing mechanism of [B] above with the present polishing mechanism. The polishing method relevant to this will be discussed later in this chapter.

[E] Non-Contact Polishing. This method corresponds to the micro-minimized processing reaction of [A] above to the order of 1/10~1/100. The particles of approximately 100 Å diameter act on the surface atoms of workpieces thereby removing several to several tens of atoms. This representative example is based on what was proposed as EEM (elastic emission machining)[38] as seen in Figure 6.24 and is applied to the non-contact processing such as

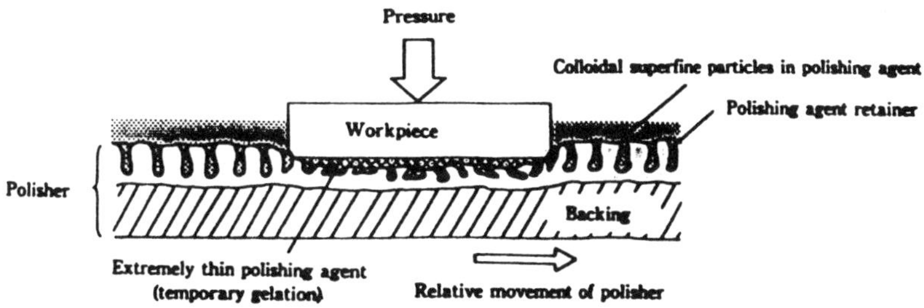

Figure 6.23: Constitutional Diagram of the Collodial Silica Polishing

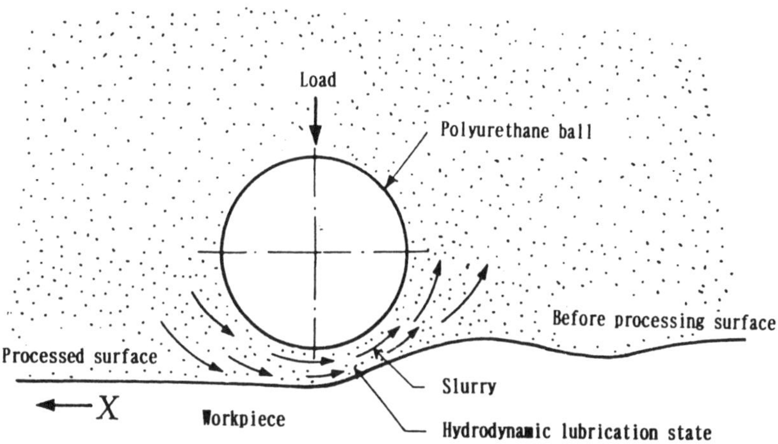

Figure 6.24: A Processing Principle of EEM (elastic emission machining)

"float polishing".[39]

When the processing unit diminishes to the atomic or molecular order, particle surface atoms join with the workpiece surface atoms without inducing plastic behavior, followed by the destruction of the joints. The polishing progresses this way. Consequently, there is almost no mechanical damage left after the atoms are removed. As regards the polishing methods referred to in [A] through [D] above, abrasives retained by a polisher or, in some cases, in a solution, act upon workpieces. On the other hand, EEM and EEM-applied non-contacting polishings by the collision of abrasives with workpieces are polishing methods that accomplish high quality and high precision polishing.

In general, processing efficiency, processing precision, and depth of damaged layers are reciprocally related. For those chemical reactions included in the processing mechanism such as mechano-chemical polishing, the higher the processing efficiency, the fewer the damaged layers. However, since the processing surface precision degrades as chemical actions become large, it is suggested that chemical reactivity be controlled in order to obtain better precision.

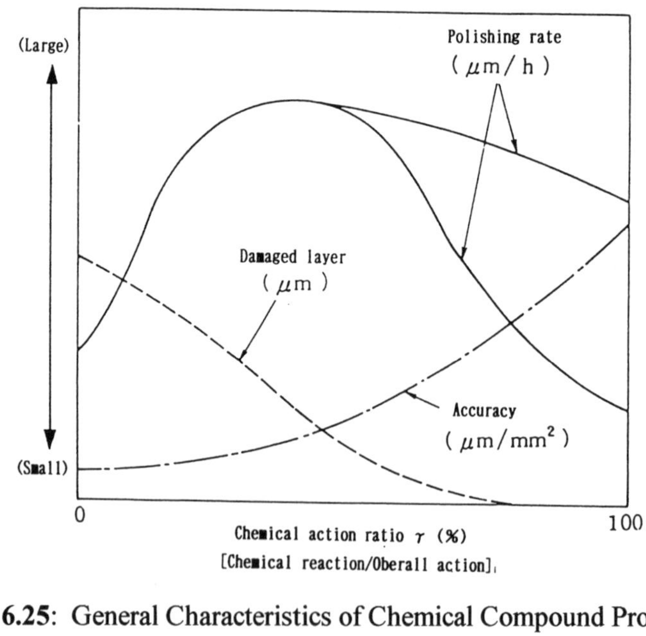

Figure 6.25: General Characteristics of Chemical Compound Processing

Figure 6.25 shows its general characteristics. With lapping, the efficiency increases approximately in proportion to the diameter of abrasives used; however, such a relation is not greatly noticeable when polishing with particles below 1 µm. For polishing with particles below 0.1 µm, the magnitude of the number of active particles has a large effect on the efficiency. In other words, high efficiency can be achieved by increasing the processing pressure with the polisher uniformly working on the surfaces of workpieces. On the other hand, since transcribing the polisher surface is a basic principle, processing precision is largely dependent upon the precision of the polisher or the plate on which workpieces are mounted. However, in the case of processing methods such as EEM which scans the area to be processed, the efficiency depends primarily upon its scanning accuracy.

Wet-Type Mechano-Chemical Polishing

Outlook

This chapter deals with chemical compound polishing, namely mechano-chemical polishing, combining mechanical removal action (mechanical polishing) of abrasives with chemical action generated by the processing reagent or abrasives.

Referring to the concepts of M. Fink,[40] T. Kubo,[41] and T. Sakurai,[42] on mechanochemistry, mechano-chemical phenomena can be defined as follows:

1) The mechanical energy (e.g., mechanical shock, grinding, rolling, tension, application of pressure, plastic deformation, etc.,) applied to a solid material causes the solid surface to change its physical and chemical properties.
2) It brings about chemical changes to the gas and liquid materials around a solid material.
3) It induces or accelerates direct reactions between the ambient gas or liquid and the solid material.

In wet-type mechano-chemical polishing, sufficient information is available that suggests that the above phenomena occurs depending on the selection of workpiece materials, abrasives, and reagent.

In accordance with the definition of the mechano-chemical phenomenon, the final finishing process of workpieces can be advantageously carried out by creating an ambient atmosphere in which chemical reactions occur in the case of conventional polishing. In other words, when viewed micro-graphically, it is presumed that abrasives mechanically act upon the work surfaces inducing high pressure and high temperature at its contacting areas. This presumption suggests a large chance of producing changes in the physical and chemical properties of the processing surfaces. As a consequence, by placing such substances in a processing atmosphere that initiates chemical reactions, polishing accompanied by mechano-chemical phenomenon, namely mechano-chemical polishing, is realized.

With regard to chemical reactions, any solid phase reaction, solid-liquid phase reaction, or solid-gas phase reaction is applicable. Dry-type mechano-chemical polishing for sapphire with $Si0_2$ fine particles is a typical example of the solid phase reactions. On the other hand, the wet-type mechano-chemical polishing for Si single crystals with alkaline reagents, demonstrates solid-liquid phase reactions. This wet-type mechano-chemical polishing has a mitigating effect on some mechanical actions and will prevent the generation of polishing defects, like scratches, if the appropriate environmental conditions are provided. However, since soft polishers are generally used to produce high-quality, strain-free, mirror-like surfaces, edge-turndown becomes predominant in comparison with dry-type mechano-chemical polishing.

Wet-type mechano-chemical polishing is diverted from the Si mechano-chemical polishing of "copper replacement type" presented by E. Mendel, et al in the latter half of the 1960s.[43]

By utilizing replacement reactions between Si and Cu, the chemical reactions are accelerated at the same instant that the Cu ion precipitates on the processing surface, using a mixed solution of NH_4F and $Cu(N0_3)_2$, and is mechanically removed by a soft polisher. Currently, this method is not applied to the polishing of a final finishing process of Si wafer for LSI due to some problems arising on the residual Cu ion and surface roughness (smoothness), although the processing efficiency is very high.

In general, processing efficiency and processing precision/depth of damaged layers are reciprocally related. For those polishings whose chemical reactions are incorporated into the processing mechanisms such as wet-type mechano-chemical polishing, the greater the chemical reactions the greater the processing efficiency with only a few damaged layers. However, the processing surface precision decreases. Therefore, good precision quality has to be achieved by controlling chemical reactions to some degree. Figure 6.25 shows the general characteristics of a wet-type mechano-chemical polishing.

The following section on wet-type mechano-chemical polishing called "mechano-chemical polishing" is based on some crystalline materials as an example.

Mechano-Chemical Polishing of Silicon Wafer for Semiconductor

[A] Characteristics of Mechano-Chemical Polishing.[35] Wet-type mechano-chemical polishing of Si wafer for LSI has proved effective as a high-efficiency, strain-free, mirror-like polishing and is technically recognized as the best polishing method. When referring to the DRAM of M bit order for ULSI, since the minimum pattern rule of the circuits becomes half submicron, a demand for higher precision for Si wafer becomes stronger. In general, Si wafers are polished using an artificial leather-made polisher and a slurry made from ultra-fine SiO_2 particles (approximately ø 100 Å) suspended in an alkaline solution of approximately pH 10. Actually, 1st, 2nd, 3rd, and even 4th polishings are carried out with a combination of slurry and polisher selected to the best of each step (Table 6.6).

Figure 6.26 illustrates the relation between atmosphere temperature of various processing surfaces (temperature of slurry, polisher surface, etc.,) and removal rate. Mechano-chemical polishing rate V_{MCP} makes an exponential increase as the temperature of the polisher surface increases; about 55 µm/h at 50°C, which is approximately 3 times that of room temperature. Similarly, a disk-type chemical polishing rate $V_{D\text{-}CP}$ without abrasives also increases as the temperature of the polisher surface increases; however, at room temperature, its polishing rate is as low as 1/20 that of V_{MCP}. We see here the big role that abrasives in slurry play on V_{MCP}.

From the above, it is clear that synergetic effects of the mechanical actions of particles and the chemical actions of reagents are distinctly effec-

Table 6.6: An example of Polishing Conditions of Si Wafers for LSI

Process	Polishing Reagent	Polisher	Processing Pressure	Stock of Removal	Main target
First Polishing	Abrasives of $Si0_2$ or $Zr0_2$: 0.1 μ m Reagent: alkaline solution (ph 9~11)	Impregnated polyurethane, Polyester non-woven fabric polisher	300~800 g/cm^2	15~20 μm	High efficiency, Mirror like surface, (Rmax 20~40Å)
Second Polishing	Abrasive of $Si0_2$:average ϕ 100~200Å Reagent: pH9~11	Forming polyurethane (artificialleather), or a soft &hard double layered polisher	100~300 g/cm^2	several~1 μm	OSF-free, Improvement of surface roughness (Rmax 10~20Å)
Third & Fourth Polishing	Abrasives of $Si0_2$:ϕ 100~24Å Reagent: ammonia or amine (pH 8~10)	Forming polyurethane (artificial leather), or a soft & hard double layered polisher	below 100 g/^{cm}	0.1~several decimal point μm	Haze-free, Low concentration of impurities

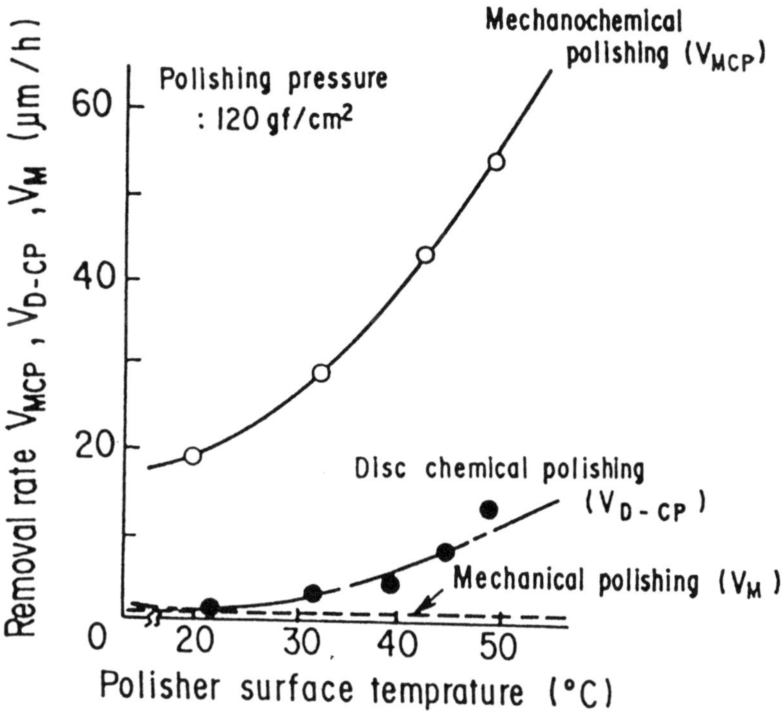

Figure 6.26: Effect of Polisher Surface Temperature on Removal Rate tive.

The Si wafers, using colloidal silica with ultra-fine particles ($Si0_2$) suspended in alkaline solution (reagent), show very high quality surface with surface roughness R_{max} of 20~10Å and flatness of about 1 μm without any resultant defect from the processing.

The afore mentioned mechano-chemical polishing method is widely applied to the strain-free, mirror-like finishing of compound semiconductors such as GaAs, InP, CdTe, etc., for high speed LSI, optical IC, and various kinds of sensors. The polishing of wafers for compound semiconductors will be discussed later.

Processing Mechanism

In the processing system of this wet-type mechano-chemical polishing, removal rate V(μm/h) excluding V_M is expressed by applying the Ahrenius' rate equation as:

$$V = V_0 \exp\left(-\frac{E}{R(T+\Delta T)}\right) \qquad (6.2)$$

$$= V_0 \exp\left(-\frac{E_0 - E_\alpha}{RT}\right) \qquad (6.3)$$

where R is the gas constant, T is the temperature (K) of the reaction system, ΔT is the temperature increased during the processing (K,$O< \Delta T/T<1$), and E_0 stands for the activation energy (kcal/mol) inherent in the reaction system of the slurry and workpieces used. E_α corresponds to the amount (kcal/mol) that the activation energy E_0 is apparently lowered by the "thermal energy" generated from the frictional heat or by the "strain energy" produced by the mechanical actions of the fine particles on the surfaces. V_0 here is the constant and can be considered as the removal rate when $E_0 = E_\alpha$ or T is ∞. According to the equation of the above (6.3), the processing efficiency improves as the temperature of the processing system increases, and as the apparent activation energy (E_0 - E_α) is lowered by the strain or friction heat produced by the increasingly strong mechanical actions of the particles.

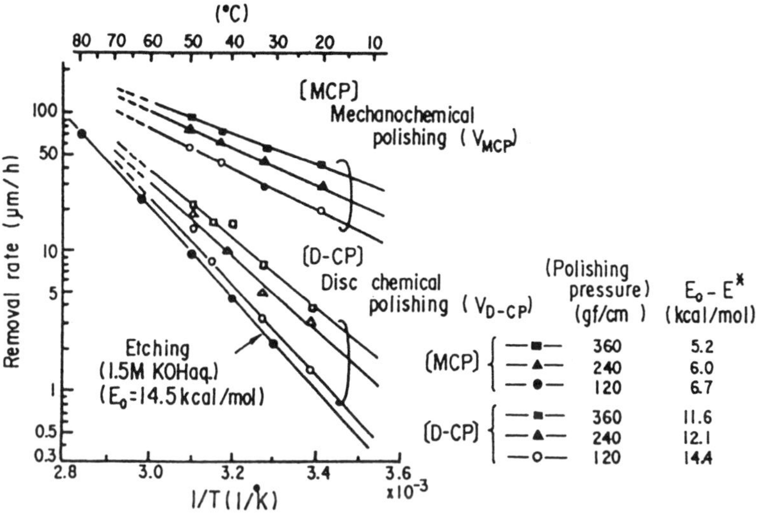

Figure 6.27: Temperature Effect on Removal Rate (Arrhenius Plots)

Figure 6.27 shows the removal rate (Arhenius-plotted) by changing the temperature of the polisher surface. The graph supports the previous discussions; it shows that the apparent activation energy decreases (inclination of straight line is reduced) as the concentration of abrasives or processing pressure increases.

We will determine by what factors a certain amount (E_α) of the apparent activation energy was dropped. Calculating the processing strain energy backwards from the experiment of etching on mechanical polishing surfaces with damaged layers, it is presumed to be below 1% of the activation energy E_0. Accordingly, E_α can be considered to be the thermal energy caused by the friction between the abrasive grains and the polisher.

Figure 6.28 illustrates the results of the investigation of the correlation between the calorific heat value Q calculated from the frictional force F (obtained by measuring the horizontal frictional resistance during the polishing process) and the thermal energy E (kcal/h) obtained based on E_α. E and Q are in the corresponding relation:

$$E = \alpha Q - \beta \qquad (6.4)$$

The coefficient α indicates the conversion rate of the calorific heat value Q which becomes thermal energy E to contribute to the processing.

Subsequently from the above, the frictional heat between abrasives and Si wafer surfaces has attracted attention. However, judging from the fact that the removal rate V_{MCP} is not dependent upon the hardness or the size of the abrasives, the cause of friction can be considered attributable not to the simple scratching actions but to the adhesion. Adhesion easily occurs when two materials' mechanical properties are similar to each other or are composed of the same kinds of material that mutually exert sliding motions. An experimental mechano-chemical polishing of Si wafer will be carried out using as abrasives the fine Si particles (approximately ø 100~200Å) of the same quality as the Si wafer sample and considered to be most easily adhered.

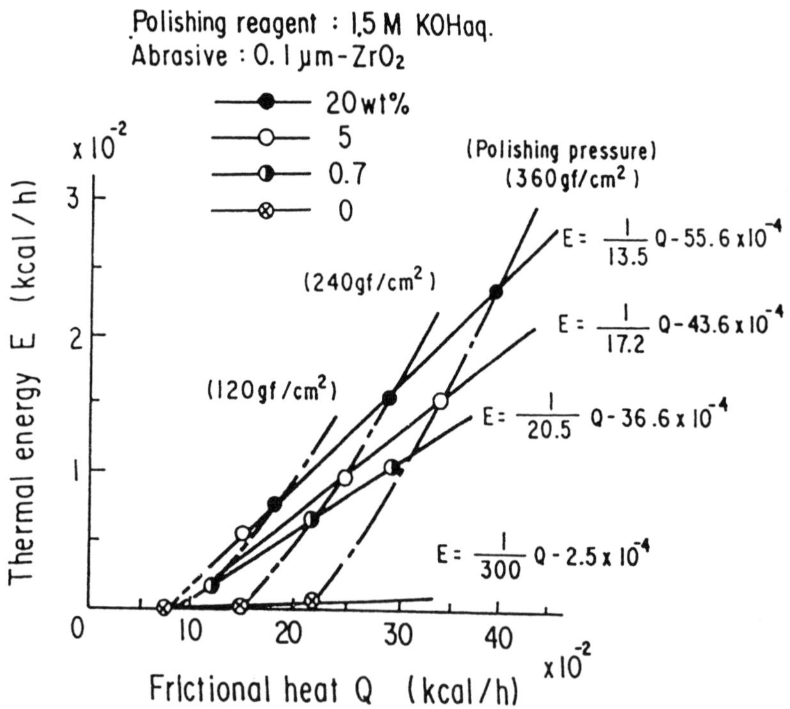

Figure 6.28: Relation between Frictional Heat Q and Thermal Energy E

Figure 6.29 shows the comparison between the removal rate V_{MCP} and V_m of various abrasive grains, each contained in the slurry of pH 12 and pure water (pH 6). Although the removal rate V_M with Si fine particles is almost none, when the chemical action ratio is compared (see footnotes of Figure 6.25), it is observed to be in the relation of γ (Si) > γ (ZrO_2) ≥ γ (SiO_2). When Si fine particles are used, V_M in particular approaches zero (i.e. $\gamma \geq 99$), which is attributed to the adhesion between the Si wafer (sample) and Si fine particles. Consequently, very few scratches are supposed to be produced on the processing surfaces.

Figure 6.30 shows the comparison of the mechano-chemical polishing rates V_{MCP} of Si wafers between the two different slurries: one, with colloidal silica (ø 100~200 Å fine particles suspended in the alkaline solution of pH11) which is used as a slurry for the final finishing process of Si wafers, and another, with colloidal silica and the afore mentioned Si fine particles. The effect is obvious from the comparison showing that by adding Si fine particles, the removal rate easily doubled over that of the slurry without Si particles.

After mechano-chemical polishing, scratch-free, mirror-like surfaces with the roughness R_{max} of some 10 Å are achieved. In order to evaluate the

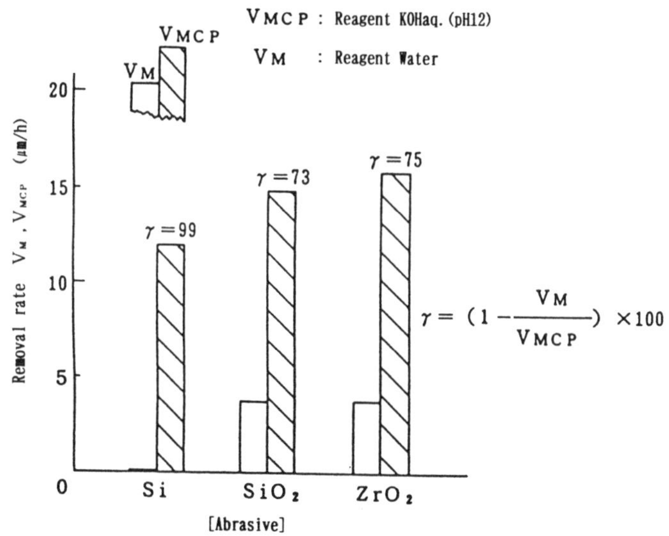

Figure 6.29: Comparison between the Removal Rate v_{mcp} and v_m of Various Abrasives

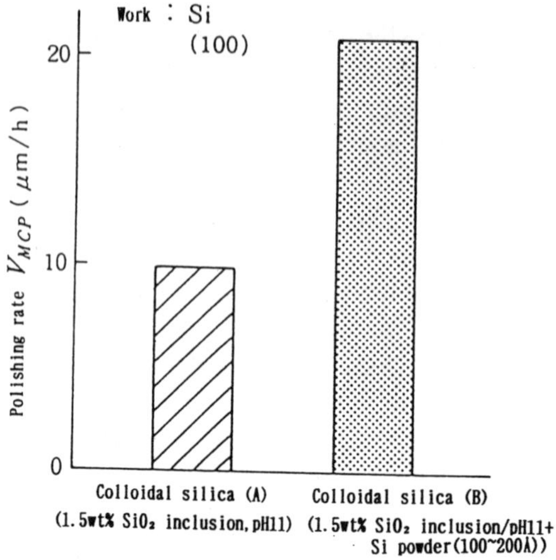

Figure 6.30: Comparison on Polishing Rate of the Slurry [A] with the Slurry [B]

crystallization of the surfaces, an OS-check (microscopic observation after Wright etching and after heat treatment at 1100°C for 2 hrs.) was carried out and the surfaces were confirmed to be free of OSF (Oxidation-induced Stacking Fault).

Figure 6.22 illustrates the activation factors of the chemical action speed V_c of the working area of abrasive grains in relation to the mechano-chemical polishing of the Si crystals mentioned in the preceding paragraphs. According to the study made so far, strain energy is approximately 1% of E_0 and almost all of E_α is thermal energy. Furthermore, it is likely that the E_α is an exothermic action produced from the adhesion and separation of the crystals (workpieces) from the abrasives.

Mechano-Chemical Polishing of Wafers for Compound Semiconductors

For the polishing of wafers for compound semiconductors such as GaAs, InP, CdTe, a mixed solution of bromine (Br) and methanol or a sodium hypochlorite (NaClO) solution have conventionally been used. These solutions have the following characteristics: a) they have a short life, b) they emit irritating odors due to their strong oxidative state, and c) they require special polishing equipment, working environments, and facilities.

Under these circumstances, the authors have been working to develop a new harmless high-efficiency slurry based on the processing mechanism of the mechano-chemical polishing. The following describes the mechano-chemical polishing of wafers for compound semiconductors by using GaAs and CdTe crystals.

[A] GaAs Crystal Wafer.[44] Considering the common points in the conventional slurries applied so far to the GaAs crystals, it has been revealed that slurries are oxidizing agents containing either Na or Br. The research has shown sodium bromite ($NaBr0_2$), which is inexpensive and used as a desizing agent for fiber processing and as a slime control agent for industrial water, does not produce a toxic gas, and its oxidization action is very mild.

When disk-type chemical polishing was carried out only with $NaBr0_2$ solution, it has been observed that reaction products partially left a "film (or stain)" on the work surface which then made the surface rough. However, when a DN agent (non-ionic solution), which has a surface penetration function, is added to the $NaBr0_2$ solution in order to remove the "films", a smooth mirror-like surface is obtained.

Furthermore, when fine particles are mixed with the above, the reaction products are positively wiped off thus achieving a high efficiency. Figure 6.31 shows the mechano-chemical polishing rate of GaAs crystals with the slurry of $NaBr0_2$ solution (0.6% of NaOH included) to which the DN agent and $Si0_2$ fine particles were added. As expected, the removal rate increases as the size of $Si0_2$ particles become large. With this mixed slurry applied to the polishing of a GaAs wafer, a surface roughness R_{max} of below 10Å was obtained as shown in Figure 6.32.

Based on these polishing characteristics, the following describes the processing mechanism.

① Both Ga and As are oxidized by the $NaBr0_2$ solution and produce gallium oxide (Ga_20_3) and diarsenic pentoxide (As_2O_5) as a reaction product respectively.

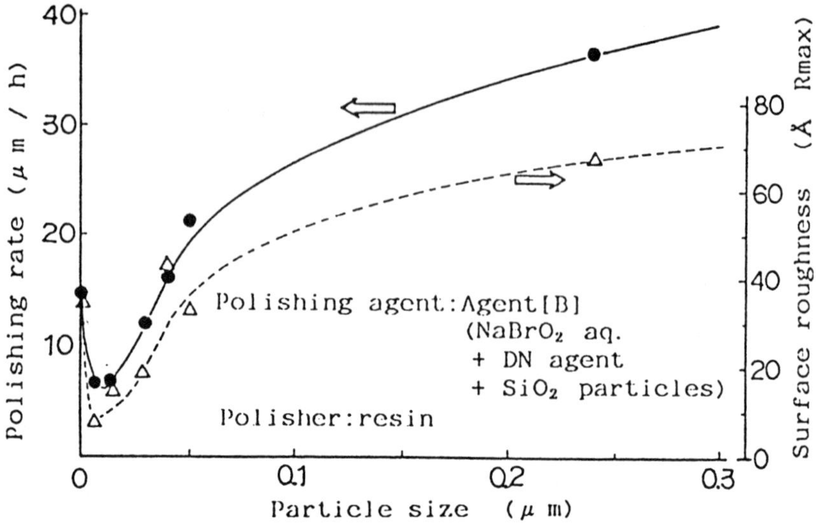

Figure 6.31: Relation between SiO_2 Particle Size, Polishing Rate, and Surface Roughness when using β -slurry (Specimen:GaAs)

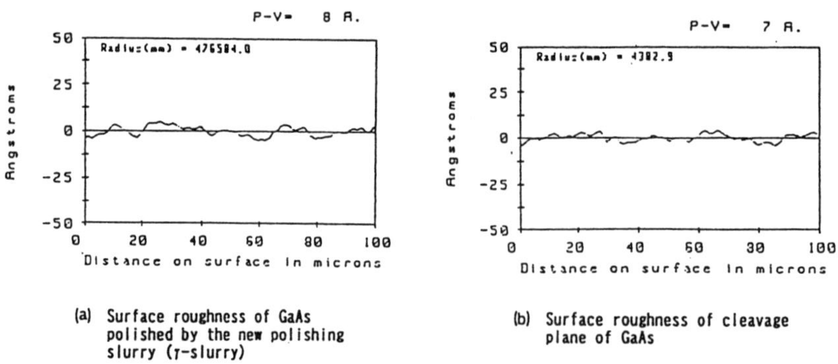

Figure 6.32: Comparison of Surface Roughness by the New Polishing Slurry γ - Slurry: $(C_4H_9)_4N \bullet BrO_2$ solution + SiO_2 (a), with cloven surface (b) specimen: GaAs

$$2GaAs + 4NaBrO_2 \rightarrow Ga_2O_3 + As_2O_5 + 4NaBr \tag{6.5}$$

② Immediately after the As_2O_5 is produced, it is converted into sodium arsenate (Na_3AsO_4) by NaOH.

$$As_2O_5 + 6NaOH \rightarrow 2Na_3AsO_4 + 3H_2O \tag{6.6}$$

③ Ga_20_3 becomes gallium hydroxide $Ga(OH)_3$ by alkali hydroxide.

$$Ga_2O_3 + 6OH \rightarrow 2Ga(OH)_3 + 3O^{2-} \tag{6.7}$$

This $Ga(OH)_3$ easily dissolves in sodium hydroxide.

④ During the reactions of the above ① ~ ③ and while surface activity penetration function is working, the above chemical actions are further accelerated by the mechanical action of the added fine particles and the frictional exoergic reaction with the processing surface.

From the above processing mechanism of GaAs, it has been proved that a slurry with an oxidant to which a very small quantity of alkali hydroxide is added is basically effective for the mechano-chemical polishing of GaAs crystals. It is certainly necessary to aim to achieve high quality surfaces by adding a DN agent with a surface penetration function allowing it to remove the reaction products (films, stains) off the surfaces.

An experimental polishing of GaAs has been made using as a slurry an oxidant H_20_2 with a very small quantity of NaOH. Figure 6.33 is an example of the processing characteristics with a H_20_2 - NaOH type solution to which a DN agent is added. Maximum removal rate is given at around pH = 10. Thus, similar removal rate and surface quality to those of $NaBr0_2$ type solutions have been obtained. However, although $Si0_2$ fine particles are added, the increase of the removal rate will not be as remarkable as that of $NaBr0_2$ solution. This is presumably because reaction products are not formed on the surfaces as solidly as $NaBr0_2$ type solutions.

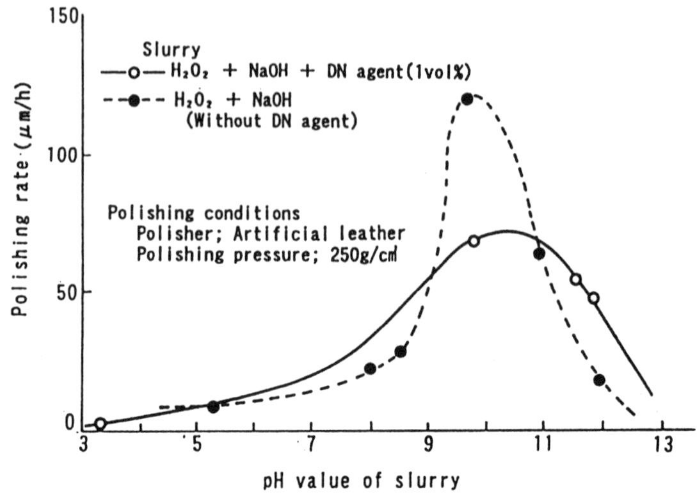

Figure 6.33: An Example of Polishing Characteristics of GaAs by using a Slurry Containing H_2O_2 (oxidizing agent), NaOH (alkali), and DN Agent (non-ionic solvent)

[B] CdTe Crystal Wafer.[45] CdTe crystals are compound semiconductors belonging to the II-VI group while the above referred GaAs crystals belong to the III-V group. CdTe has come into focus particularly as an infrared/ radiation detector, light emitting diode, and a solar battery substrate. CdTe crystals, like GaAs crystals, have a marked cleavage and are very soft (Moh's hardness of GaAs and CdTe is 4.5 and 2.8 respectively).

When establishing a selection guide of polishing slurries for CdTe crystals based on the processing mechanism of CaAs crystals, this requires processing solutions that oxidize CdTe surfaces. An assumption was made to remove (with a polisher or fine particles) the oxide, like $CdTe0_3$, formed by such a processing solution.

Figure 6.34 indicates the removal rate of the CdTe crystals when the oxidizing bleaching agent that is harmless and commonly used at home is primarily used; it clearly shows the effect of the oxidizing bleaching agent and fine particles ($Si0_2$). Through the microscopic observation of the processed surfaces, it has been found that more uniform surfaces were produced when DN added-polishing slurry was used as in the mechano-chemical polishing of

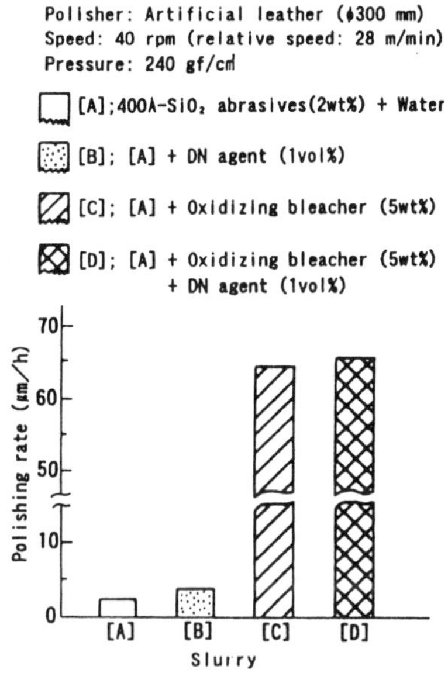

Figure 6.34: Polishing Rate by Various Slurries/Workpiece: CdTe

GaAs wafers. However, its effect is not so obvious as that of GaAs crystals. The surface roughness R_{max} was below 20Å, which means that better surface roughness has been efficiently obtained when compared to the conventional polishing slurry.

From the processing experiments of CdTe crystals and the results of its chemical analysis, the following will be drawn as a processing mechanism of the mechano-chemical polishing of CdTe crystals.

First, by creating an atmosphere that makes the CdTe surface easily oxidized by a DN agent, the substrate surface of CdTe becomes oxidized by an oxidizing bleaching agent, and finally some kind of reaction product is deposited on the surface. Then, oxide films or reaction products deposited on the surface are mechanically removed by abrasive grains or a polisher. By repeating these processes, a high-quality surface is produced.

We have discussed the processing mechanism of GaAs and CdTe crystals for compound semiconductors and drew a conclusion that its processing mechanism is basically the same as that of the mechano-chemical polishing of Si crystals.

Colloidal Silica Polishing

Basic Processing Characteristics

Table 6.6 shows an example of the composition and properties of colloidal silica. This is an example of the alkaline solution (pH 9.5 ~ 10.5) in which $Si0_2$ fine particles (of ø 100 ~ 200Å) are suspended by 38 wt%. In general, gelling of colloidal silica is controlled by the addition of alcohol. By the addition of salt, like sodium sulfate (Na_2SO_4), gelling is accelerated.

The relationship between the processing pressures and the removal rates of polishing are shown in Figure 6.35, using colloidal silica on a variety of single crystals such as GGG, $LiTa0_3$, Si, and sapphire which are chemically and mechanically very stable. Removal rates of all crystals increase in proportion to the increase of processing pressures; particularly high pressures were needed for the hard sapphire single crystals.

Referring to the removal rate shown in Figure 6.35 as a parameter, the relation between Vickers hardness of several crystals and processing pressures necessary to obtain a given constant removal rate is illustrated in Figure 6.36. In the case of Si single crystal recognized clearly as affected by the chemical action of the alkaline solution, the stock of removal in the figure is considerably off the straight line, achieving as a consequence a high removal rate at low polishing pressure. For the rest of the crystals, the polishing pressure P(kg/ mm^2) necessary to obtain the removal rates V(μm/h) is expressed in the following equation.

$$P = 10^4 \exp(K_1 \cdot H_v) \tag{6.8}$$

Table 6.7: Composition of Colloidal Silica and its Characteristics (for the first polishing)

SiO_2 content	50 wt%
NaO_2 content	below 0.4
Hydrogen ion exponent (pH)	10.8~11.2
Particle diameter (at 20°C)	below 25 CP
Specific gravity (at 20°C)	1.36~1.40
Appearance	transparent milky colloidal solution
Freezing point	0°C
Stability	semi-permanent
TEM photo of colloidal silica (Transmission Electron Microscope)	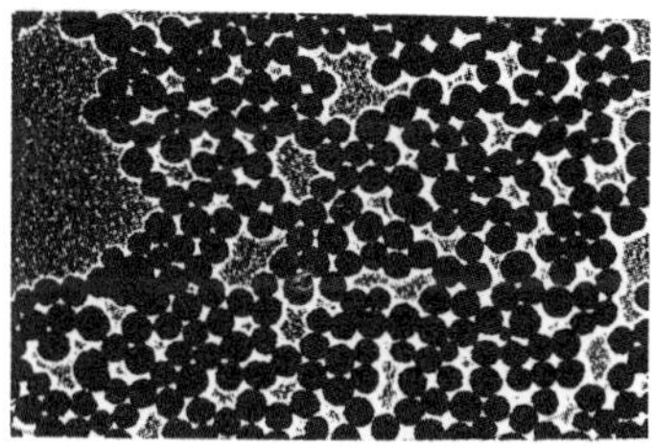

where K_1 is a constant determined by the processing conditions and H_v is Vickers hardness (kg/mm²). The constant K_1 can be obtained with the equation, $K_1 = 2 \times 10^{-4} V + 1.4 \times 10^{-3}$, and when its relation is substituted for the equation (6.8), it will be expressed as:

$$V = 5 \times 10^{-3} \left(\frac{1}{H_v} \ln 10^4 P - 1.4 \times 10^{-3}\right) \tag{6.9}$$

Consequently, the removal rate V of various crystals using colloidal silica is inversely proportional to the Vickers hardness H_v of crystals. This is quite similar to the tendency of abrasion loss produced by cohesion or ploughing abrasion which is proportional to the inverse number of the hardness H_v.

In the case of polishing with colloidal silica, almost no mechano-chemical effect has been recognized with the exception of Si single crystal;

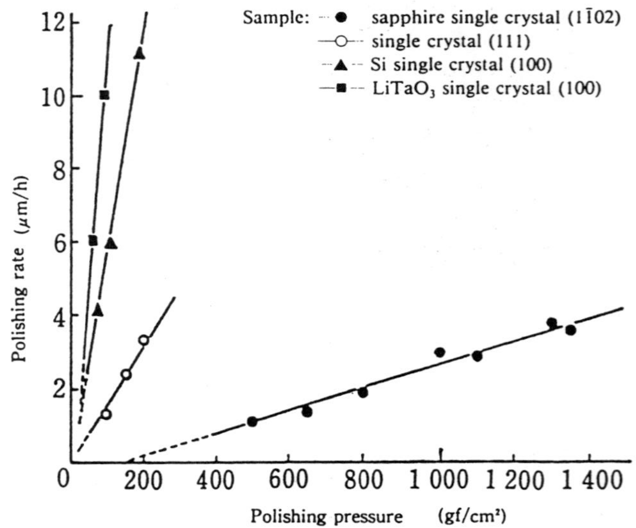

Figure 6.35: Relation between Polishing Pressure and Polishing Rate of Various Single Crystals using Colloidal Silica. Polishing Agent: colloidal silica. Polisher: artificial leather (300 Mm Dives.), 60 rpm.

- • Sapphire Single Crystal (1102);
- ○ Single Crystal (111),
- ▲ Si Single Crystal (100), and
- ■ $LiTaO_3$ Single Crystal (100)

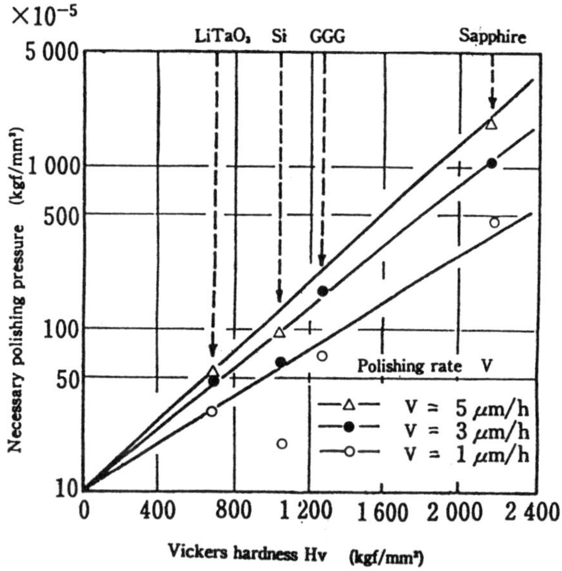

Figure 6.36: Relation between Necessary Processing Pressure and Vickers Hardness of Various Kinds of Crystals.

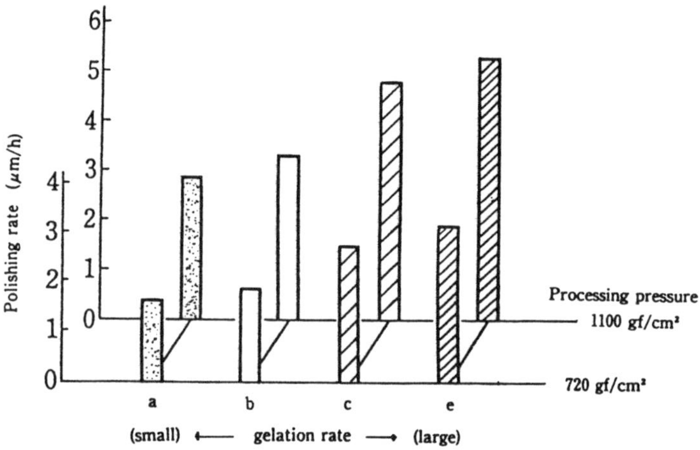

Figure 6.37: Comparative Processing Rates of Sapphire Single Crystals at Different Gelation Rates of Colloidal Silica. a: Ethylene Glycol (2.5 wt.%) + colloidal silica; b: colloidal silica (no additive); c: $NaSO_4$ (1 wt.%) + colloidal silica; d: Na_2SO_4 (1.3 wt.%) + colloidal silica

however, it has been understood that with micro-removal actions mechanically induced by soft SiO_2 fine particles, strain-free polishing is performed as described in the previous chapter.

Based on the processing mechanism, we have also studied high-efficiency polishing. During polishing, frictional heat or a strong shearing stress is imposed under high pressure on an ultra-thin film-like colloidal silica found between a polisher and crystal substrate (sample). If such situations are created, it is likely that a reversible gelling phenomenon (dilatancy) temporarily appears as is often seen in colloids. If so, we think that the retaining force of SiO_2 fine particles becomes intensified as if SiO_2 fine particles were retained or fixed in the slurry which will help to efficiently and effectively promote polishing.

Figure 6.36 compares the processing characteristics after adjusting the gelling speed of the colloidal silica as slurry by adding sodium sulfate ($Na2SO_4$) or ethylene glycol. When sodium sulfate is added to colloidal silica, the removal rate increases as the ratio of the addition increases. To the contrary, when ethylene glycol is added, the removal rate decreases by 10 to 20%.

Applications to Functional Materials

[A] Silicon Wafer for ULSI. Colloidal silica polishing is typically used and highly ranked for the finishing process of the silicon wafers for ULSI. This process produces excellent, high-efficiency, and strain-free mirror-like surfaces. As the design-rule of circuits in LSI has become submicroscopic, requirements for the processing precision of silicon wafers has become more strict. As shown in Table 6.6, 1st to 3rd and sometimes 4th polishes are currently performed for ø 8~12" wafers aiming to accomplish surface roughness R_{max} of several Å to 10Å, TTV (total thickness variation) of 1~2μm, and flatness of 1~3μm.

Since a silicon wafer is active against alkali, processing is promoted with the superimposing actions of a colloidial effect combined with the mechano-chemical effect.

[B] ***Sapphire Single Crystal Substrate for Silicon on Sapphire (SOS).*** The following is an example of the evaluations made on the crystallization of the surfaces of a sapphire single crystal substrate for SOS processed by colloidal silica polishing.

The surface roughness R_{max} of the sapphire single crystal polished with colloidal silica was found to be around 10Å, which proved to have been processed in the order of an atom or molecule. Figure 6.38 shows the results of an electronic microscopic observation of the surfaces polished with colloidal silica and conventional diamond grains (2/6 μm), on which a silicon thin film of 3000 Å was epitaxially grown. The surface polished with colloidal silica showed no defect from the processing when compared with the diamond polished surface. As per the resistivity measurement of the Si thin film deposited on the sapphire substrate, an excellent polished surface of the base substrate was obtained with a uniform film thickness of 10 Ω • cm of p-type.

[C] Application to the Shaping Process of LaB_6 Single Crystal for Electronic Gun. LaB_6 single crystals are used as a cathode material due to the thermal radiation characteristics inherent to its structural properties. In order to use this kind of single crystal as a cathode for electron beam analyzers such as TEM and AES or electron beam exposure equipment, finishing the tips to a specified shape and to a smooth mirror-like surface without damaged layers is indispensable.

LaB_6 single crystals are dissolved by a strong oxidative acid such as nitric acid but are not dissolved by hydrochloric acid or dilute sulfuric acid. It is stable also against alkali. Although the cleavage is noticeable, LaB_6 single crystal is mechanically hard with a hardness of $2660/mm^2$ (microhardness).

Figure 6.39 is one of the electron microscopic photos showing the tips of LaB_6 single crystals (0.65 x 0.6 x 2.2mm chip) finished to a variety of shapes with colloidal silica using a relatively hard plastic polisher. Compared to the surfaces processed by conventional diamond polishing, it has scratch-free, smooth surfaces.

We have also investigated the processing defects of LaB_6 crystal surfaces finished with colloidal silica through etching (dilute $HN0_3$ solution); we detected no defects (like latent scratches) arising from the processing. The polished surfaces were also found to be strain-free and of very high quality.

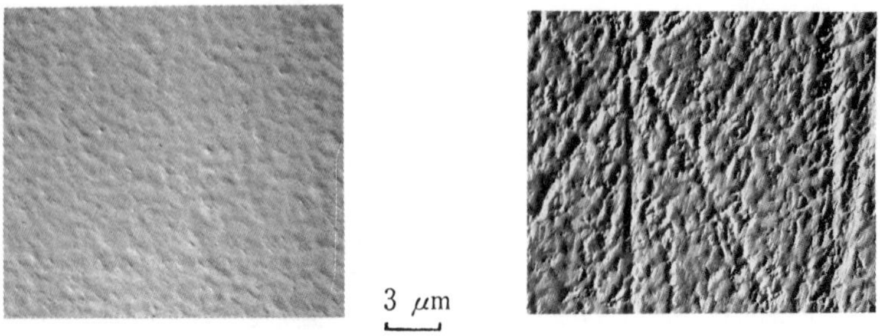

(a) Epitaxial film on surface polished by colloidal silica

(b) Epitaxial film on surface polished by diamond abrasives

Figure 6.38: Example of Electron Microscope Photographs of Epitaxially Grown Si Thin Films on Sapphire Single Crystals treated in Various Ways (Si Thin Film: about 3000Å, Substrate: Sapphire Single Crystal (1102)).

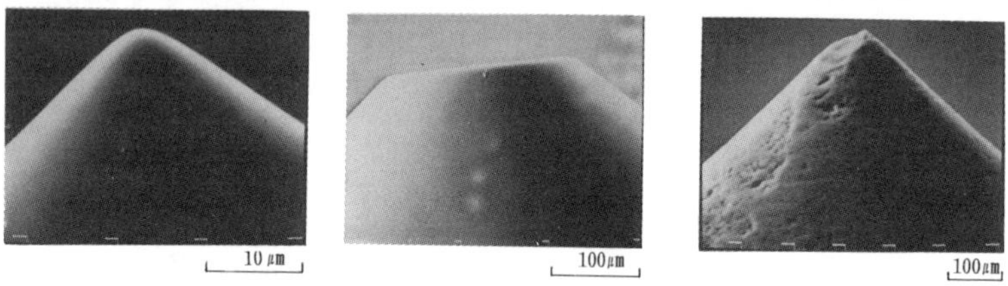

(a) Example finished with top curvature radius R=2 μm

(b) Example finished with flat top

(c) Example finished by conventional processing method

Figure 6.39: Example of the Strain-free Mirror-like Surface finished by the Colloidal Silica Polishing Technique and Conventional Processing Technique of LaB_6 Single-crystal Chip for an Electron Gun (SEM Photograph).

An electron gun was experimentally manufactured applying this method (Figure 6.40), which was afterwards subjected to the evaluation of its characteristics and confirmed to have excellent electron radiation characteristics.

[D] Strain-free, Mirror-like Polishing of $LiTaO_3$ Single Crystal Substrates for Epitaxial Growth. Microscopic photos of the surfaces of $LiTaO_3$ single crystals for optical components: each polished by colloidal silica and conventional pitch-polishing and another set of surfaces similarly polished on which films were epitaxially grown are shown in Figure 6.41. As seen from Figure 6.41(a), the surface polished with this polishing method demonstrates crystalline defect-free, high-quality epitaxial film while in the case of the conventional method, a large number of crystalline defects were produced as shown in Figure 6.41(b).

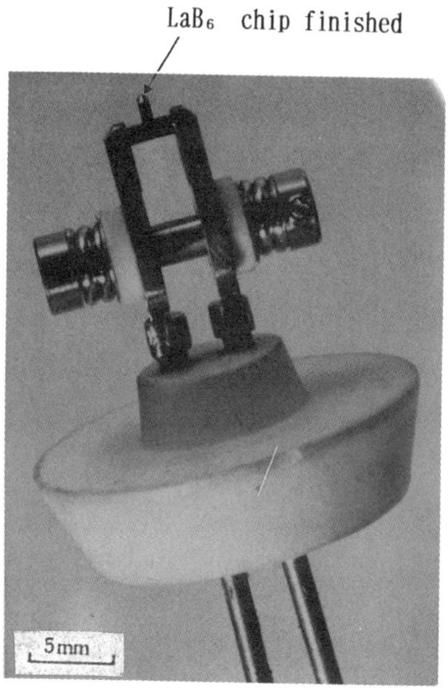

Figure 6.40: An Example of an Electron Gun Experimentally Manufactured applying Colloidal Silica Polishing (LaB_6 assembled Vogel Type Electron Gun)

With regard to the surfaces polished using the conventional method, it has been shown that only a few μm etching can produce variation Δα according to the relationship between the depth etched by melt back and the variation amount Δα to the lattice constant (α =5.153Å) of a bulk. On the other hand, the polished surfaces with colloidal silica are capable of obtaining a strain-free (Δα = 0) epitaxial film even without undergoing etching. This proves that the surfaces polished with colloidal silica polishing are completely strain-free and without damage.

6.0 ULTRASONIC LAPPING

Hans K. Tonshoff

Ultrasonic lapping is an abrasive process that uses the suspension of hard grains and the liquid phase as the active material removal medium.[46] The abrasion is performed by the grains alone. They are driven by a form tool and a sonotrode. The sonotrode vibrates with frequencies in the ultrasonic domain, i.e., 20 kHz and more. For the abrasive action of the grains, there are two possible principles:

1) The grains in suspension are accelerated by the form tool and hammer onto the workpiece; the material is removed by a dynamic action.
2) The grains are pressed against the workpiece by the form tool and the static action of forces applied to the grains: material is removed.

It could be determined that the hammering effect is dominant.[47][48][49][50] In any case, the abrasive grains act by indentation in the work material. They generate microcracks in normal and transversal directions and thus remove particles of the work.[51] That means that hard and brittle materials can be machined with comparably high efficiency because with the indenting grain, the material is taken off mainly by cracking, whereas ductile materials tend to deform plastically. The process is applied to glass, ceramics, metallic carbides, ferrites, semiconductive materials, and to jewelry as work materials. The principle of ultrasonic lapping is shown in Figure 6.42. The suspension is supplied either externally as shown in the figure or internally

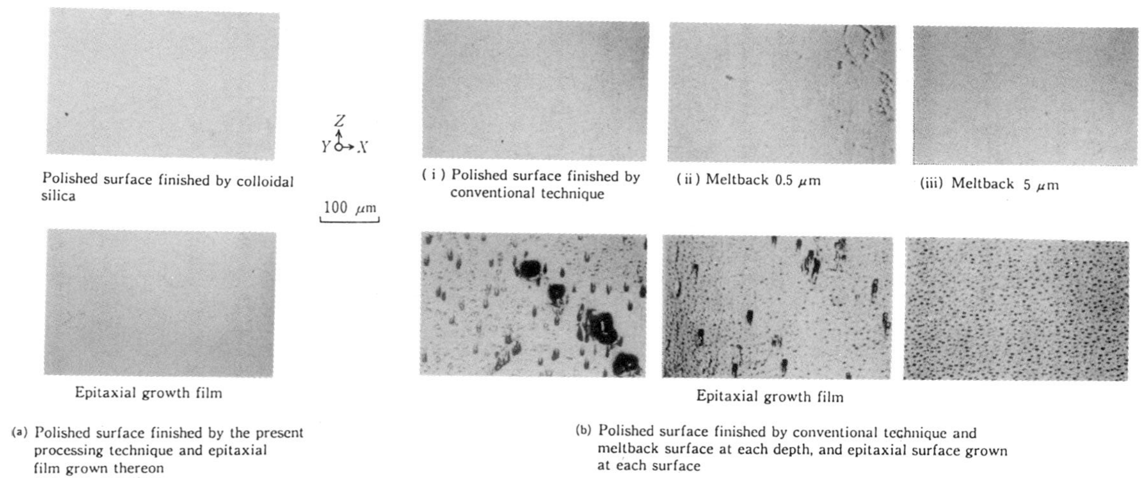

Figure 6.41: Photographs of Surface and Epitaxial Growth Film of $LiTaO_3$ Single Crystal Polished by the Present Processing Technique and the Conventional Processing Technique.

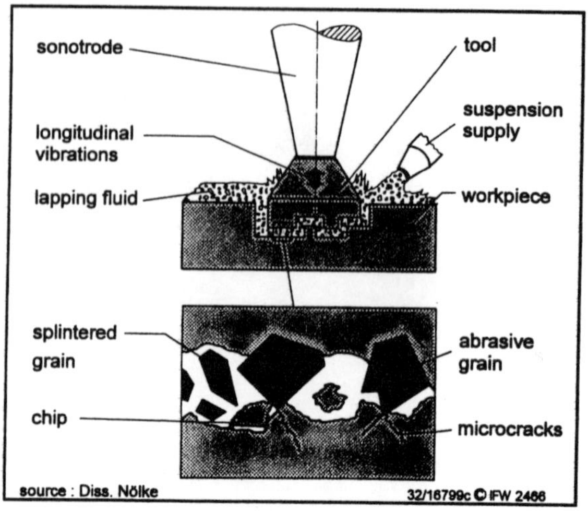

Figure 6.42: Principle of Ultrasonic Lapping

from the form tool side or - if possible - from the workpiece. The suspension flow can be effectively supported by suction through the form tool. The conventional application of the process is for sinking operations. This means that the shape of the form tool is equidistantly introduced into the work material. The gap between the form tool and the workpiece is due to the material removal mechanisms explained above, about the maximal grain size. For electrically non-conductive ceramics, ultrasonic lapping (besides the laser caving, which is still a process in development) is the only method to generate complex shapes.

There are basically three different variations of ultrasonic lapping. They are distinguished by their individual tool geometry or feed direction.

Shape Sinking

This process reproduces the 3-dimensional tool shape at its bottom. By this process, carvings in ceramics, glasses, or other brittle materials are generated. It is also used in the jewelry industry to manufacture engravings.

Boring

This process reproduces a plane cross-sectional profile of the tool. The surface lines of the prismatic holes are straight. Clearance holes in any cross-sectional form can be manufactured. Electronic and optical components made of glass, semiconductors, ferrites, and other brittle materials are machined this way. This process is also used for extracting any geometries in plates.

Shaping by Geometrical Generation

When, in addition to the axial feed of conventional ultrasonic lapping, a translational or rotational feed of the tool is available, it is possible to achieve a workpiece shape which is largely independent on the tool shape. This is called ultrasonic milling or ultrasonic turning and is analogous to conventional production processes.[46][52]

When working according to the first two modes, it is obvious that the active surface of the form tool is its bottom. In mode 3, there are principally two removal mechanisms. The tool may be introduced in the workpiece by sinking it to the full depth. It is then moved with a low feed transversely, comparable to a creep-feed grinding process. The other possibility is to move the form tool with a comparatively high feed speed and to generate the necessary depth of caving by a reciprocating movement. It is possibly shown[52] that the material removal rate is much higher in the latter case. This is due to the active bottom surface of the form tool. In the creep-feed lapping process, the hammering effect cannot take place. Terms for a description of geometry and feeds differ in the literature. The cutting direction corresponds to the longitudinal tool vibrations. The vibration amplitude X_a and frequency describe the vibrations as follows:

$$X_c = X_a \sin\omega t \tag{6.10}$$

From this equation, the cuffing speed follows

$$V_c = \omega X_a \, \mathrm{Cos}\omega t \tag{6.11}$$

Technologically meaningful is the perpendicular to the work-piece surface oriented speed vector. The tool feed in profile counter boring and in conventional ultrasonic boring is usually the motion caused by the axial force or static boring pressure. The feed direction is parallel to the cutting direction. In contrast to conventional production processes, the axial feed has no constant feed speed but a constant force. This is required by the specific process conditions. Otherwise, an excessive increase of static axial forces in combination with a collapse of the resonance vibration can occur. The axial feed speed is related to the achieved material removal and can be used as a parameter for the machining speed. In an enhanced process configuration, a cross feed can be overlaid to the axial feed. The cross feed has a constant feed speed like feed speeds in conventional production processes. Here, the workpiece material is also removed by the tool flank. The setting points for defined chip removal in ultrasonic milling or turning are the cross feed f_q and engagement e. Axial feed f_a and depth of cut result from the material removal rate. The ease of machining in ultrasonic lapping can be evaluated by the following criteria:

- material removal rate,
- tool wear,
- workpiece form accuracy, and
- workpiece surface quality.

Figure 6.43 shows main process parameters and working results.

In Figure 6.44, the schematic design of an ultrasonic generation unit is shown as follows:[50]

- high frequency generator,
- electromechanical ultrasonic converter,
- ultrasonic transformer,
- sonotrode, and
- form tool.

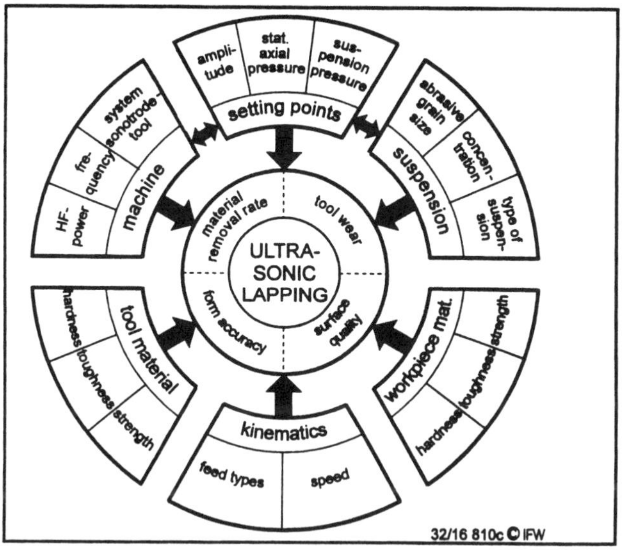

Figure 6.43: Process Parameters in Ultrasonic Lapping

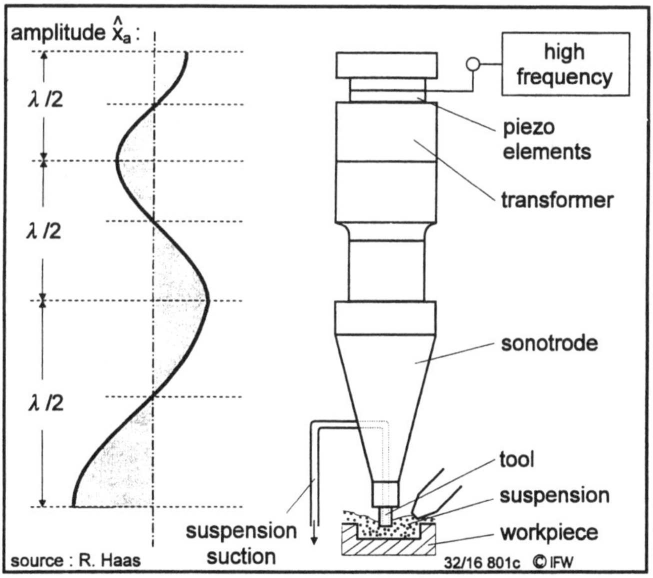

Figure 6.44: Ultrasonic Generator

The form tool vibrations are generated mechanically by an electromechanical source. Normally, this movement is introduced into a sonotrode that acts as a mechanical amplifier (Figure 6.44). The generator can be based on piezoelectric actuators that reach high frequency vibrations at small amplitudes. Amplifiers serve the transformer and the sonotrode together.[53] The sonotrode fixes the tool. Useful geometries for the sonotrode are exponential or hyperbolic forms, cones, and reduced cylindrical shapes. The whole system operates in a resonance state.

These vibrational movements are the reason for some process limits. If the form tool is not symmetric, the sonotrode may be stimulated in considerable bending or torsion amplitudes as shown in Figure 6.45. This results in reduced material removal rates, because a part of the vibrational power is "wasted" for such non-effective movements, form deviations, additional tool wear, and also tool or sonotrode fracture. Due to false movements as well as longitudinal vibrations, the interface between the sonotrode and the form tool is often critical; the joints may break.

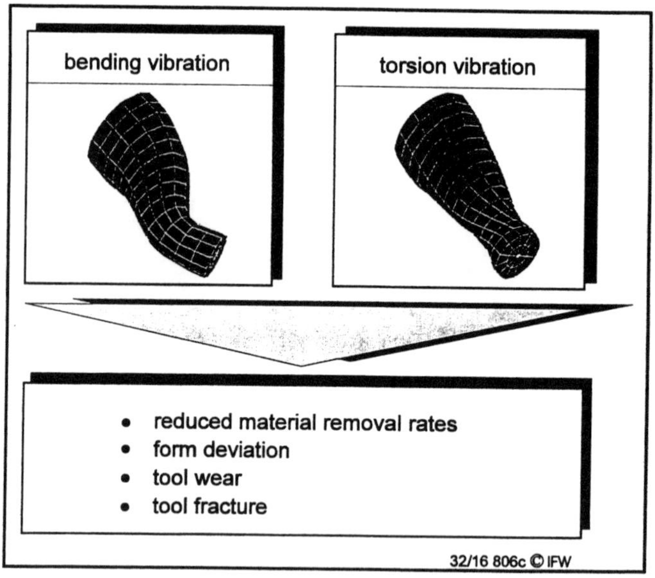

Figure 6.45: Forms of Interfering Vibrations and their Effect on the Process

For a proper layout of a sonotrode, this aspect must be considered. Calculations can be performed by the finite element method (FEW). It is possible to determine various resonant frequencies with their forms of vibration and hence to predict process stability which depends on the distance between longitudinal resonance frequency to other natural frequencies of the sonotrode.

The tool amplitude has a major influence on the material removal rate. As demonstrated in Figure 6.46, an increase in axial speed is observed up to a maximum. This is reversed at higher amplitudes. The optimal amplitude depends on the size of abrasive particles added to the suspension fluid.

An essential problem of ultrasonic lapping is the tool wear. Tool wear can be as large as the removed work material. Using tools with a layer of diamond (PCD tools), the wear can be reduced sharply. Abrasive wear and shock wear during machining is drastically lower. In an ultrasonic milling long term trial, a 12.6 mm diameter PCD tool was used for at least 1010 min to machine a ring groove of 43.5 mm (outer diameter) in silicon nitride (Figure 6.47).[52]

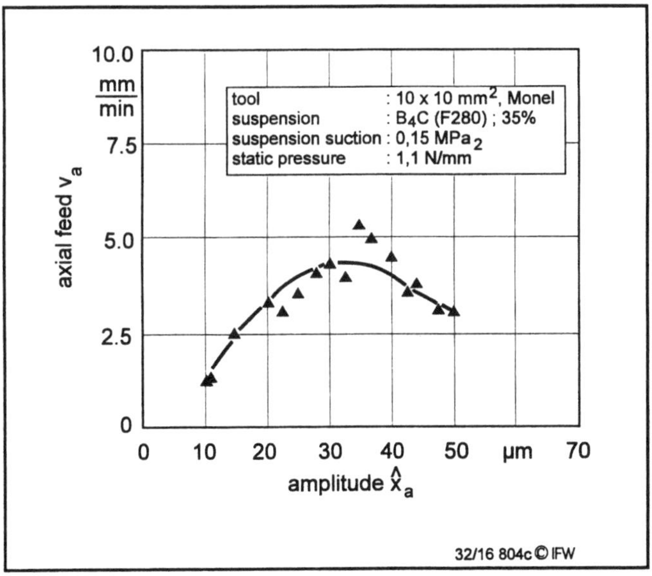

Figure 6.46: Influence of Tool Amplitude on Axial Feed

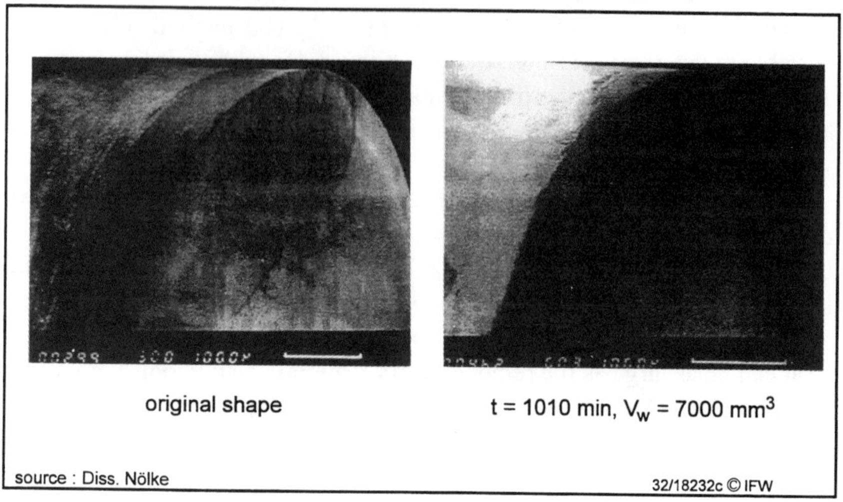

Figure 6.47: Wear Behavior of PCD Tool after 1010 Min Ultrasonic Milling

The PCD layer shows a medium wear of 0.25 mm on the bottom. The wheel speed of the tool bottom is 0.032 mm^3/min. Ultrasonic boring with conventional tools has wear speeds of 5 mm^3/min. This would reduce the tool length by 40 mm (Figure 6.48). The ratio between tool wear and workpiece material removal is given for PCD tools in comparison to conventional tools. When machining silicon nitride, the order of two magnitudes difference shows the potential of PCD tools. Concerning the cross feed speed in ultrasonic milling and thus the material removal rate, it is possible to achieve values that are used for conventional tools.

The price of PCD tools is about 100 times higher; this will partially compensate the advantages of low wear speeds. With its equivalent material removal rate, an implementation of these tools seems to be useful when close tolerances of shape and form of the ceramic components are demanded. Higher tool life avoids multiple replacing, especially when higher amounts of material must be removed.

The drastic wear reduction of PCD tools leads to the assumption that abrasion is the dominant wear mechanism. The use of tool materials with

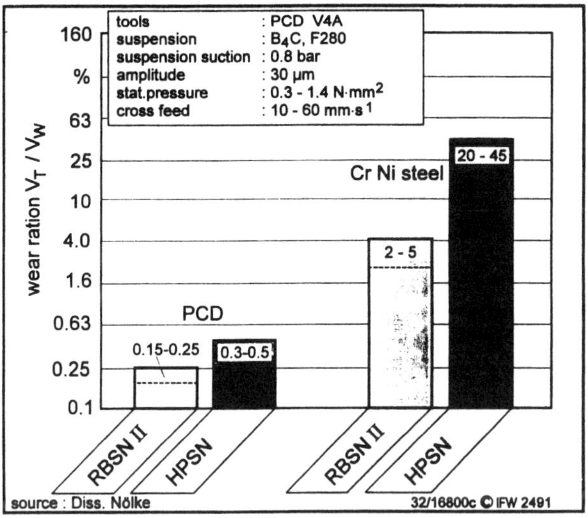

Figure 6.48: Wear of PCD Tools and Conventional Tools in Ultrasonic Milling.

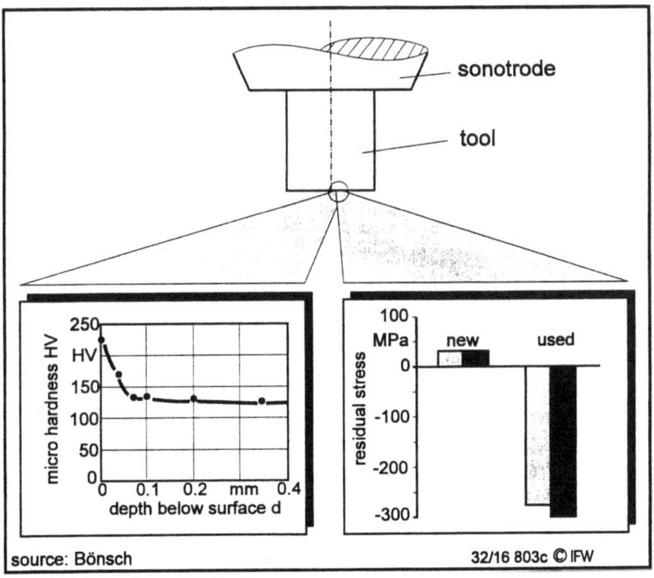

Figure 6.49: Strain Hardening of Tool during Machining

high hardness will therefore show better resistance to abrasion. Micro-hardness measurements of tools show remarkably increased values after machining without an influence on the sub-surface microstructure. Residual stress measurements determine changes of the surface integrity (Figure 6.49). Higher hardness and compressive residual stress point to a strain hardened surface layer. Obviously, this effect can be exploited by using materials that tend to strain hardening. A good compromise between brazing ability for mounting and wear behavior is austenitic steel.

The suspension is the medium which fills the gap between the tool and workpiece. It contains abrasive grains and a liquid phase as a carrier. In ultrasonic lapping, the abrasive grains of the suspension can be taken as a tool. Hence, process results can be influenced by varying the suspension.

The basic demand on abrasive grains is a hardness superior to the workpiece material. The compressive strength must be sufficient. Sharp edges avoid plane contact with the workpiece surface. This requires good fracture behavior.

In most applications, the abrasive material is boron carbide with a grit size of F280. The concentration is in the range of 25-30 weight%. This choice allows high removal rates at high surface quality. Figure 6.50 shows the influence of grit size on material removal rate and workpiece roughness. Of course, larger grains increase removal rates. Due to the improved surface roughness compared to F180, it is obvious why F280 is the standard grit size. The fine grain of 400 mesh produces the best surface quality at a significantly reduced removal rate. The abrasive grain material also affects the removal rate. Due to the superior hardness, diamond yields the best results. Silicon carbide is applied for US-lapping of glass, but the hardness is not sufficient when machining advanced ceramics. Boron carbide compensates the lower material removal rates in comparison with diamond by lower costs.

The process reliability and high machining accuracy depend on a sufficient suspension supply of the gap between the form tool and the workpiece. An implementation of a suspension suction inside tool and sonotrode can ensure this. A suspension supply under pressure is also possible. Figure 6.51 demonstrates the influence of the suspension supply pressure or suction pressure under the variation of axial force on material removal rate.

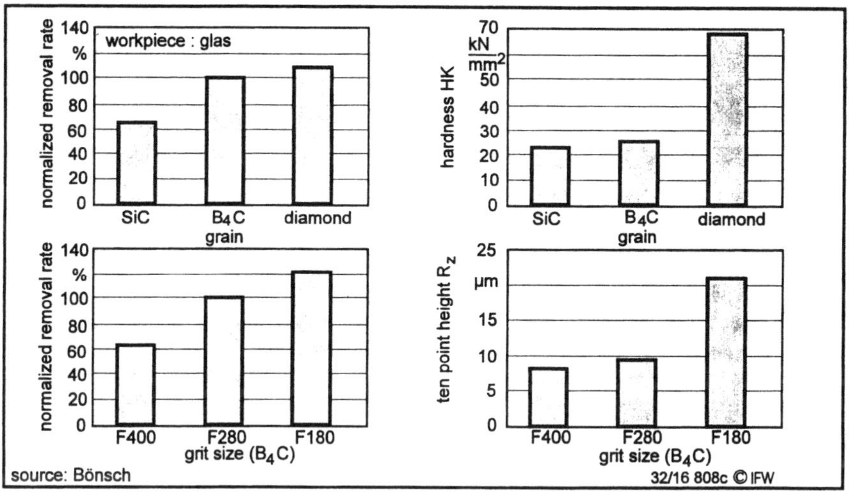

Figure 6.50: Influence of Abrasive Grain Material and Grit Size on Work Result

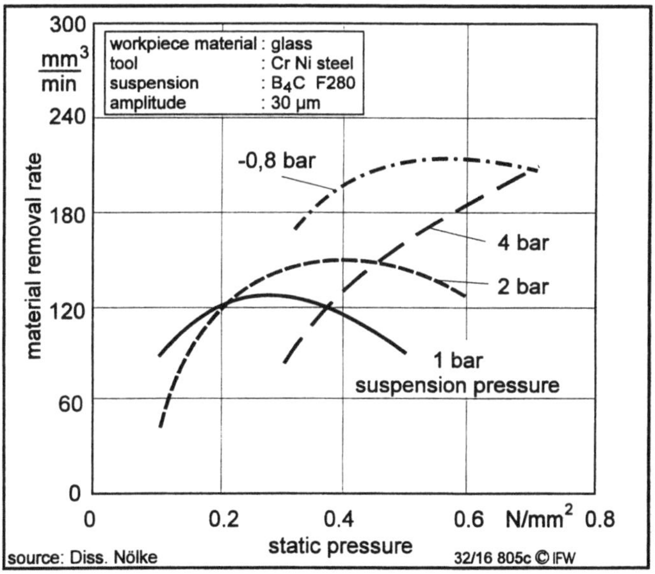

Figure 6.51: Influence of Suspension Pressure and Suction on Removal Rate

Generally, the curves increase to a maximum for discrete axial force values and decrease at higher axial forces. Overly high forces overload the abrasive grains and allow them to crack rapidly. The number of active grains changes; this has an adverse influence on the process stability.

A suspension suction device increases the removal rate significantly. Additionally, the removal rates are nearly constant in a wider range of axial pressures. The same efficiency requires a pressure of 4 bars, but this will influence the vibrational behavior of the sonotrode system negatively. Higher power supply for equal amplitudes is necessary, and a more sensitive reaction with process variations is observed.

Besides the abrasive grain, the carrier medium also influence the process. Variations of the suspension fluid have shown the best removal rates for water, compared to alcohol, benzene, or glycol. In Figure 6.52, obtained material removal rates for different fluids in relation to water are illustrated.

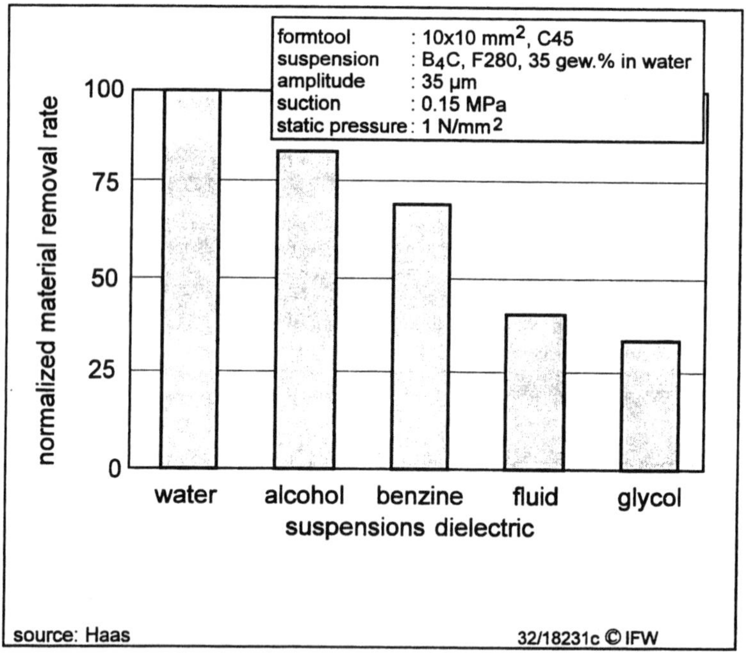

Figure 6.52: Material Removal Rate under Variation of Suspension Fluid

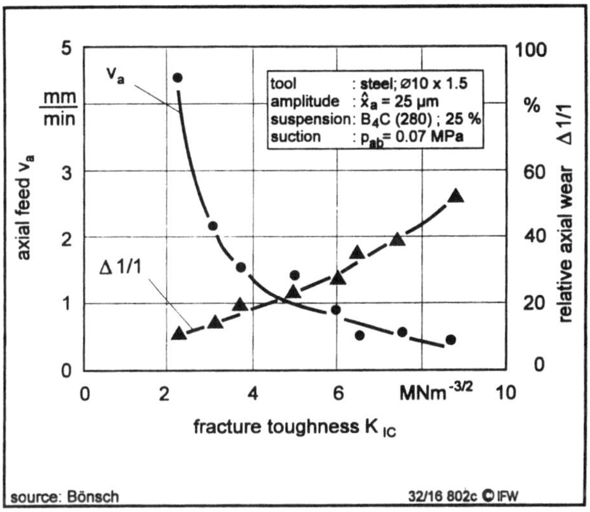

Figure 6.53: Material Removal Rate and Tool Wear in Relation to Fracture Toughness K_{ic}

Experimental investigations proved the dependence of machinability of work materials towards the critical fracture toughness K_{IC}. This relation results from Figure 6.53. Materials with a low fracture toughness have a lower resistance against crack initialization by abrasive grains. The removal mechanism is well supported, and process efficiency is stressed by low values for tool wear. The limited reproducibility of ultrasonic lapping can be confirmed in principle. But absolute values can spread in a wide range, mainly due to different ways of process optimization. Due to the material removal mechanism by crack generation in the workpiece surface layer, the surface roughness can be influenced only slightly. Reduced local mechanical load (which is achieved by finer abrasive grains) leads to smaller cracks. Investigations in grinding of ceramics have shown a major influence of the workpiece microstructure on the surface quality. Similar results have been observed for ultrasonic lapping (Figure 6.54). The average sizes of sintered workpiece material grains correlate to measured surface roughness. Ceramic grains tend to splinter. This has an effect on the surface roughness, especially when the microstructure contains huge grains. Therefore, the best surface

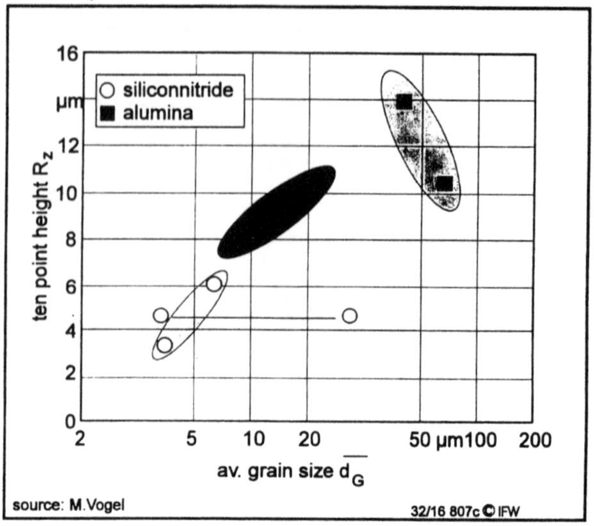

Figure 6.54: Influence of Average Grain Size of Ceramic Materials on Surface Roughness.

roughness can be achieved for materials with small grain size.

Ultrasonic lapping, like grinding, has numerous input parameters that govern the process. Their influence on the output and the machining results has been investigated by experimental and theoretical approaches in the 60s and again in the 80s. Of major influence on the material removal rate, i.e., on the process efficiency, are the amplitudes of the form tool and the suspension supply. Suction supports the removal rate considerably; this goes together with minimal form tool wear. The machining parameters influence the surface roughness only slightly, whereas the grain size of the ceramics to be machined is dominant. An interesting process variant is proposed by using transverse feed; this allows US milling or turning. Under these conditions, tool wear is important. PCD layers can decrease volumetric wear considerably. The ultrasonic lapping process is well suited for ceramic machining due to the material removal process that supports brittle behavior. The main advantages are derived from possible geometric forms that can be produced. Furthermore, the low process forces allow the manufacturing of very fine structures on ceramic surfaces. Due to the low material removal rate, this process can hardly compete with conventional ceramic grinding; it shall be viewed as a complementary process. So far, ultrasonic machined components have expanded the variety of products.

REFERENCES

1. Kasai, T., 4. Ultraprecision polishing, Foundation, Ultraprecision Manufacturing Technology (in Japanese), Fuji-technosystem Ltd., p.266 (1995)
2. Caseate, Horio,K. and Doy,K.T., Lapping and Polishing Characteristics of Soft Material Tool Use, Sensors and Materials, MYU 2:301-312 (1989)
3. Caseate, 6. Lapping and Polishing, Ultraprecision Machining Technology, Kumagaya Memorial Committee Meeting, Japan Society of Grinding Engineers (in Japanese) p.451 (1984)
4. Taniguchi,N., The state of the art of nanotechnology for processing of ultraprecision and ultra fine products, 1993 ASPE distinguished lecture, Precision Engineering, 16 (1)pp.5- 24 (1994)
5. Nakamura,N., 1-4 Grinding - Lapping & Polishing, Optical Production Technology '93 (in Japanese), Japan Optoelectro- Mechanics Association, pp.58-123 (1993)
6. Ywyman, F., Prism and lens making, 2nd ed., Hilger & Watts Ltd., Co., p.363 (1952)
7. Rumsayj., The Effect of Eccentric Loading of Optical Flat during Grinding or Polishing, J. Sci. Instrum., 32 (9) pp.338- 339 (1955)
8. 'Kasai,T. and Noda, J., Improvement of parallelism in lapping or polishing using eccentrically distributed loading, Study on free abrasives machining flat and parallel planes (3rd report) (in Japanese) J. of Japan Society for Precision Engineering, 44 (9) pp. 1064- 1060 (1968)
9. Noda, J., Caseate and Maj., Devices for preparing oriented crystal surfaces by lapping. Review of the Electric Communication Laboratories, NIT Public Corporation, 21 (1,2)pp.69- 66 (1963)
10. Shinmura,T. and Aizawa,t., Study on a New Finishing Process of Fine Ceramic by Magnetic Abrasive Machining Development of Diamond Magnetic Abrasives and its Finishing Performances (in Japanese), J. of Japan Society of Grinding Engineers, 36 (2) pp.91- 98 (1992)
11. Yaniaguchi,H. and Shinmura,T., Study on a New Internal Finishing Process by the Application of Magnetic Abrasive Machining, Discussion of the Cylindoricity (in Japanese), J. of Japan Society for Precision Engineering, 61 (6) pp.996- 1000 (1995)

12. Caseate, Noda.J. and Suzuki, J., Lapping characteristics of LiTaO 3 single crystal, Study on precision machining of octo- electronic crystals (1st report), (in Japanese) J. of Japan Society for Precision Engineering, 44(11) pp.1360-1366 (1968)
13. Caseate, Study on Precision Machining of Flat Planes, Electric Communication Laboratories Report No. 13634, NTT Public Corporation, (in Japanese) p. 165 (1969)
14. Ikeda,M., Study on Lapping Mechanism of Non- metallic Crystals (in Japanese), J. of Japan Society of Grinding Engineers, 14 (12) pp,1- 153 (1960)
15. Caseate, Horio,K., Doy,K.T. and Kobayashi, A., Improvement of Conventional Polishing Conditions for Obtaining Super Smooth surfaces of Glass and Metal Works, Annals of the CIRP, 39/1 pp.321-323 (1990)
16. Izunitani,T. and Adachi,S., Polishing Mechanism of Fused Silica Glass, Topical Meeting on the Science of Polishing, Technical Digest, OSA TuB- Al- I - 3 (1984)
17. Brown,Nj., Some Speculations on the Mechanisms of Abrasive Grinding and Polishing, Precision Engineering, 9 (3) p.129 (1986.6)
18. Doy,K.T. and Watanabe, J., Effect of Frictional Heat on Removal Rate in Mechanochemical Polishing of Crystals Used in Electronics, Wear of Materials, ASME pp.226- 234 (1983)
19. Caseate, Orioka,T., On the high speed polishing of glass using felt plate (5), On the polishing of glass, (in Japanese) J. of Japan Society of Precision Engineering 33, 5 (1966) pp.306-311
20. Browm, J.N., Baker,C.P. and Maney, T.R., The optical polishing of metals, SPIE's 25th Annual International Symposium, San Diego, California, Aug. 24-28 (1981)
21. Caseate, Study on Precision Machining of Flat Planes, Electric Communication Laboratories Report No.13634, NTT Public Corporation, (in Japanese) p.45 (1969)
22. Caseate, Horio,K., Yamazaki,T., Komoda,M., Doy,K.D. and Kubo, N., Polishing to Reveal Micro- defects on Glass, J. Non- Crystalline Solids, 166, pp.396-404 (1994)
23. Caseate, Horio,K., Doy,K.T. and Kobayashi, A., Improvement of Conventional Polishing Conditions for Obtaining Super Smooth Surfaces of Glass and Metal Works, Annals of the CIRP 39/1, pp.321- 324, (1990)
24. Caseate, Horio,K. and Doy,K.T., Polishing Characteristics of Soft Glass (Laser Glass) The 3rd International Conference on Nanotechnology, Program and Abstracts, pp. 16- 16 (1992)
25. Otte,G., An Improved Method for the Production of Optically Flat Surfaces, J. Sci. Instrum., 42 p.911 (1965)

26. Lerstner, A., Progress Report on Teflon Polishing for Precision Optical Flats, Applied Optics, 11 (4) p-960 (1962)
27. Bajuk, D. J. , Computer Controlled Generation of Rotationally Symmetric Aspheric Surfaces Optical Engg., 15 (5) p.401 (1966)
28. Becker,K. and Beckstette,K., M400 and P400-A Pair of Machines for Computer Controlled Fine Correction of Optical Surfaces, Proceedings of the International Congress for Ultraprecision Technology in Aachen/FRG pp.212- 223 (1988, 5)
29. Mori, Y., Yamauchi, K. and Endo,K., Elastic emission machining, Precision Engineering, 9 (3) pp.123128 (1986)
30. Namba,Y., Mechanism of Float Polishing, Technical Digest at Topical Meeting on Science of Polishing, OSA Tub- A (1984,4)
31. Sato,Y., Knight, J. D. and Somerday, B., Observation and Prediction of Cavities in a Lubricant Between Non- Parallel Disks in Relative Rotation, Tribology Transactions, 35(1) pp.160-166 (1992)
32. Gormley,J.V., Manfra, M.J., and Carawa,A.R., Hydroplane Polishing of Semiconductor Crystals, Rev. Sci. Instrum., 52(8) pp.1256- 1259 (1981)
33. Caseate, Kobayashi, A., Progressive Mechanical and Chemical Polishing, Technical Digestat Topical Meeting on Science of Polishing, OSA, TuB- A (1984, 4)
34. Caseate, Matsumoto, F. and Kobayasi, A., Newly Developed Fully Automatic Polishing Machines for Obtainable Super- smooth Surfaces of Compound Semiconductor Wafers, Annals of CARP, 36/1 pp-536- 540 (1988)
35. T.Karaki: et al: Wear of Materials "83 (1983) 226-234
36. N.Yasunaga: et al: Proc. 1st Int. Conf. Production Engy., (1964) 34
37. T. Karaki-Doy: Sensors & Materials, 3 (1988) 153-166
38. Y.Meri: et al: Proc. 3rd Int. Conf. Production Engy., (1961) 336
39. Y. Namba: et al: Proc. 4th Int. Conf. Production Engy., (1980) 1016
40. M. Fink: et al: ASLE-ASME Lubricantin Conf., 6 (1965) 65
41. T. Kube: Science of Machine, 23, 11 (1961) 1500 (in Japanese)
42. T. Sakurai: J. Jap. Soc. Of Lubrication Engy., 11, 1 (1966) 26 (in Japanese)
43. E. Mendel: et al: Proc. Of IEEE, Sep. (1969) 1466
44. T. Karaki-Doy: et alL Int. J.JSPE, 30, 1 (1996)
45. T. Karaki-Doy: et al: Advanced Materials in Optics, Electro-Optics and Communication technologies (1995), 541-548
46. Tönshoff, H.K., H.H.Nölke: Spanende Bearbeitung von Siliziumnitrid durch Ultraschallschwingläppen. IDR 16 (1982) 1, pp.41-49.
47. Shaw, M.C.: Das Schleifen mit Ultraschall. Mikrotechnik X (1956) 6, pp. 265-265.
48. Rozenberg, L.D.:An Investigation of the Mechanism of Ultrasonic Cutting by Means of High-peed Cinematography. Stanki I Instrument, 5 (1959).

49. Adithan, M., V.C. Venkatesh: Tool Wear Mechanism in Ultrasonic Drilling. Wear 34 (1965), pp. 449-453.
50. Haas, R.: Technologie zur Leistungssteigerung beim Ultraschalischwingläppen. Dr.-Ing. Diss. TH Aachen 1991.
51. Lawn,B.R., A.G. Evans, D.B. Marshall: Elastic Plastic Indentation Damage in Ceramics. Journal American Ceramic Soc. 63 (1980).
52. Nölke, H.H.: Spanende Bearbeitung von Siliziumnitridwerkstoffen durch Ultraschalischwingläppen. Dr.-Ing. Diss. Universität Hannover 1980.
53. Cronjäger, L.: Einsenken unter Ultraschall. Dr.-Ing. Diss. TU Hannover 1961.

Appendix

Manufacturing of Ceramic Components: Case Studies

G. Warnecke, L. Shäfer, K. Eichgrün, C. Barth, and W. Pfeiffer

1.0 INTRODUCTION

Due to their superior properties, advanced ceramics have great potential for industrial applications. However, the high failure sensitivity and brittle fracture behavior of ceramics still limit their application. Complex influence-cause-effect-relations, not only within single processes of the process chain but also between different processes, must be considered when manufacturing components made of advanced ceramics. It is recognized that each of the processes can induce failures of components that can hardly be equalized by subsequent manufacturing operations. These failures can add up to deficient reproducibility of grinding and finishing operations, or even to complete failure of the component. For this reason, sensitive points within the process chain must be identified and their relevance on end-product features evaluated. These sensitive parameters must be kept within close tolerances and under statistical control. Other less sensitive parameters may be kept within wide, cost saving tolerances.

Three exemplary components have been chosen in a joint project to demonstrate the possibility of manufacturing high quality ceramic components economically: ceramic engine valves made of silicon nitride, as an example for a mass-production component, cylinder liners made of silicon carbide (medium scale production) and cutters made of zirconia (small-lot production). In the following discussion, some aspects of the manufacture of the three components will be presented as case studies focusing on the grinding operations.

The common approach in these three cases was to analyze the whole process chain shown in Figure A-1, with the example of the cylinder liners. Starting with a systematic analysis of the application conditions, the requirements of the component must be defined. The component design should be suitable for ceramics, and the process chain must meet the manufacturing requirements of ceramic materials. Existing process chains have been examined using the following methods of comprehensive quality management:

- qualitative and quantitative analysis with a high measurement involvement recording the features of intermediate products and process condition,

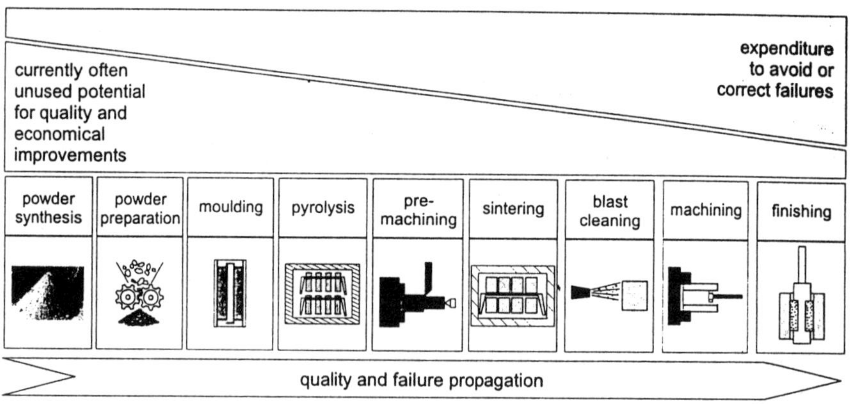

Figure A.1: Process Chain for the Manufacture of Ceramic Components

- statistic design of experiments intergrating several single processes,
- component-specific documentation to enable tracing of the production path in all manufacturing steps,
- statistic analysis of production and workpiece data,
- optimization of process chains and manufacturing methods using modeling and simulation,
- proof of component capability and reliability in the application followed by laboratory analysis.

Close cooperation in a technology group consisting of companies for all steps of the process chains, and of research institutes leads to an information exchange on further advanced design, manufacturing and testing technologies as well as on examination results and the holistic adaption of the processes to the end-product requirements. Due to this, the potentials of the already existing process chains could be further optimized.

2.0 CASE STUDY GRINDING OF CERAMIC ENGINE VALVES IN MASS - PRODUCTION

In the following, the process chain of ceramic automotive engine valves made of silicon nitride is taken as an example to show the capability of the integrated approach. To prove the technical feasibility of replacing metal engine components with components of advanced ceramics and to test the reliability of the components under real application conditions, the manufactured valves are currently used in a fault-free field trial on 1250 Mercedes-Benz cars.

As a consequence of the high temperatures in the combustion chamber and the dynamics of the valve drive, the engine valves' functional elements are subjected to high thermal and mechanical strain. To fulfill the technical requirements of this application, such as component strength and surface roughness as well as economical boundary conditions, cost efficient manufacturing processes with high process reliability and reproducibility within the entire process chain are required. From the manufacturing point of view, the feasibility of manufacturing reproducibility including functioning quality assurance had to be proved under conditions of mass production.

Requirements of the Component

The requirements for geometric accuracy and surface roughness of inlet and exhaust valves are shown in Figure A.2. To achieve high air-tightness of the valve in the combustion chamber, the roundness and surface roughness of the valve face must be highly accurate. The cyclic movement of the valves leads to tribological load on the valve stem. High requirements of cylindricality, concentricity, roundness and surface roughness on the stem are the consequence. The connection between the valve spring and the valve groove requires not only high positional accuracy but also precise line contour and high quality of roundness and surface roughness.

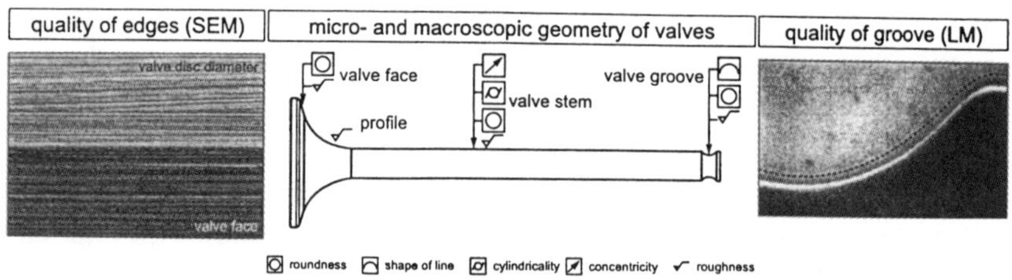

Figure A.2: Macro- and Micro- Geometrical Requirements of Engine Valves

Process Chain and Grinding Process of Ceramic Valves

The process chain in the manufacture of automotive engine valves of silicon nitride consists of powder synthesis, powder preparation, the moulding process, sintering and the grinding process. Finally a non-destructive ultrasonic material test is performed on the workpieces.[1] The grinding process is performed on a four-spindle grinding lathe using the Quick-Point grinding method as a high performance process. This method was developed to increase the rate of material removal without increasing the process forces and wear rate of tools.[2] Ideally, the circumferential surfaces of the grinding wheel and of the rotationally symmetrical workpiece contact each other only at one point. This is achieved by skew arrangement of the tool axis to the axis of the rotating workpiece. The advantages of this grinding method are as follow:

- optimum feeding of cooling lubricant to the contact zone, improving surface quality,
- grinding of complex geometries due to the two-axis feed of the contact point,
- increased tool life due to small grinding forces and continuous regeneration of the engaging circumferential tool edge, which is shifted over the axial profile of the grinding wheel.

The machine system used for the grinding operation is a machining center which allows the entire grinding process to be done on one machine. This Junker grinding machine type Valve Center (4001) was partly adapted to ceramic specific requirements on the machine system, for example, concerning the spindle driving power, the stiffness of the spindle, grinding wheels, dressing tools, back rest material etc. With one exception (infeed grinding of the valve groove) all operations comprise traverse grinding.

Starting Situation

The starting situation of the valve manufacture was characterized as follows:

- tool wear in grinding was too high, causing high costs of grinding wheels and set-up operations,
- due to tool wear, the surface roughness was in a critical tolerance range,
- the number and size of break-outs on the ground edges was partly too high.

In order to improve the grinding process, extensive examinations were carried out along the process chain and led to results that enabled optimization of the process chain and the grinding process. Some of the main results are discussed in the following.

Influences on End-product Quality Features, Tool Wear and Process Parameters

All the properties of silicon nitride which are suitable from the application point of view make the processes of final machining more difficult. The process chain for sintered ceramic valves generates the workpiece specific input quantities of the grinding process. Due to the complex influence-caus-effect-relations,[3] the characteristic features of powder synthesis and powder preparation already influence features such as strength and geometrical accuracy of the sintered compacts. Depending on the degree of reproducibility and process reliability of powder synthesis, powder preparation and of the moulding process, the green compacts show mainly oval and trigonal circumferential profiles(Figure A.3). The sintering process intensifies these preferred shapes of the green compacts. Thus, geometrical accuracy of the ceramic valves is normally distributed within components manufactured from one powder batch, but may differ widely if another powder batch is used or if the pressing mould is changed.

Analogously, the strength of the valve depends on several influential quantities of the processes of manufacture of sintered compacts. Due to the dependence of the strength of ceramic parts on microstructure components, varying material strength results from different batches of powder preparation, for instance. Thus, varying component strengths influence the machinability of the valves in the subsequent grinding process.

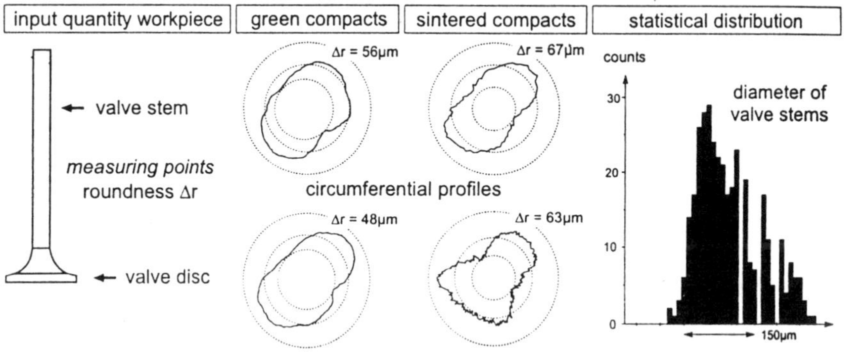

Figure A.3: Geometrical Features of Ceramic Valves of Different Intermediate Products

The necessity of taking all previous processes of the process chain into account is also shown by the influence of the geometrical shape of components. Due to the distortion of the sintered compacts resulting in statistical distribution of stem deflection, it is difficult to achieve high clamping accuracy of ceramic compared with metal components because the applied collet chucks are developed for metal parts with higher geometric accuracy and lower stiffness. To assess the clamping condition, the contact surface between the collet chuck and workpiece was examined with a touch-sensitive lacquer. The inprints show that the ceramic valve is fixed only at a few points (Figure A.4). This leads to undefined clamping conditions and to radial run-outs varying in a range of $100\ \mu m < \Delta r < 600\ \mu m$ of the valve disc during rotation of the clamping device. The run-out of the rotating workpieces leads to dynamic exitations of the grinding process, thus increasing the mechanical load on the workpiece and grinding wheel. Within the contact zone of the workpiece and tool, a great variation in depths of cut results from the clamping condition. Concerning the grinding wheel, dynamic exitation mechanisms increase the rate of wear.

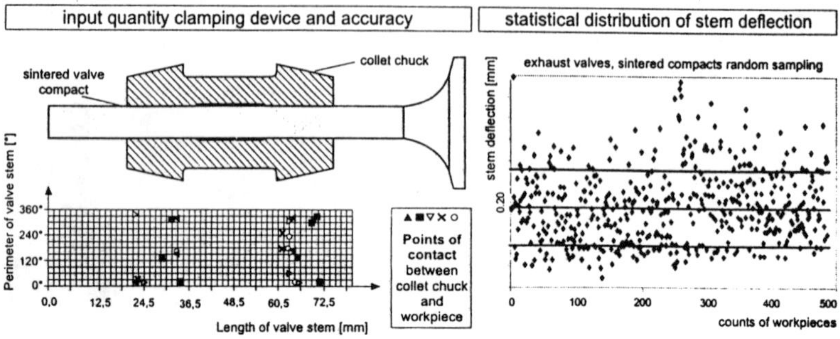

Figure A.4: Influence of Valve Stem Deflection and Clamping Device on Clamping Accuracy

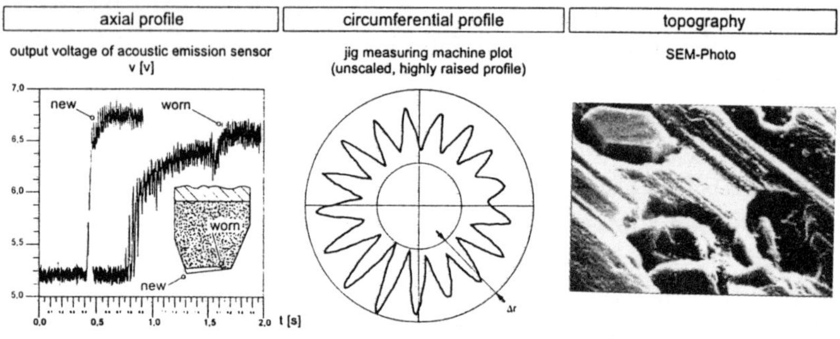

Figure A.5: Wear of Grinding Wheel Machining Ceramic Component in Mass Production (measurements with support of grinding wheel manufacturer Tesch, Ludwigsburg, Germany)

Due to the engagement conditions typical of the Quick-point process, the cutting edge of the grinding wheel is shifted axially over the profile (Figure A.5). While the defined conditions of tool engagement of an unworn axial tool profile are displayed by a sharp rise in the output voltage of an acoustic emission sensor, a worn profile leads to an undefined course of signal before the steady state cutting phase. Dynamic exitations intensify this wear mechanism and also the wear mechanisms leading to periodic waviness on the circumferential wheel profile (Figure A.5 middle). Once such a periodical circumferential profile is generated on the grinding wheel after several thousands of ground workpieces, high dynamic process forces lead to intensified removal and splintering of diamond grits of the grinding (Figure A.4 right) wheel and to surface and sub-surface damage of the workpiece.

The effect of high stress peaks on components of advanced ceramics during the grinding process is shown in Figure A.6. Due to the brittleness, cracks are induced within the material. In particular, functional edges of the valve disc are damaged by punctionally effecting peaks of stress. On the one hand, this results from the lack of surrounding and supporting material at the

Figure A.6: Workpiece Quality of Functional Edges of the Valve

edges. On the other hand, this damage results from the succession of grinding operations. For instance, grinding of the valve disc diameter induces lateral and radial cracks on the machined surface. Grinding of the valve disc chamfer generates the edge between the machined surfaces and again induces cracks. The joining of lateral cracks leads to break-outs on edges (Figure A.7).

Derived Measures for Optimization and Improvement of Work Result

All processes of the process chain offer a potential for optimization of process reliability by adapting design and control of the individual processes. The measures of optimization introduced in the following are focused on set-up parameters and grinding wheels. As a first step to optimize the application of grinding wheels, the accuracy of tool mounting was improved. If the hub flange of the driving spindle is cleaned of cooling lubricant and the machine operator fits the grinding wheel with care, axial run-out can be reduced by a factor of 4. Due to the higher degree of defined tool engagement, dynamic exitations and process forces can be reduced.

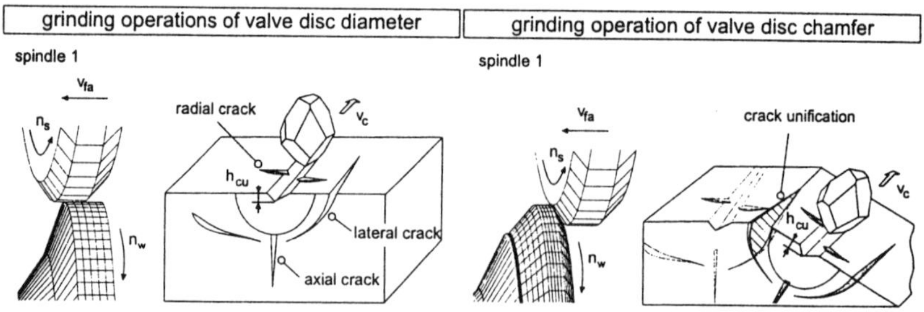

Figure A.7: Generation of Break-outs by Unification of Cracks

The specifications of grinding wheels have been modified to reduce the rate of wear due to breakouts of grits and the generation of waviness of the circumferential profile. A further objective in changing the specification was to increase the surface quality of machined valves. Thus, the grit size was reduced and the concentration of grits increased. Bonding systems have been modified by increasing the capability of grit fixing and improved damping properties. On the one hand, single grits of the engaging edge of the grinding wheel profile withdraw higher process forces this way. On the other hand, improved damping behavior of the bonding system reduces the effect of dynamic exitation mechanisms of the grinding process to generate waviness on the circumferential profile.

Setup parameters were improved using a comprehensive simulation technique for grinding processes called kinematic simulation. The principle of kinematic simulation, which allows reconstruction of the physical chain of effects in the grinding processes, and the obtained simulation results are shown in Figure A.8. Kinematic simulation computes the penetration between the grits on the enveloping profile of the grinding wheel and the workpiece.

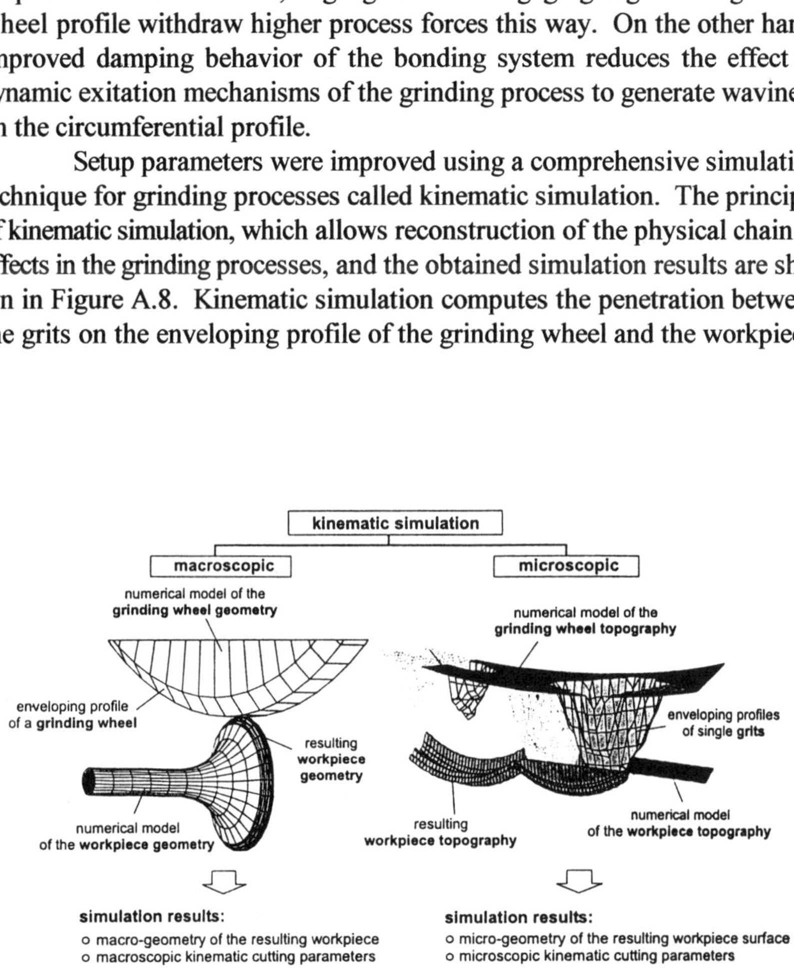

Figure A.8: Principle of the Kinematic Simulation[4]

The resulting macro-geometry and ideal surface roughness of the workpiece as well as undeformed kinematic cutting parameters can be calculated by the cumulation of multiple grit engagements. At the microscopic level, among other characteristic values, the undeformed chip thickness can be calculated. Thus the influence of grinding technique and process parameters can be analyzed.

Due to the high demands on micro and macro geometrical accuracy of the valves and wear rate of the grinding wheels, the given insufficient set-up parameters had to be adapted to grinding advanced ceramics. Starting from given process parameters, simulations were carried out with varying combinations for the axial feed rate and angular velocities of the workpiece, while maintaining cutting speed and depth of cut. Besides further kinematic cutting parameters, such as chip cross section, length of engagement of the kinematic grits etc., the resulting maximum surface-relevant undeformed chip thicknesses were calculated in the simulation and analyzed. Afterwards, the optimum process parameters from the simulation results were transferred to real grinding operations on ceramic valves.

In addition to the set-up parameters, the sequence of cuts and the depth of cut were changed. Investigations with a scanning electron microscope show that higher surface quality results from increased depth of cut. This can be explained by higher process temperatures that shift the properties of the work material to higher ductility. The number of grinding operations can be reduced this way, if the grinding wheels are adapted to this requirement.

With improved geometrical accuracy of unmachined valves and fitting accuracy of grinding wheels, modification of tool specification, and application of the simulation results on set-up parameters, a significant reduction in break-outs on the functional edges was achieved, whereas the cycle time of the grinding process could be kept within the limitations of mass-production process requirements. Figure A.9 shows a comparison of the simulated surface-relevant undeformed chip thickness for different parameter variations and the number of experimentally identified break-outs distinguished between defined size classes of the functional edge 2 of ceramic valve discs. The investigations show a clear correlation between experimentally determined size and number of break-outs on the analyzed functional edges and the simulated surface-relevant undeformed chip thickness. If the sum of the mean value and standard deviation of the simulated surface-relevant undeformed chip thickness remains below the experimentally determined critical surface-relevant

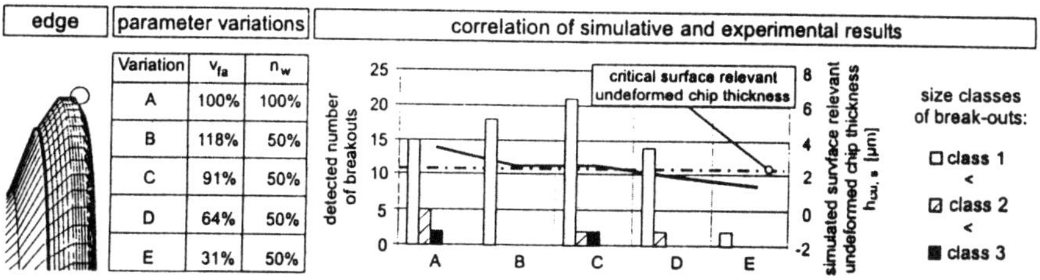

Figure A.9: Comparison of Simulated Surface-relevant Undeformed Chip Thickness and Number of Breakouts of Different Process Parameters

undeformed chip thickness, the number of break-outs is fundamentally minimized.

To fulfill the requirements of a ceramic specific clamping device within the given system of the grinding machine, we developed the following concept: a clamping device to clamp the valves at the stem that is self adjusting to the shape and deflection of the stem, in order to obtain defined contact conditions between collet chuck and workpiece. A prototype of the developed clamping device will be tested within 1999.

To summarize, optimization of the entire process chain led to the following results for the grinding process:

- improved macro- and micro- geometrical accuracy of finished ceramic valves by reduced breakouts on edges and higher quality of surface,
- improved component strength of finished valves and
- increased tool life of grinding wheels and thus a significant reduction in tool and setup costs.

Thus, mass-production of ceramic valves has been improved with regard to process reliability, reproducibility and economic efficiency based on an integrated approach. It was proved that ceramics are capable of replacing metal components in automotive applications, leading to improved engine characteristics such as reduced fuel consumption and noise emission.[5]

3.0 CASE STUDY - HIGH PRECISION GRINDING OF CYLINDER LINERS IN MIDDLE SERIES

Under application conditions with extreme tribological load, such as dry-run, metal components often reach the end of their lifetime very quickly, due to wear of surface and fretting. Particularly under such conditions, advanced ceramics offer an optimum profile of characteristics, due to their high hardness, great wear resistance and low tendency to adhese. In the following, the high precision manufacture of cylinder liners made of advanced ceramics

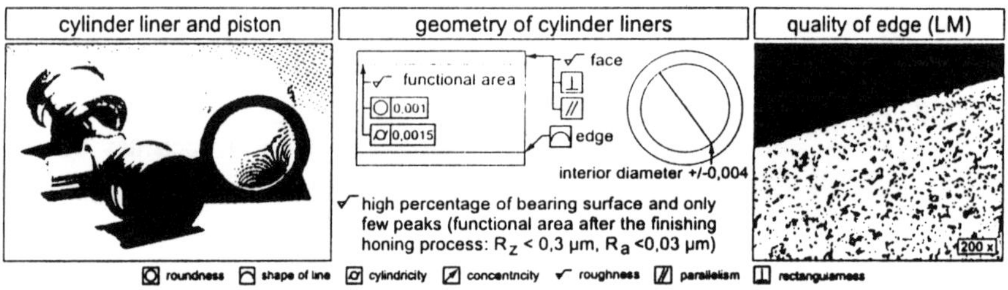

Figure A.10: Piston and Cylinder for Dry-run Compressor of Advanced Ceramics and Requirements of the Cylinder Liner

is presented as an example for a component designed for lasting dry-run in a compressor. Such compressors are suitable for applications in medicine or foodstuff technology. Figure A.10 shows the components' piston and cylinder liner for the ceramic compressor. The example demonstrates that under tribological load, ceramics are not only capable of replacing metal materials, but also of realizing superior technical solutions. Reliable manufacture of these components requires high reproducibility in all stages of the process chain and statistical supervision of the processes.

Process Chain and Grinding Process of Cylinder Liners

To achieve the capability of dry-run over the desired lifetime of the component, a number of high requirements must be fulfilled, as shown in Figure A.10. Besides measurable surface characteristics, the microscopic structure of the machined surface must be even and closed (see Figure A.14), but must keep the quality of low adhesion at the same time. The whole workpiece must be free of break-outs and splinters, since ceramic particles on the running surface lead to immediate fret of the tribo pairing. This leads to the requirement for the lowest possible induction of cracks in the subsurface.

To meet these requirements, silicon infiltrated silicon carbide (Si-SiC) was chosen as the workpiece material for the cylinder liners, due to its favorable tribologic characteristics, high wear resistance, and hardness. The process chain for the manufacture of the cylinder liners consists of powder preparation, moulding process, pyrolysis, a white machining process, sintering, internal and external cylindrical grinding and the final honing process. To have the grinding processes suitably designed for machining of ceramics and for the given requirements, a number of adjustments concerning process design, machine system and tools were carried out (see Figure A.11).

process design

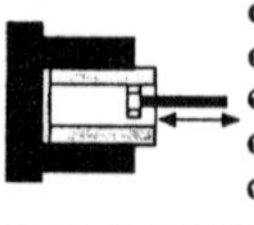

- o external cylindrical grinding for high clamping accuracy
- o low force pendulum grinding on the inside face for high roundness and cylindricity
- o sparc-out cycles
- o radiusses instead of edges to avoid break-outs
- o honing process to improve roundness, cylindricity and surface

machine system

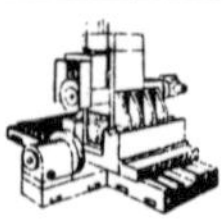

- o thermal stabilisation by cooling of the machine bed
- o application of shoe nozzles for improved coolant supply
- o central high performance coolant cleaning-system for microscopic particle filtering
- o low but sufficient clamping pressure
- o precisely adjusted clamping devices

tools

- o diamond grinding wheels with small grits for finish grinding of functional surface
- o machine integrated dressing processes for low radial deviation, thus a constant number of engaging grits
- o small amount of uncut chip thickness

Figure A.11: Adjustment of Process Design, Machine System and Tools on High Precision Grinding of Advanced Ceramics

Influences on End-product Quality Features, Tool Wear and Process Parameters

Examination of the process chain leads to results that are meaningful for the understanding of cause-effect relations and for the reproducibility of grinding processes. The characteristics of the end product develop over the process chain and are temporarily or finally influenced by single processes. A starting point for the assessment of the development of end product features is the examination of the so-called "tolerance channels". Figure A.12 shows exemplary tolerance channels for the development of a geometry and a surface characteristic value for the examined cylinder liners.

As the examples show, the assumption that tolerance channels narrow down like a funnel with every single process is not always applicable. Characteristics may be influenced in a negative way and may worsen in mean value and deviation during the process, since the influencing of certain single

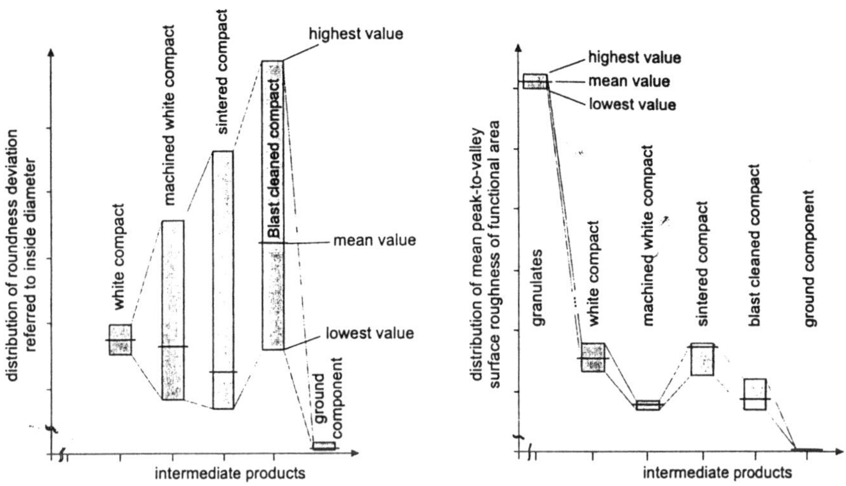

Figure A.12: Tolerance Channels of Surface Roughness and Roundness Deviation of Inside Diameter of Intermediate Stages of Cylinder Liners

characteristics by a single process can have a negative impact on other characteristics. Examining the tolerance channels more closely, it could not be observed in this process chain that workpieces which lay on the edge of the tolerance in one single process were found on the edge of tolerance again in the following process. Due to low grinding forces and the more extensive process strategy of the middle series, effects of geometry reproduction and other reproduction effects were avoided.

The example of tolerance channels shows that a strategy for keeping all tolerances of quality characteristics in all single processes as tight as possible is not always advisable. Rather, those characteristics which have a considerable influence on the finished product must be determined. To enable good reproducibilty of the finished product, these features must be specially supervised and kept within lowest possible variation. All other features can be kept within a wider tolerance to save process and measurement costs. Regarding the process chain of the cylinder liners, the exact composition of the powder batch has been identified to be particularly influential. To examine the influence of the powder batch on the end-product geometry, the

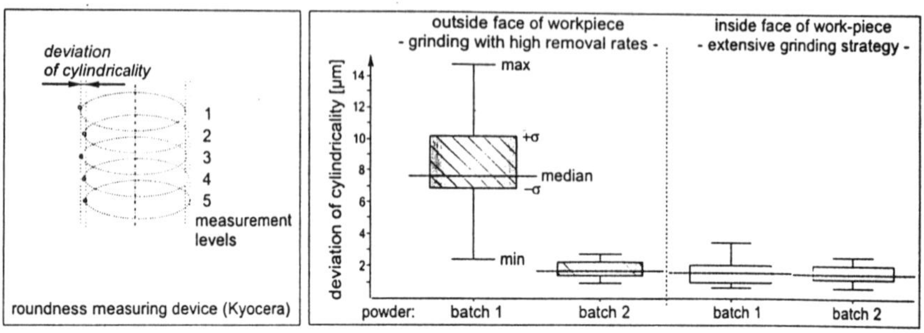

Figure A.13: Interior and Exterior Cylindricality of Ground Components depending on Batch of Ceramic Powder and Grinding Strategy

distribution of cylindricity on the outside and inside of the face of the workpiece is displayed by boxplots that show the median, the spread and the 2σ-interval (Figure A.13).

To exclude systematic faults, e.g. due to tool wear, sintering positions or other influences, all workpieces were well mixed before the examination of each process. This showed that variations within the powder preparation process lead to varying properties of the powder batch that influence the workpiece right up to the grinding process for the generation of its final geometric features. The distribution of cylindricality on the outside of workpieces made from batch 2 is (without any other systematic variations) significantly better than the distribution of cylindricality of workpieces made from batch 1. This was observed analogously for comparable properties such as the deviation of roundness of the workpieces. On the other hand, the example shows that effects like this can be compensated by an adapted process design. Machining of the outside face is carried out with a significantly higher material removal rate than the pendulum-grinding process of the functional area on the inside of the cylinder. Lower forces during the internal grinding process due to the extensive grinding strategy lead to reproducible manufacture of the functional area within the given tolerances, even if properties of the workpiece materials are varied.

Differences between the two batches with respect to the surface and subsurface structure could not be proved. Due to the high requirements on the surface of the functional area, SEM-examinations were carried out in order to enable visual assessment of the ground surfaces. Figure A.14 shows the surface of the inside face of a ground (a) and honed (b) cylinder liner.

Both pictures reveal a surface structure typical of the workpiece material. Among larger flat areas, surface defects and grain break-outs can be observed. This leads to the conclusion that varying holding loads within the bonding matrix result in occasional removal of SiSiC-grains, but basically, transcrystalline material removal is present. Comparing ground and honed surface, it can be observed that the honing process only partially leads to the generation of a new surface. Mainly, the flat areas are further extended, which explains the identical course of the roughness values for measured samples before and after honing (Figure A.14 below). The honing process

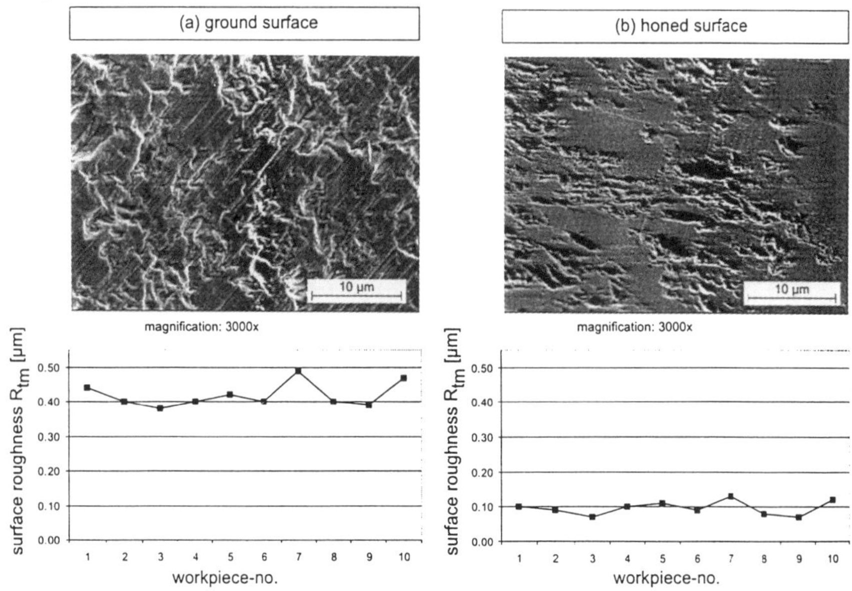

Figure A.14: Surface on the Functional Area of Ground (a) and honed (b) cylinder liners and readings of surface roughness of 10 samples before and after honing (Data: Pöschl-Werke Gmbh, Artem, Germany)

reduces the surface roughness significantly but is obviously not capable of completely correcting deficient grinding results. This indicates the consequence that the grinding process must be carried out so that the desired surface characteristics are already obtained after grinding. Since the understanding of generation and assessment of surfaces is of great importance for the reliable manufacture of ceramics, this subject is more closely discussed in the next section.

The results of the process chain analysis of the cylinder liners show that the reasons for failure and lack of reproducibility of machined ceramic components can often be traced back to previous processes, and do not necessarily result from the grinding or honing process itself. Due to the high sensitivity of ceramics, failures in previous stages of the process chain can often not be compensated at all by machining operations. In other cases, they are only compensated with great effort.

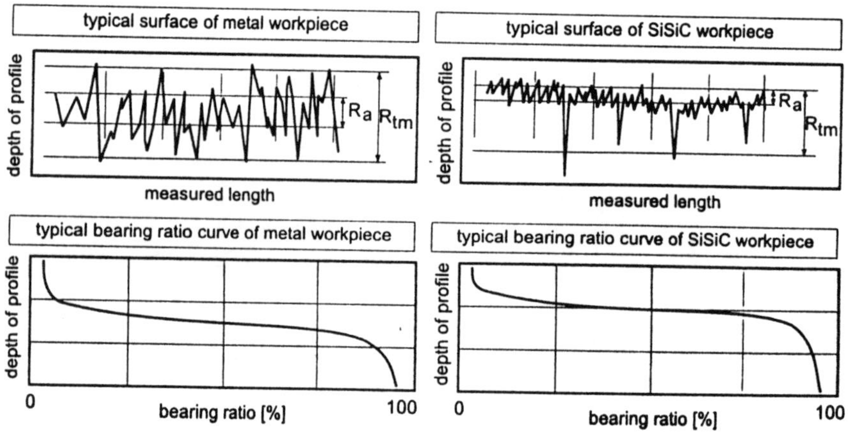

Figure A.15: Typical Surfaces of Ground Metal and Si-SiC Workpieces (schematically)

Considerations of Surface Assessment

Regarding the surface shown in Figure A.14 for cylinder liners designed for high tribological load, it becomes clear that the surface structure of ground Si-SiC components must be considered for the assessment of the surface with characteristic values. With regard to the profile recording, occasional hollows in the surface result in deep valleys that interrupt the otherwise even profile stochastically. The commonly used surface roughness values R_a and R_{tm} (= R_z, according to German DIN) were basically designed for the typical roughness profiles of metal materials. Figure A.15 compares typical surfaces of ground metal and ground Si-SiC workpieces.

As the figure shows, the basis for the calculation of the surface roughness value R_{tm} is significantly different from those of Si-SiC materials. Regarding Si-SiC, the mean peak-to-valley-height is by few, deep valleys, where as with metal materials the mean peak-to-valley-height is calculated based on a balanced relation of many peaks and valleys. One of the consequences is that the ratio of surface roughness R_{tm} to the average surface roughness R_a is significant higher than the same ratio using metal materials. The differences in distribution of peaks and valleys become more evident by comparison of the courses of the bearing ratio curve. In comparison to metal workpieces, the bearing ratio curve runs significantly earlier into bearing sections. This is equivalent to small R_{pk}-readings. Due to the dominating fracture of valleys for the bearing ratio curve, the value R_{vk} that describes the depths of the valleys is significantly higher for the same surface roughness. In quality inspection of surfaces of Si-SiC components, this means that the characteristic values of the bearing ratio curve R_{pk} (peak height), R_k (core roughness) and R_{vk} (valley depth) should be considered for surface assessment. Measurements on ceramic surfaces should be ceramic-specific. Based on extensive measurements, a group of scientists in Germany developed a guideline for carrying out roughness measurements on ceramic surfaces, as shown in Table A.1.[6]

Table A.1: Recommendations for Surface Roughness Measurements on Ceramic Surfaces according to [6] in Addition to ISO 4287/1

Surface roughness measurements on ceramic surfaces	
measuring system	vibration isolated reference surface system
tracing pin geometry	angle of tracing pin: 90° radius of tracing pin: 5µm for high quality surfaces ($R_{tm} < 1\,\mu m$): 2µm
preparation of measurements	cleaning of sample (fat and dust free) microscopic visual inspection of tracing pin calibration and check of results with roughness normal
performance of measurement	performance of at least 15 single measurements per surface uniform distribution of measurements over the entire testing surface documentation of average surface roughness (R_a), mean peak-to-valley-height (R_z/R_{tm}) and the characteristic values of the bearing ratio curve
evaluation	calculation of the mean value from the 15 single measurements and the confidence interval (95%)

4.0 CASE STUDY - CERAMIC CUTTERS FOR PAPER INDUSTRY

Cutting tools are applied in many branches of industry for manufacturing paper, cardboard, textile and leather products. In particular, cutters for the paper industry are subjected to high operating strain and wear strain as a consequence of paper ingredients such as non wood-free recycled paper, fillers and pigments (kaolin, titanium oxide, etc.). Wear resisting advanced ceramics appear to be suitable for this application. In the following, cutters made of partly stabilized zirconia (TZP) currently examined in applications of the paper industry to replace hard metal cutters, are introduced (Figure A.16 left).

Requirements of the Components

To achieve long and reliable tool service life, cutters must fulfill high requirements. The requirement profile is constituted by micro- and macro-geometrical requirements such as surface roughness, radius of curvature and straightness of cutting edge as well as material specific component requirements such as stability of cutting edges and wear resistance (Figure A.16). In particular, the design of manufacturing technologies and strategies has to be oriented to the application to manufacture defined cutting edges. The radius

Figure A.16: Industrial Application of Ceramic Cutters in the Paper Industry (Kolbus, Rahden, Germany) and Macro- and Micro- Geometrical Requirements of Cutters (Schematically, SEM Photography)

of curvature of cutting edge should be machined without material-weakening surface and subsurface damage to preserve highly defined sharp edges enabling low cutting forces.

State of the Art Grinding Process of Ceramic Cutters

The final machining of the ceramic cutter is structured in a sequence of surface-side traverse grinding operations with cup grinding wheels of graded specifications, followed by a manual lapping process. The state-of-the-art approach using set-up parameters of current surface grinding operations was taken to enable high material removal rates. Graded specifications of diamond grinding wheels were used to achieve high surface qualities of ceramic cutters.

Figure A.16 (right) shows the geometry of a machined cutting edge. The single marks and multiple breakouts on the cutting edge plane and on the edge result from deficient dressing of the tool and excessive uncut chip thicknesses of single grits engaging the workpiece. This damage of cutting edges leads to low reliability of operating performance. However, certain ceramic cutters applied in face cutting machines clearly exceeded the service life of hard metal cutters. Thus, measures of optimization within the entire process chain were required to improve reproducibility of component properties

Derived Measures for Optimization and Improvement of Work Result

To increase the reliability of ceramic cutters, the entire process chain was investigated, with the main emphasis on the principle of component design, material quality and finishing process. The end of service life of cutters resulted more from micro break-outs on the cutting edge and less from abrasive wear due to operating conditions (Figure A.17 left). To increase stability of the cutting edge, the material quality has been changed to a homogeneous and fine grained quality of TZP. If the strain induced phase transformation typical for TZP occurs during application, compressive residual stress fields in the vicinity of possibly induced micro-cracks improve stability of the cut-

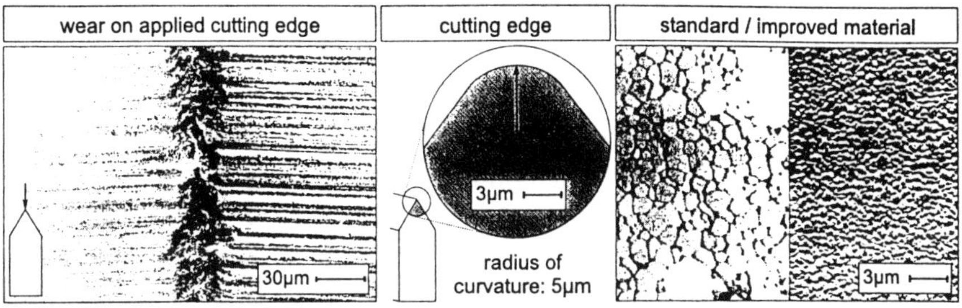

Figure A.17: Geometry of Ceramic Cutting Edges Worn (SEM-photography), basic shape schematically and material structures (SEM-photography)

ting edge. Another advantage of improved material quality is the higher process reliability within grinding operations on the edge. The desired reading of the radius of curvature is r = 5 μm. Machining of this edge is impossible if the micro structure of the ceramic component is characterized by grains with sizes of comparable scale which are located close to the edge (Figure A.17 right). During the grinding process, engaging diamond grits may remove this material grain from the workpiece. Selecting a fine grained material reduces occurrence of this removal mechanism to a great extent.

To reduce fabrication costs, the principle of component design has also been changed. Following the approach of function sharing, only highly strained parts of the cutters during application were designed with ceramic material. Thus, cutters are designed with metallic holders with specific TZP-inserts as ceramic cutting edges. Due to reduced machining operations on the ceramic inserts, this concept is advantageous compared to full-ceramic cutters.

With particular focus on the cutting edge, all finishing operations were optimized to improve process reliability and reproducibility by the following measures:

- Specification of cup grinding wheels: Soft resin bonding system and reduced diamond grit concentration to reduce cutting forces. The graduation of diamond grit sizes of each grinding operation has been fine tuned to each other by reduced grit sizes and additional grinding operations.
- Tool preparation: Reduction of the diamond grit protrusion to avoid grits with great protrusion from removing grains and clusters of ceramic material, particularly at the cutting edge.
- Machining strategy: To reduce the uncut chip thickness, the cutters are pendulum-ground with low feed rates. A grinding operation within this small-lot production is finished when all grinding marks, as visible surface influences of the previous process, are removed. Finishing each grinding operation with spark-out pendulum motions guarantees high geometrical accuracy. During the machining of cutting edges, an additional rest supports the edge of the workpieces to avoid breakouts.
- Process automation: In order to improve reproducibility, the final manual lapping process has been automated. Thus, cutting edges can be machined with higher reproducibility.

Parallel to optimization of the technological process chain, an effective method for testing work results has been developed to assess the performance of different final machining processes and setup parameters. Due to the extremely thin subsurface zone influenced by finishing processes, current four-point bending tests are not suitable to control this damage. Improved sensitivity to the structural condition of the subsurface was proved by examinations of the developed ball indenter tests compared with the four-point bending test (Figure A.18). Using a commercial edge tester with acoustic detection of fracture initiation, different variants of machining cutters could be clearly distinguished while four-point bending test results show only slight differences. Thus, the ball indenter test is suitable for testing surface relevant material strength quickly and is applicable without time consuming preparation of the ceramic components.

Due to the results of the ball indenter test, the ceramic cutters machined after optimization by variant 4 were chosen for the application test

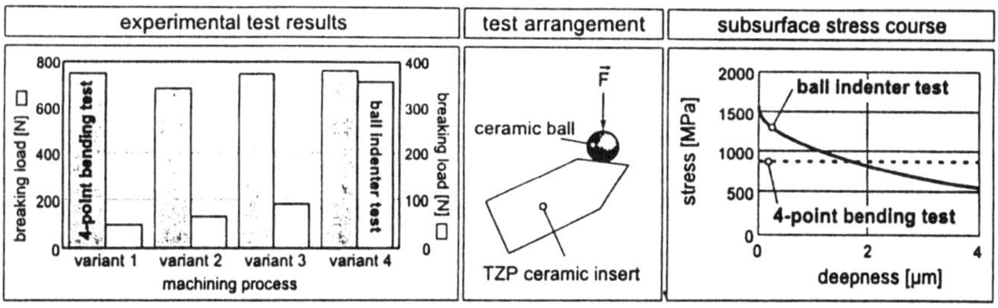

Figure A.18: Principle and Experimental Results of Ball Indenter Test for Testing Stability of Subsurface Zones of Machined Ceramic Cutters

(Figure A.18).[7] These ceramic cutters have succeeded in application tests in the paper industry. The service tool life of the ceramic cutter was more than sixteen times longer than the average service tool life of currently applied hard metal cutters. Further improvement of the duration of application is expected.

5.0 CONCLUSIONS

The three presented case studies demonstrate the feasibility of replacing metal parts with ceramics in certain applications. In all cases, ceramic material and component reliability was improved during the research project and manufacturing cost was concurrently reduced. The prerequisite of improvement was the integrated optimization approach, considering all processes within the process chain with regard to ceramic-specific material behavior.

The results of the investigations are summarized as follow: work results of grinding processes may often be greatly influenced by varying quality features of intermediate products of previous processes. This may lead to expensive and time consuming strategies for grinding processes to achieve ceramic components of the required quality and may lower reproducibility of the component properties . For instance, it is not only the geometry of the sintered component that influences geometrical properties of the end-product, such as roundness and shape deviation, but also the powder preparation and other processes which influence the end-product geometry. Quality and economical improvements in manufacture of components made of advanced ceramics are possible by identifying these sensitive parameters over the entire process chain and maintaining tight control over their variation. It can be assumed that this potential of the manufacture of ceramic components is currently not often fully utilized.

REFERENCES

1. Gugel, E., Woetting G., Claeys, P., Woditsch, P.: ‚Silicon Nitritde Valves available for Automotive Engines" , 29th International Symposium on Automotive Technology & Automation, Florence/ Italy; Ed. D. Roller, Stuttgart (1996), Vol 1, pp. 677-684
2. N.N.: Leaflet Fa. Junker Maschinenfabrik GmbH, Nordrach, Germany, 1992-1996
3. Warnecke, GL, EichgrQn, K., Schdfer, L.: ‚Manufacture Optimization of Process Chains of Ceramic Components", Proceedings of the Ultrahard Materials, Technical Conference, Windsor, Canada, May 28-30 1998, pp. 297-307
4. Warnecke, G., Zitt, U.: ‚Kinematic Simulation for Analyzing and Predicting High-Performance Grinding Processes", Annals of the CIRP 47 (1998) 1, pp. 265-270
5. Nickel, K.-G.: Ventile aus Siliziumnitrid. ‚Hoch leistu ngs kerami k in der Motortechnik", Technische Rundschau Transfer 86 (1994), Vol. 21, pp. 38-40

6. Ardelt, T., EichgrOn, K., Friemuth, T., Gdbler, J., Gerent, 0., Holl, S., Kreis, R., Lierse, T., Schippers, C., Zapp, M.: ,Rauheitsmessungen an keramischen Oberf lAchen", Werkstattstechnik 88 H. 5 (1998), pp. 239-243
7. Pfeiffer, W., Hollstein, T.: jnfluence of Grinding Parameter on Strength-Dominating Near-Surface Characteristics of Silicone Nitride Ceramics", J. European Ceramic Society 17, Elsevie Science Limited (1997), pp. 487-494

Index